MEMS Packaging

WSPC Series in Advanced Integration and Packaging

ISSN: 2315-473X

In the 21st century, innovative integration and packaging drives success in the marketplace, establishing the "look and feel" of electronic products and their price point, and enabling differentiation and branding. The rapid development and proliferation of design methodologies, assembly techniques, and thermal solutions for components ranging from integrated circuits to photonics, and from sensors and MEMS to mixed signals and power amplifiers, drive the often frenetic pace of exploration, discovery, and innovation characteristic of the electronic industry.

This recently introduced WSPC book series on "Advanced Integration and Packaging of Electronic Systems" (AIPack), with a particular focus on 2.5D and 3D form factors, will provide a "technology commons" for the packaging community, a place where ideas can be articulated, examined, and disseminated rapidly and efficiently through both print and electronic media. It is our hope that product developers will find AIPack a convenient and reliable source of technical information; that researchers and students will be guided to the challenges and barriers facing key technologies, and that technology managers and investors will find the tools and data needed to make broad technology assessments and comparisons for emerging packaging and integration technologies.

Published

Vol. 5: *MEMS Packaging*
edited by Y. C. Lee, Yu-Ting Cheng and Ramesh Ramadoss

Vol. 4: *Cost Analysis of Electronic Systems (Second Edition)*
by Peter Sandborn

Vol. 3: *Cooling of Microelectronic and Nanoelectronic Equipment: Advances and Emerging Research*
edited by Madhusudan Iyengar, Karl J. L. Geisler and Bahgat Sammakia

Vol. 2: *Design and Modeling for 3D ICs and Interposers*
by Madhavan Swaminathan and Ki Jin Han

Vol. 1: *Cost Analysis of Electronic Systems*
by Peter Sandborn

More information on this series can also be found at http://www.worldscientific.com/series/wssaip

WSPC Series in Advanced Integration and Packaging Vol. 5

MEMS Packaging

Editors

Y C Lee
University of Colorado, Boulder, USA

Yu-Ting Cheng
National Chiao Tung University, Taiwan

Ramesh Ramadoss
IoT Technologies Inc., USA

World Scientific

NEW JERSEY • LONDON • SINGAPORE • BEIJING • SHANGHAI • HONG KONG • TAIPEI • CHENNAI • TOKYO

Published by

World Scientific Publishing Co. Pte. Ltd.
5 Toh Tuck Link, Singapore 596224
USA office: 27 Warren Street, Suite 401-402, Hackensack, NJ 07601
UK office: 57 Shelton Street, Covent Garden, London WC2H 9HE

British Library Cataloguing-in-Publication Data
A catalogue record for this book is available from the British Library.

WSPC Series in Advanced Integration and Packaging — Vol. 5
MEMS PACKAGING

ISBN 978-981-3229-35-8

For any available supplementary material, please visit
http://www.worldscientific.com/worldscibooks/10.1142/10692#t=suppl

Typeset by Stallion Press
Email: enquiries@stallionpress.com

Printed in Singapore

Preface

MEMS sensors and actuators are enabling components for smartphones, augmented reality/virtual reality (AR/VR) and wearable electronics. MEMS packaging is recognized as one of the most critical activities to design and manufacture reliable MEMS. A unique challenge to MEMS packaging is how to protect moving MEMS devices during manufacturing and operation. With the introduction of wafer level capping and encapsulation processes, this barrier is removed successfully. In addition, MEMS devices should be integrated with their electronic chips with the smallest footprint possible. As a result, 3D packaging is applied to connect the devices vertically for the most effective integration. Such 3D packaging also paves the way for further heterogeneous integration of MEMS devices, electronics and other functional devices.

This book consists of chapters written by leaders developing products in a MEMS industrial setting and faculty members conducting research in an academic setting. After Chapter 1 on introduction to MEMS packaging Chapters 2 to 6 are written by industry professionals to address practical issues on through silicon vias, vertical interconnects, wafer level packaging, motion sensor-to-CMOS bonding and printed circuit board technology for MEMS. Then, Chapters 7 to 13 are written by leading academic researchers on fundamental issues and future technologies. The topics include: encapsulation of MEMS, heterogeneous integration, microfluidics, solder bonding, localized sealing, microsprings and reliability.

This book is the first book covering MEMS packaging by leading professionals and researchers in the field. It is going to be a major resource for those who participate in R&D and manufacturing of MEMS products. In addition, it is also a good resource for every professional associated with the integration for systems represented by smartphones, AR/VR and wearable electronics. MEMS sensors and actuators are enabling components for such "smart" systems, and these professionals can learn how to integrate MEMS devices with other functional components.

Y. C. Lee, University of Colorado Boulder
Yu-Ting Cheng, National Chiao Tung University
Ramesh Ramadoss, IoT Technologies Inc.
May 2017

List of Contributors

Carmen Aracil
Dpto. Ing. Electronica
Universidad de Sevilla
Avda. de los Descubrimientos sn
E-41092 Sevilla, Spain
Email: caracil@gte.esi.us.es

Tomas Bauer
Box 595,
Bruttovägen 1,
SE-175 26 Järfälla,
Sweden
Email: tomas.bauer@silex.se

Yu-Ting Cheng
Department of Electronics Engineering
National Chiao Tung University
ED640, 1001 Ta-Hsueh Road, Hsinchu, 300, Taiwan
Email: ytcheng@g2.nctu.edu.tw

Eugene M. Chow
Microsystems Area Manager and Principal Scientist
PARC, a Xerox Company
3333 Coyote Hill Road, Palo Alto, CA 94306, USA
Email: Eugene.Chow@parc.com

Christopher L. Chua
Principal Scientist
PARC, a Xerox Company
3333 Coyote Hill Road, Palo Alto, CA 94306, USA
Email: Christopher.Chua@parc.com

Ravi Doraiswami
CALCE Center for Advanced Life Cycle Engineering
1103 Engineering Lab Building
University of Maryland
College Park, MD 20742, USA
Email: washravi@gmail.com

Thorbjorn Ebefors
Box 595,
Bruttovägen 1,
SE-175 26 Järfälla,
Sweden
Email: thorbjorn.ebefors@spinverse.com

Masayoshi Esashi
Micro System Integration Center (μSIC), Professor, Tohoku University,
Center of Innovation (COI), Research fellow
519-1176 Aramaki-Aza-Aoba, Aoba-ku, Sendai 980-0845, Japan
Email: esashi@mems.mech.tohoku.ac.jp

Weileun Fang
Department of Power Mechanical Engineering and NEMS Institute
National Tsing Hua University,
No. 101, Section 2, Kuang-Fu Road, Hsinchu, 30013, Taiwan, R.O.C.
Email: fang@pme.nthu.edu.tw

Nils Høivik
Professor
Faculty of Technology and Maritime Sciences (TNM)
Department of MicroSystems (IMS)
Campus Bakkenteigen, Vestfold, Norway
Email: nils.hoivik@usn.no

Chien-Ming Huang
CALCE Center for Advanced Life Cycle Engineering
1103 Engineering Lab Building
University of Maryland
College Park, MD 20742, USA
Email: cmhuang@umd.edu

Yunhan Huang
CALCE Center for Advanced Life Cycle Engineering
1103 Engineering Lab Building
University of Maryland
College Park, MD 20742, USA
Email: hyh@umd.edu

Thomas W. Kenny
Department of Mechanical Engineering
440 Escondido Mall, Bldg 530/Rm 224
Stanford University
Stanford, CA 94305-3030
USA
Email: kenny@cdr.stanford.edu

Y. C. Lee
S.J. Archuleta Professor
Department of Mechanical Engineering
1111 Engineering Drive
University of Colorado, Boulder, CO 80309-0427, USA
Email: leeyc@colorado.edu

Chiung-Wen Lin
c/o Prof. Weileun Fang
Department of Power Mechanical Engineering and NEMS Institute
National Tsing Hua University,
No. 101, Section 2, Kuang-Fu Road, Hsinchu, 30013, Taiwan, R.O.C.
Email: fang@pme.nthu.edu.tw

Liwei Lin
Department of Mechanical Engineering
5135 Etcheverry Hall MC#1740
University of California at Berkeley, Berkeley, CA 94720-1740, USA
Email: lwlin@me.berkeley.edu

Antonio Luque
Dpto. Ing. Electronica
Universidad de Sevilla
Avda. de los Descubrimientos sn
E-41092 Sevilla, Spain
Email: aluque@gte.esi.us.es

Sigurd Moe
SINTEF Digital, Minalab
Gaustadalléen 23C
0373 Oslo
Norway
Email: Sigurd.Moe@sintef.no

Steven Nasiri
Nasiri Ventures LLC
Los Altos Hills, CA, USA
Email: snasiri@nasiriventures.com

Michael Osterman
CALCE Center for Advanced Life Cycle Engineering
1103 Engineering Lab Building
University of Maryland
College Park, MD 20742, USA
Email: osterman@calce.umd.edu

Michael Pecht
CALCE Center for Advanced Life Cycle Engineering
1103 Engineering Lab Building
University of Maryland
College Park, MD 20742, USA
Email: pecht@calce.umd.edu

Paul Pickering
Micralyne Inc.
1911-94 Street
Edmonton, AB
Canada T6N 1E6
Email: paulp@micralyne.com

Ramesh Ramadoss
414 Chestnut St
Brentwood, CA 94513, USA
Email: Dr.Ramesh.Ramadoss@ieee.org

Janna Rodriguez
Department of Mechanical Engineering
440 Escondido Mall, Bldg 530/Rm 224
Stanford University
Stanford, CA 94305-3030
USA
Email: jannar@stanford.edu

Dean Spicer
Micralyne Inc.
1911-94 Street
Edmonton, AB
Canada T6N 1E6

Yu-Chuan Su
Department of Engineering and System Science
National Tsing Hua University
101, Section 2, Kuang-Fu Road, Hsinchu, 30013, Taiwan
Email: ycsu@ess.nthu.edu.tw

Maaike M. Visser Taklo
SINTEF Digital
Forskningsveien 1
0373 Oslo
Norway
Email: maaikemargrete@gmail.com

Shuji Tanaka
Laboratory for Smart System Integration, Chair of Nanosystem,
Department of Robotics, Tohoku University
Division of Mechanical Engineering,
6-6-01 Aramaki aza Aoba, Aoba-ku, Sendai, Miyagi 980-8579, Japan
Email: tanaka@mems.mech.tohoku.ac.jp

Colin Twanow
Micralyne Inc.
1911-94 Street
Edmonton, AB
Canada T6N 1E6

Arvind Vasan
900 Pepper Tree Ln, Apt 523
Santa Clara, CA 95051, USA
Email: arvind@empowermicro.com

Sandra Winkler
586 Lakeview Way,
Emerald Hills, CA 94062, USA
Email: s.l.winkler@ieee.org

Hsueh-An Yang
Department of Power Mechanical Engineering
National Tsing Hua University
No. 101, Section 2, Kuang-Fu Road, Hsinchu, Taiwan 30013, R.O.C.
Email: d917706@alumni.nthu.edu.tw

Contents

Preface v

List of Contributors vii

Chapter 1. Introduction to MEMS Packaging 1
Y. C. Lee, Ramesh Ramadoss and Nils Hoivik

Chapter 2. Silex's TSV Technology: Overview of Processes and MEMS Applications 31
Tomas Bauer and Thorbjörn Ebefors

Chapter 3. Vertical Interconnects for High-End MEMS 45
Maaike M. Visser Taklo and Sigurd Moe

Chapter 4. Using Wafer-Level Packaging to Improve Sensor Manufacturability and Cost 53
Paul Pickering, Collin Twanow and Dean Spicer

Chapter 5. Nasiri Fabrication Process for Low-Cost Motion Sensors in the Consumer Market 73
Steven Nasiri, Sandra Winkler and Ramesh Ramadoss

Chapter 6. PCB Based MEMS and Microfluidics 85
Ramesh Ramadoss, Antonio Luque and Carmen Aracil

Chapter 7. Single Wafer Encapsulation of MEMS Resonators 117
Janna Rodriguez and Thomas W. Kenny

Chapter 8. Heterogeneous Integration and Wafer-Level Packaging of MEMS 141

Masayoshi Esashi and Shuji Tanaka

Chapter 9. Packaging of Membrane-Based Polymer Microfluidic Systems 171

Yu-Chuan Su

Chapter 10. Wafer-Level Solder Bonding by Using Localized Induction Heating 197

Hsueh-An Yang, Chiung-Wen Lin and Weileun Fang

Chapter 11. Localized Sealing Schemes for MEMS Packaging 223

Y. T. Cheng, Y. C. Su and Liwei Lin

Chapter 12. Microsprings for High-Density Flip-Chip Packaging 251

Eugene M. Chow and Christopher L. Chua

Chapter 13. MEMS Reliability 279

Chien-Ming Huang, Arvind Sai SarathiVasan, Yunhan Huang, Ravi Doraiswami, Michael Osterman and Michael Pecht

Index 345

Chapter 1

Introduction to MEMS Packaging

Y. C. Lee, Ramesh Ramadoss and Nils Hoivik

1.1 Introduction

Electronic packaging and semiconductor manufacturing are two major activities to produce microsystems including microelectronics, optoelectronics, microwave/RF, biomedical devices, sensors, actuators, micro-scaled energy systems, etc. Electronic packaging interconnects "chips" and other components into a microsystem. Figure 1.1 shows an iPhone 7 Plus [1]. The processes to assemble such a total system are categorized as the third level of packaging. The logic board, battery, antenna, camera, display, casing, and many other components are assembled in this packaging level. Figure 1.2 shows the logic board taken out of the iPhone 7 Plus. The processes to assemble such a logic board are categorized as the second level of packaging. Many chips and discrete components are interconnected on a multilayer printed circuit board (PCB). For example, A10 microprocessor, LTE modem, three power amplifier modules, touchscreen controller, and barometric pressure sensor as marked by colors are some major components on this side of the board.

Figure 1.2 also shows a 6-axis accelerometer and gyroscope chip with details exposed. In fact, it is not a single chip; it consists of a MEMS chip (2.64 mm × 1.67 mm) and an application-specific integrated circuit (ASIC) chip (2.64 mm × 1.92 mm) stacked inside the plastic package (3.00 mm × 3.00 mm × 0.76 mm) [2]. The processes to package these two chips are categorized as the first level of packaging. This first level packaging for MEMS device is to be introduced in this chapter.

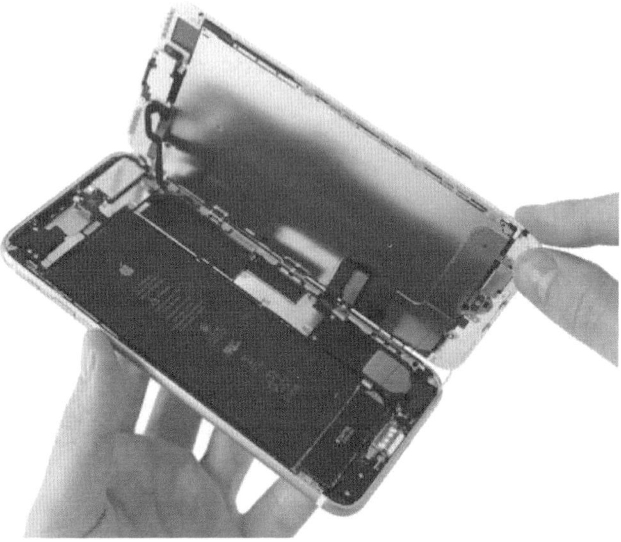

Fig. 1.1. iPhone 7 Plus with major components exposed.

Fig. 1.2. Logic board of iPhone 7 Plus with A10, 6-axis accelerometer and gyroscope chip and quite a few other chips and discrete components interconnected on a PCB.

Figure 1.3(a) illustrates MEMS packaging with MEMS, ASIC and other components integrated in a single package. A unique challenge to packaging is how to protect moving MEMS devices as represented by an accelerometer shown in Fig. 1.3(b). Such a device has to be enclosed by a cap; otherwise,

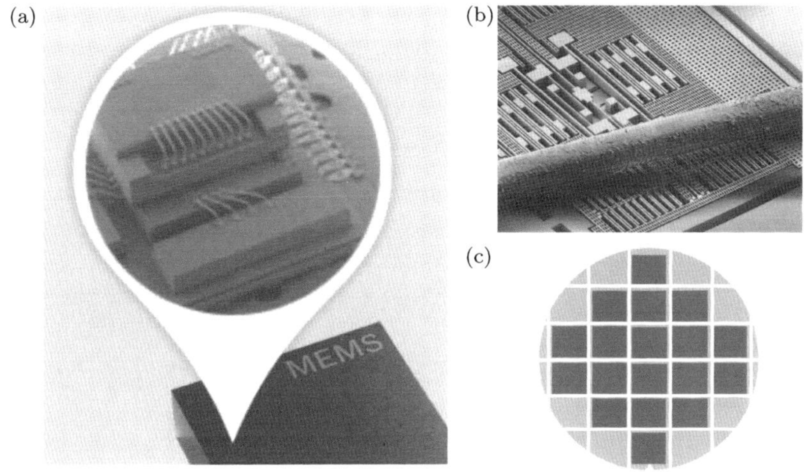

Fig. 1.3. MEMS packaging illustrated by: (a) a MEMS device, an ASIC and other components integrated in a single package; (b) the moving parts in a MEMS device require special care and unique processing steps; and (c) packaging processes represented by wafer dicing that requires protection of moving devices. (Courtesy of Bosch for (a) and (b))

it would be damaged during packaging such as a dicing process as shown in Fig. 1.3(c). It is also very important to reduce the size, and overall footprint of the package. As a result, the device and the package are often designed and manufactured at the same time. This introduction will present different MEMS packaging configurations and processes. MEMS technology enables sensing and actuating devices, which are used in a wide spectrum of applications such as microelectronics, optoelectronics, radio frequency (RF), biomedical lab-on-a-chip, and micro-energy systems. Therefore, there are many other opportunities, which will be illustrated with a few examples: packaging technology to fabricate microelectromechanical systems (MEMS) devices, MEMS devices to enhance packaging technology, and micro/nano systems.

1.2 MEMS Packaging: Configurations

In general, MEMS packages can be differentiated into ceramic/metal and plastic packages. Wafer-level capping and other protection means are enabling technologies for the plastic packaging. In each package, there are various MEMS-to-package connection schemes: single die wire bonded or flip-chip bonded; multiple dies wire bonded or flip-chip bonded; and multiple dies stacked with wire bonds or flip-chip bonding. These configurations are to be reviewed as follows:

1.2.1 *Ceramic/Metal Package*

A conventional ceramic package consists of a base and a lid (see Fig. 1.4). The base is also referred to as a header. One or more dies are attached to the base through adhesive die attach and electrical interconnection between the metal pads on the die and the base are formed through wire bonding. Alternatively, flip-chip solder bonding process may be employed to provide mechanical and electrical interconnection between the metal pads on the die and base. Then, the package is capped and sealed with a lid. Materials typically used for making TO-can packages as shown in Fig. 1.4(a) include Kovar, copper and different metal alloys. Materials typically used for making ceramic packages as shown in Fig. 1.4(b) include alumina (Al_2O_3), aluminum nitride (AlN), beryllium oxide (BeO), and silicon carbide (SiC). Lids are commonly made of glass, metal, ceramic or plastic and the choice of material depends on the end-use or application. Figure 1.5(a–d) shows a pin-grid-array (PGA) ceramic package used for comb-drive MEMS relay. The MEMS relays with electroplated nickel as the structural layer are fabricated using MetalMUMPs™ process.

The metal, TO-can, package for MEMS is very robust and simple and can be applied to a limited number of products. The ceramic/metal packaging for MEMS is an expensive approach. It demands a class-10 packaging environment to avoid any damaging contamination inside the package. The package itself is also expensive. It is widely used to package the first generation of MEMS products. For example, the DMD™ from Texas Instruments

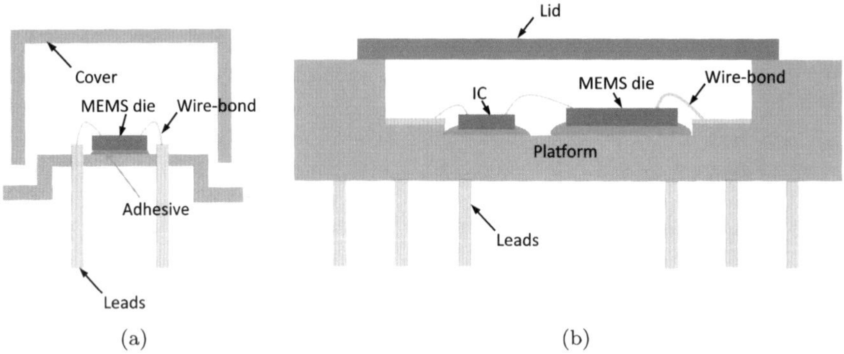

Fig. 1.4. (a) A TO-can metal package and (b) a ceramic package for MEMS.

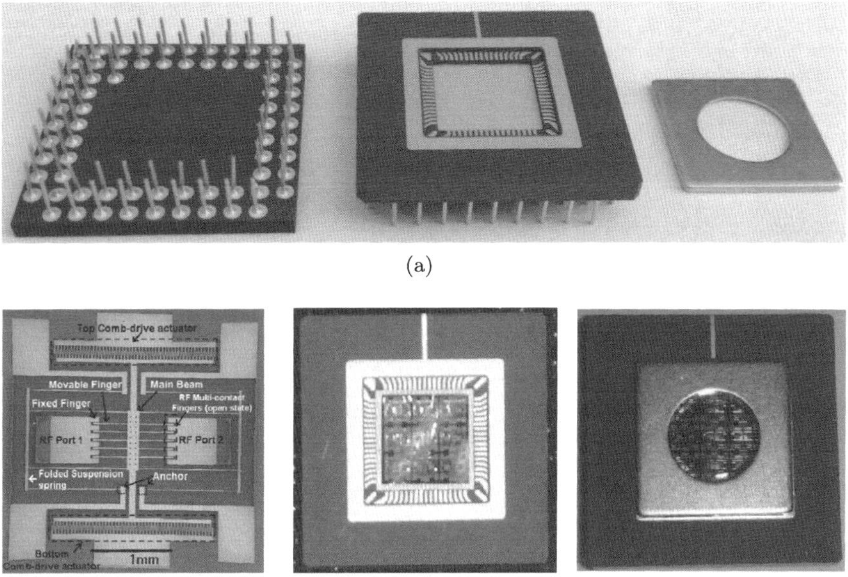

(a) (b) (c) (d)

Fig. 1.5. Packaging of MEMS comb-drive relays in a PGA ceramic package: (a) bottom view of the header, top view of the header, and top view of the glass lid, (b) a single MEMS comb-drive relay, (c) wire bonding of the die with 9 MEMS relays (3×3 array) to the package, and (d) header capped with a sealing lid.

used a ceramic package with a transparent glass lid suitable for optical MEMS application. Most ceramic/metal packages are now replaced by low-cost plastic packages.

1.2.2 *Plastic Package*

Plastic packages offer low-cost solutions for packaging MEMS devices. However, the fragile and moving MEMS devices must be protected before packaging processes. The wafer-level capping technique is an enabling technology for plastic packaging of MEMS. As illustrated in Fig. 1.6, a glass wafer, with pre-fabricated cavities, can be bonded to a silicon wafer of MEMS devices through anodic bonding. The cap protects the MEMS device throughout the entire packaging processes. The anodic bonding process also provides a hermetically sealed clean environment inside the package to assure device's reliable operation. In addition to the glass cap, there are many alternative approaches, such as metal capping, polysilicon encapsulation and polymer/metal cap as illustrated in Fig. 1.6 [3].

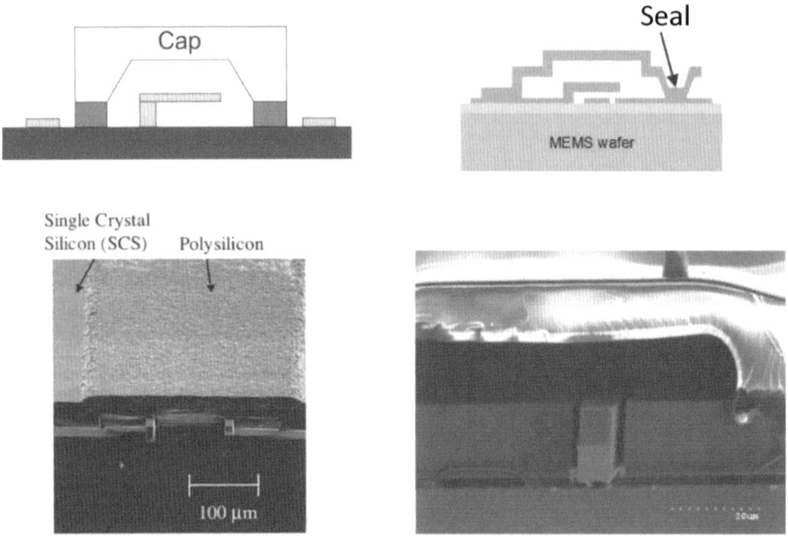

Fig. 1.6. Wafer-level capping using glass cap, metal cap, polysilicon encapsulation or polymer/metal cap for protecting MEMS during the plastic packaging processes [3–5].

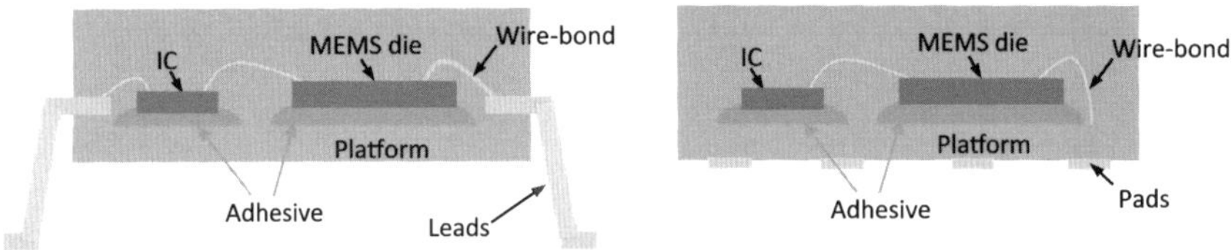

(a) (b)

Fig. 1.7. Surface mount packages: (a) peripheral type and (b) area array (or BGA) type.

The capped MEMS device can be further packaged with one of the two approaches: pre-molding and post-molding. In the pre-molding approach, a plastic package is pre-molded with a cavity and lead frame for housing the die. The die is attached to the bottom surface of the cavity and the pads on the die are connected to the leads through wire-bonding. In the post-molding approach, a plastic molding material is used to encapsulate the die and lead frame assembly (see Fig. 1.7). Most commonly, epoxy is used as the plastic molding material. Typically, fillers such as glass, silica, or alumina powder are used to reduce thermal expansion mismatch and improve thermal conductivity of the plastic material. The final finish of plastic packages is typically black in color due to carbon and organic dye ingredients. Inside the package, there are different MEMS-to-package connection schemes, which are to be reviewed next.

1.2.3 *MEMS-to-Package Connections*

After dicing, each individual die can be attached onto a platform inside a package using adhesives (organic or inorganic) or solder alloys as illustrated by Fig. 1.8(a). Typical adhesives include epoxies, silicones, and polyimides. The choice of bonding material is critical as it should be capable of handling stresses between the silicon die and package due to a mismatch of thermal expansion between the two components over the designed operating temperature range. Excessive thermal stress may lead to cracking of the bond and/or die. The choice of a solder alloy for a particular application is made based on its melting temperature and mechanical properties. Hard solders melt at temperatures near or above 500°C. Soft solders melt at lower temperatures (between 90°C and 450°C) depending on their composition. Eutectic alloys transition from liquid-to-solid phase directly (melting point is the same as freezing point) and thus have the lowest melting points of alloys sharing the same composition. For die attachment using solders, the bottom of the die must be coated with a thin metal film such as gold, platinum or palladium. After the die attachment, the capped device is wire bonded to the package.

Alternatively, the capped MEMS die can be flip-chip bonded to the platform as shown in Fig. 1.8(b). In these applications, there are different bonding technologies such as soldering, thermocompression bonding, and thermosonic bonding. However, this flip-chip configuration is rarely used for a single MEMS device connected to its package platform.

Since an ASIC is usually required for low-power operation, signal processing, etc., several packaging schemes integrates both a MEMS and an ASIC assembled into the same package. Figure 1.9(a) illustrates the approach using wire bonds to connect the MEMS and ASIC dies attached

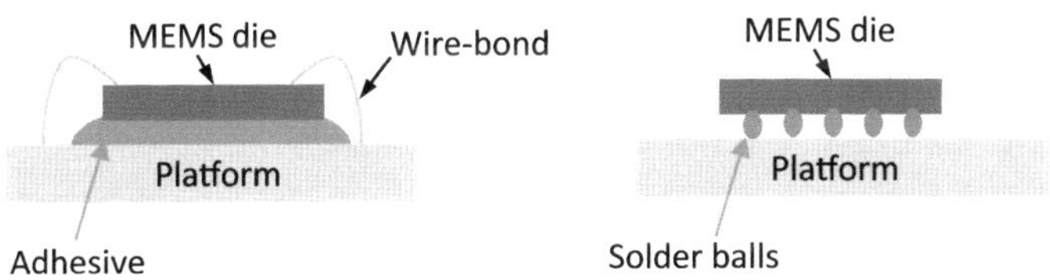

(a) (b)

Fig. 1.8. Illustration of MEMS connection to the package: (a) die-attachment with wire bonding and (b) flip-chip bonding.

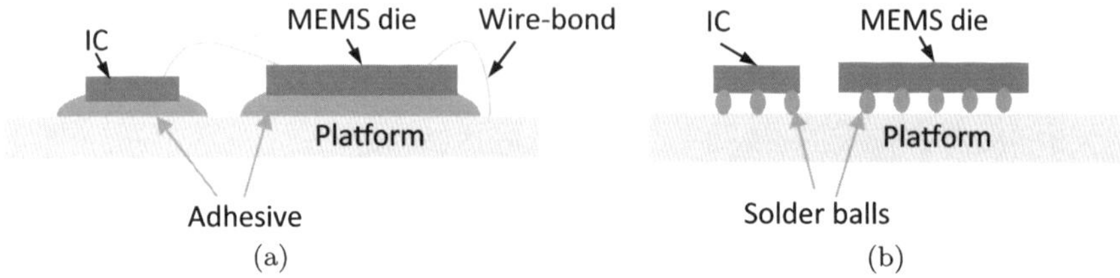

Fig. 1.9. Illustration of horizontal side-by-side integration of MEMS and IC inside a package: (a) die-attachment and wire bonding process and (b) flip-chip bonding process.

Fig. 1.10. A MEMS product with an ASIC and a capped MEMS device wire bonded onto a package.

on the package platform. Figure 1.10 shows the image of a real product commercialized. This approach is more popular than the flip-chip configuration as shown in Fig. 1.9(b).

In order to reduce the footprint of the package, the most popular approach is to stack an ASIC on a MEMS device as illustrated in Fig. 1.11(a) with a real product shown in Fig. 1.12 [6]. With the advancement of flip-chip bonding and the use of through silicon vias (TSVs) technologies, other configurations shown in Figures 1.11(b) and (c) can be manufactured. See Section 1.3.3, 3D Integration, for details.

In addition to these configurations shown, there are other approaches. For example, a monolithic approach can fabricate MEMS device on top of the ASIC [7]. For bioMEMS applications, microfluidic coupling has to

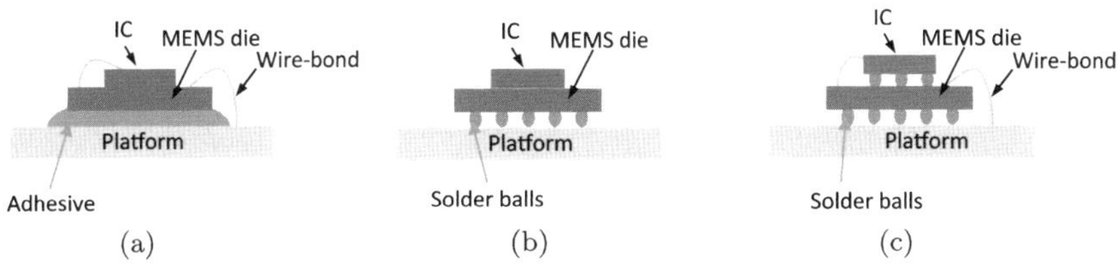

Fig. 1.11. Illustration of vertical stacked up integration MEMS and IC: (a) die-attachment and wire bonding, (b) flip-chip bonding, and (c) hybrid approach.

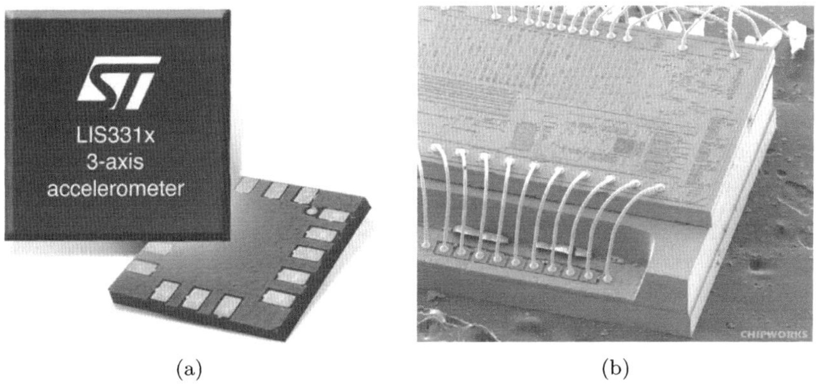

Fig. 1.12. Plastic packaging of ST microelectronics 3-axis accelerometer: (a) top and bottom view of the package, and (b) stack up ASIC and a MEMS device.

be considered to meet the specific requirements resulting from bio functions [8]. For photonic applications, photonic and MEMS interactions would be the main consideration [9]. A comprehensive review of different configurations is beyond the scope of this chapter. Some examples are illustrated in Section 1.4.

For each configuration, there are quite a few fabrication or assembly processes involved. In the following section, we will introduce some of these processes.

1.3 MEMS Packaging: Processes

The major packaging processes to be introduced include monolithic packaging and different bonding processes for packaging of MEMS at wafer level. In addition, 3D integration is essential to the advancement of MEMS applications by reducing its footprint.

1.3.1 *Packaging of MEMS at Wafer-Level: Monolithic*

Monolithic packaging refers to sealing of MEMS devices, directly on the wafer of which it is fabricated, during completion of the micro-machining process. This approach is quite appealing, as it reduces fabrication cost and the package can be made to have very little impact on the device performance. The downside to monolithic packaging is that if the MEMS packaging or fabrication yield is low, the overall production cost will quickly increase. Compared to the traditional system in a package where only known-good dies (KGDs) are used, it is less practical and cost-efficient to test MEMS devices before the monolithic packaging is completed. Throughout the last decade, several monolithic approaches have been demonstrated, but as of today two technologies are used in commercially available systems. As shown in Fig. 1.13, "MEMS-first": where MEMS devices are fabricated using single crystalline silicon before CMOS on the wafers are made [10], or "MEMS-last": using the back-end materials on the wafers on top of the IC circuits [11]. The main difference between "MEMS-first" and "MEMS-last" is the temperatures that can be used in the micro-machining process. Most IC circuits have a temperature limit of 400°C which rules out some

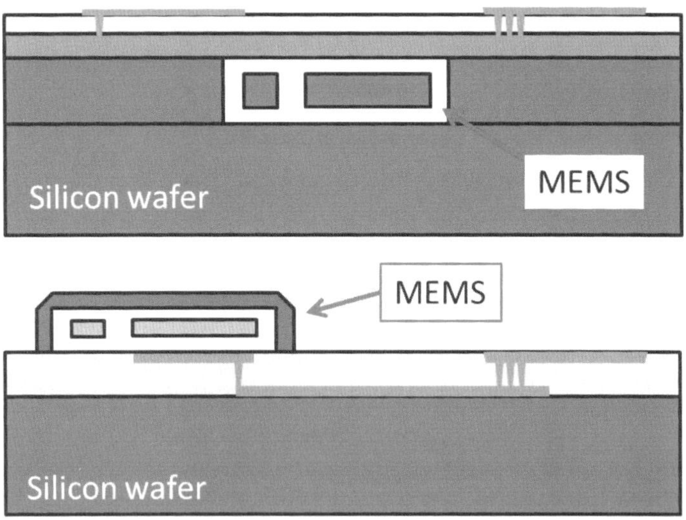

Fig. 1.13. Illustration of monolithic packaging of MEMS at wafer-level. MEMS-first: Devices are usually fabricated using single crystalline silicon structures on SOI devices. CMOS and other electronic circuits are fabricated afterwards. MEMS-last: Devices are fabricated using the dielectric and metal layers available in the back-end on top of the IC components. The device is capped, or sealed during the last fabrication steps.

deposition and annealing processes for MEMS-last, whereas MEMS-first permits high temperature processing to create the device.

Fabricating MEMS devices before building the CMOS devices on top is the success behind the Si-Time MEMS first process used to fabricate very high performance commercially available MEMS oscillators [10, 12]. Looking closer at the MEMS-first approach illustrated in Fig. 1.14, the structures are first fabricated using single-crystalline silicon on an SOI wafer, which is later patterned by deep reactive ion etching (D-RIE). Subsequently, the silicon structure is oxidized, and the Si structure is covered by deposition of a polysilicon layer that is later patterned with fine vents. The vents enable removal of the oxide layer surrounding the resonator and define the volume of the enclosed vacuum cavity. After sealing, a thick durable encapsulation layer of polysilicon is grown on top. Since this process is performed at very high temperatures in an ultraclean vacuum environment, the volume surrounding the resonators maintain ultraclean vacuum for the lifetime of the device.

Discera, another MEMS oscillator company, employs the MEMS-last approach with its oscillator as shown in Fig. 1.15 [13]. The packaged resonator is on an ASIC chip fabricated first. Monolithic packaging is

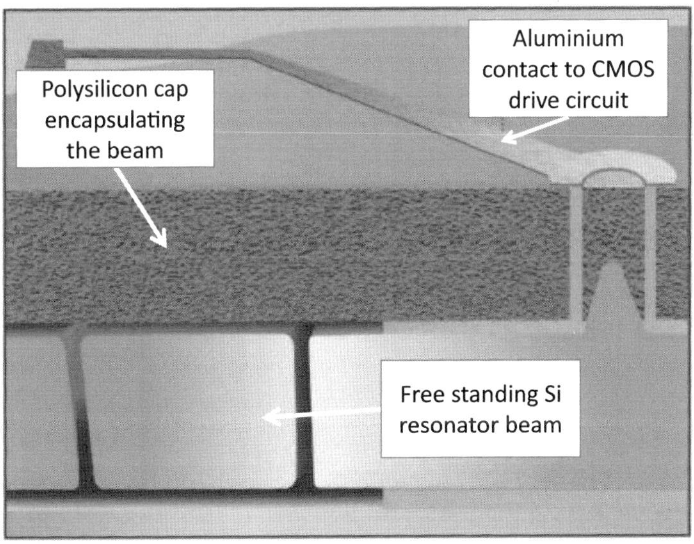

Fig. 1.14. Illustration of MEMS first process used to fabricate free-standing single crystalline silicon resonator beams encapsulated by a polysilicon cap. The volume surrounding the MEMS structure maintains ultraclean vacuum conditions.

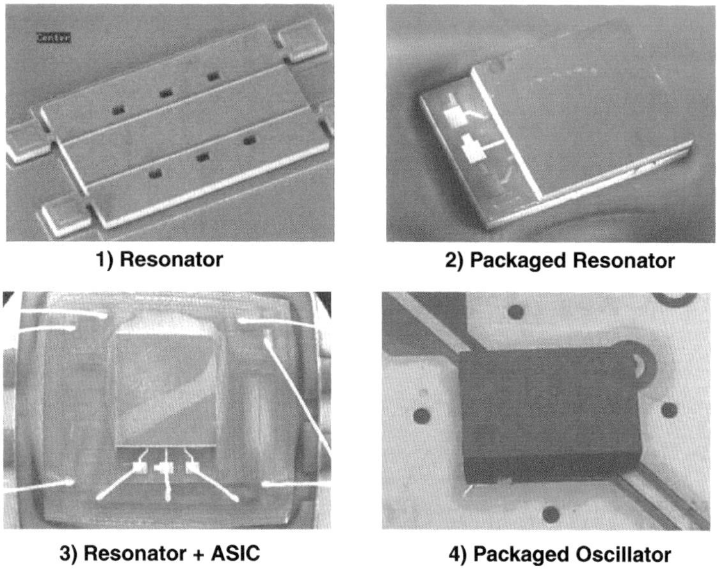

Fig. 1.15. MEMS oscillator fabricated with MEMS-last approach.

essential to the commercialization of MEMS oscillators that demands high volume manufacturing of extremely small and stable products.

Some of the benefits of monolithic packaging are:

- No seal ring takes up space around the MEMS device — which means more devices per unit chip area.
- No large packaging lid required — reduces parasitic coupling effects and no buried feed-through signal lines.
- Short interconnect distances between the MEMS device and electronic circuit.
- No alignment required for wafer bonding — standard photolithography processing only.
- Very clean process — which means that the MEMS devices are packaged and protected as soon as possible after the release process is completed.

1.3.2 *Packaging of MEMS at Wafer-Level Bonding*

Monolithic packaging introduced as above is an excellent approach to packaging MEMS devices. However, it is not applicable to complicated MEMS devices. Wafer-level capping using different bonding technologies is the primary approach for MEMS packaging. Three bonding technologies, glass-frit, anodic and metallic bonds, are to be introduced as follows.

1.3.2.1 *Glass-frit*

Glass-frit bonding is one of the most common technologies used in volume production to package MEMS devices (see Fig. 1.16). It is a universal bonding technique and can be used to bond several different surfaces and materials commonly used in microelectronic and MEMS. The principle behind glass-frit bonding is to use a low-temperature melting point glass (between 400°C and 450°C) between two wafers that melt and wet the surfaces and follows any contour or surface roughness. As a result, a perfect sealing is achieved by the liquid glass, and when it cools down, a mechanically strong and hermetic bond frame is obtained. The process is very often done in vacuum conditions to package and seal off gyroscopes and other resonating MEMS devices. Some of the benefits of glass-frit bonding are:

- Low-cost screen printing of glass-frit bond frames on the wafer.
- No electrical potential is required for bonding — good for electrostatically sensitive MEMS devices.
- Metallic feed-through is permissible for contacting the MEMS device inside the package.
- CTE of glass-frit is well matched to silicon, which reduces stress in the package.
- Final bond is very reliable, mechanically strong and hermetic.

The challenge with glass-frit packaging is that the area used on the wafer for sealing is relatively large. Typical screen printed bonding frames vary between 250 μm and 500 μm in width. The thickness after bonding is

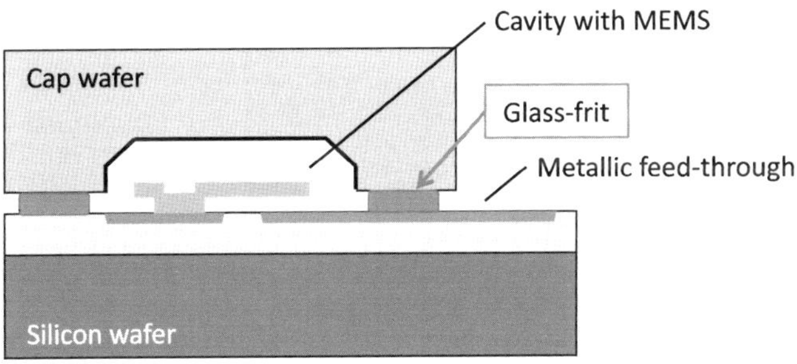

Fig. 1.16. Illustration of glass-frit used to package a MEMS device. The glass-frit seal can cover the signal lines going into the cavity.

usually 5 to 15 μm, and would normally not be used to define functional gaps in micro-devices [14].

1.3.2.2 *Anodic*

For MEMS devices which require interaction with a media, such as pressure sensors, the main challenge to overcome is to protect the moving and sensitive elements of the sensor, while at the same time permitting the media to be in contact with the active parts of the sensor. A typical example of this is a pressure sensor where a diaphragm, which is manufactured by micro-machining, must be in contact with the medium for pressure sensing. To enable this, Sensonor developed a highly successful pressure sensor using triple-stack wafer-level bonding with extremely good long term properties [15]. The device is manufactured by wafer-level bonding of a bulk micro-machined silicon sandwiched between two glass wafers. This forms a so-called triple-stack, as illustrated in Fig. 1.17. The sandwich design provides excellent media compatibility since the inlet is located on the backside of the sensor. The piezo resistors will therefore not come in contact with the measurement media. Anodic bonding of silicon to glass, which is very strong and hermetic, does not permit the electrical connections from the sensing element to run along the surface of the MEMS substrate. Sensonor

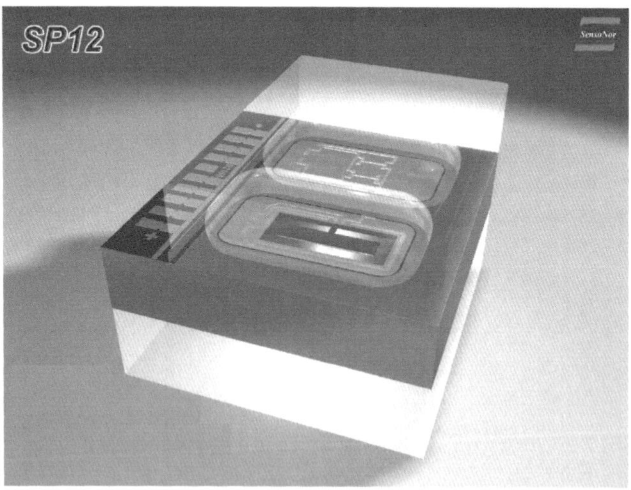

Fig. 1.17. Illustration of the triple-stack design. The bulk micro-machined silicon and the sensing elements are sandwiched between two micro-machined glass wafers.

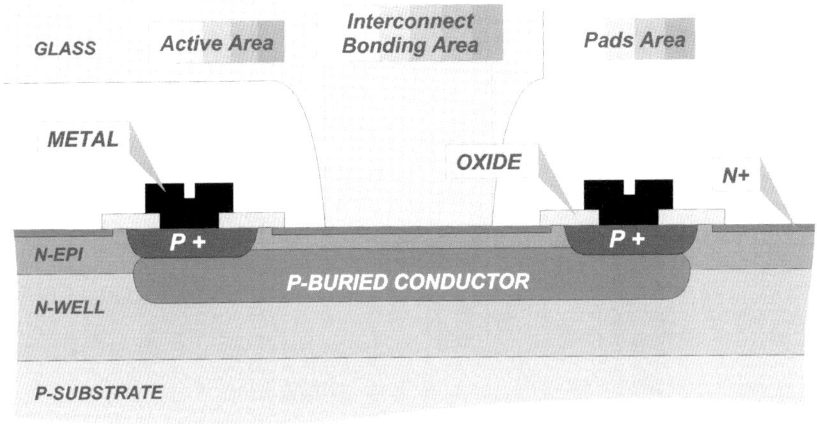

Fig. 1.18. Illustration of the buried conductor under a layer of epitaxial silicon. This conductor allows the top silicon layer to be anodically bonded to the cap glass wafer to create a very reliable and hermetic seal.

solved this challenge by integrating buried conductors created by diffusion under a layer of epitaxial silicon, as illustrated in Fig. 1.18. The glass wafer on top is micro-machined to create a cavity above the pressure sensor while the bottom glass wafer is anistropically etched from both sides to create the opening to the media to be sensed.

With anodic bonding and a triple-stack, very sophisticated MEMS devices can be created. One example is the Sensonor SW412 where an uncompensated piezoresistive pressure and acceleration sensor die is combined into one device as shown in Fig. 1.19. The design and performance of SW412 makes it ideal for high accuracy measurements, also in harsh environments. The long-term stability is outstanding and has been proven in applications during a period for more than 10 years. The pads along the one side of the die permits the sensor die to be connected to passive compensation and/or signal conditioning as required for a given application.

1.3.2.3 *Metallic*

Using metallic bonding is appealing for MEMS packaging since the bond frames, or seals, can be significantly smaller than for glass-frit or anodic bonding. This means that more devices can be fabricated on the wafer and thus reducing the overall cost. Metallic bonds are hermetic and temperature

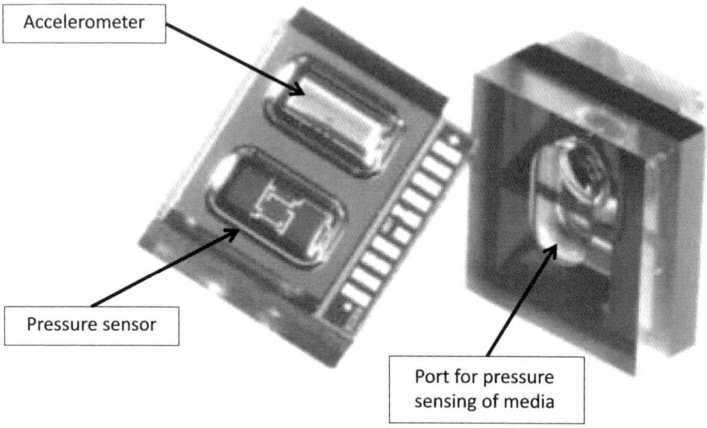

Fig. 1.19. Triple-stack anodically bonded MEMS device with a piezoresistive pressure sensor and acceleration sensor combined into one package. The metallic bond pads along the one of the edges are used to connect the die depending on the given application (Courtesy of Sensonor AS).

stable without significant outgassing into the package. Furthermore, additional processing of the device can be carried out without affecting the metallic seal, and subsequent stacking or bonding of the packaged dies will enable 3D integration. Among the more common metal bonding scenarios demonstrated are Au–Sn, Au–Au, Cu–Sn and more recently Au–In. In addition, Cu–Cu bonding and similar direct metal bonding have been demonstrated for 3D–IC applications, while eutectic and solid liquid interdiffusion (SLID) bonding is becoming an attractive approach for MEMS encapsulation and packaging. The benefits of SLID bonding are that one can use electroplated metals with high bond strength, and lower processing temperatures compared to traditional encapsulation techniques, such as glass-frit, direct metal (thermocompression), and anodic bonding [16]. SLID bonding is based on rapid formation of intermetallic compounds (IMCs) between two metal components, one metal with high melting temperature (Au, Cu, Ag) and one metal with low melting temperature (In, Sn). The bond is formed at low temperature, just above the lowest melting point. At the bond temperature, the low melting component melts, and IMCs solidify isothermally. The reaction is brought to thermal equilibrium, and the resulting bond-line consists only of the high-melting metal and IMCs with elevated melting temperatures. This allows the joint to survive repeated processing at the same process temperature without re-melting.

Material used in bonding process	Bonding temperatures used [°C]	Re-melting temperature of intermetallic [°C]
Au–Sn	280–350	278 (eutectic)
Ni–Sn	300	>400
Cu–Sn	260–300	>415
Ag–Sn	250	>600
Au–In	>156	>495

Source: [17–20]

Aside from noble metals (Au), metal surfaces oxidize at various rates. Removing the oxide from the metal surface of the bond frame using reducing gases or other chemicals is not compatible for wafers which contain released and fragile MEMS devices. This is one reason why Au–Au thermocompression bonding is so appealing for packaging MEMS devices. For SLID bonding relying on Sn with Cu, one can either cap the Sn with a thin layer of Au, or cap the Cu with thicker layer of Sn in a so-called symmetric bonding scheme, as shown in Fig. 1.20. In the latter case, the Sn layers

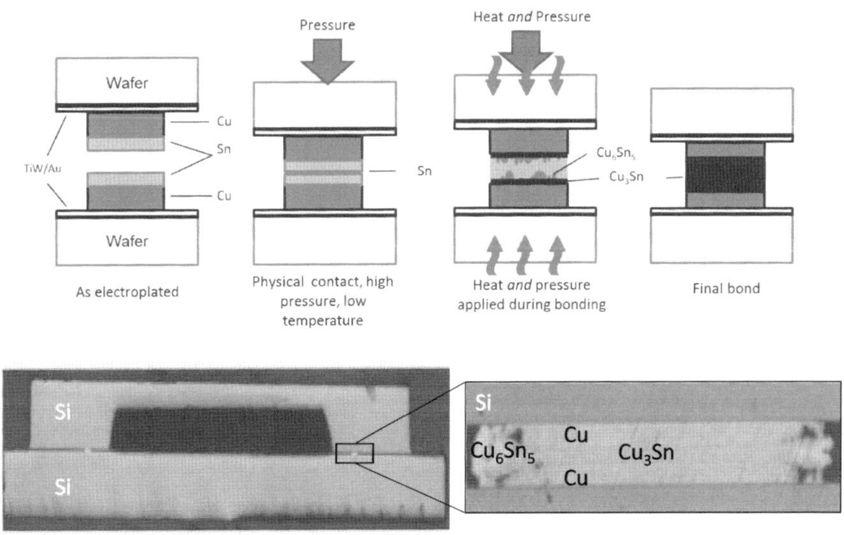

Fig. 1.20. Top: Illustration of symmetric Cu–Sn bonding used to create a hermetic seal for wafer-level bonding. Below: Photo of a wafer-level assembled package with an 80 μm wide Cu–Sn bond frame highlighted.

must be sufficiently thick to both prohibit oxidation of the Cu surface and to form a pure Sn–Sn intimate interface when the wafers are brought into contact. This minimizes oxidation and makes flux-less bonding possible that is critical for MEMS applications [21–23].

1.3.3 *3D Integration*

3D integration with the use of TSVs and flip-chip assembly is an emerging technology for microelectronics systems [24–26]. 3D integration of MEMS is more challenging than 3D integration for microelectronics mostly due to the mechanical restrictions working with a sensing or moving device. MEMS devices typically have specific requirements regarding mechanical issues like stiffness, volume, and mass in addition to temperature limitations. One of the more critical challenges is stacking in the third dimension since the underlying layers or entities have to sustain both subsequent high temperatures and sometimes experience forces from bonding tools, etc. Furthermore, MEMS devices can be more fragile with special handling requirements that may require special tooling in terms of touching and holding MEMS dies.

For most 3D MEMS integration applications, the goal is to combine both a processor and logic (ASIC) with memory and MEMS into one final device. Adding new functionalities to existing technology platforms, such as motion sensing is important for many consumer products. This requires some unique features such as TSVs and a cleverly integrated and adaptable interposer. Wafer-level bonding for 3D integration of MEMS is challenging since various technologies are fabricated on wafers of different size. MEMS devices are often fabricated on six-inch wafers, whereas memory and controllers are fabricated on 12-inch wafers, which make wafer-bonding challenging.

1.3.3.1 *TSV technology*

Over the past few years, TSV technology initially developed for 3D IC has made an entry to the MEMS world. By creating conductive vias in the silicon wafer, one can directly contact the MEMS devices inside a package without buried or feed-through conductors on the surface of the die. It also eliminates wire bonding of MEMS devices. Figure 1.21 illustrates two approaches where a MEMS device is directly bonded and packaged to an ASIC or sealed using a conventional cap wafer. Note that in the illustration shown, both the interconnects and the bond frame are fabricated from the

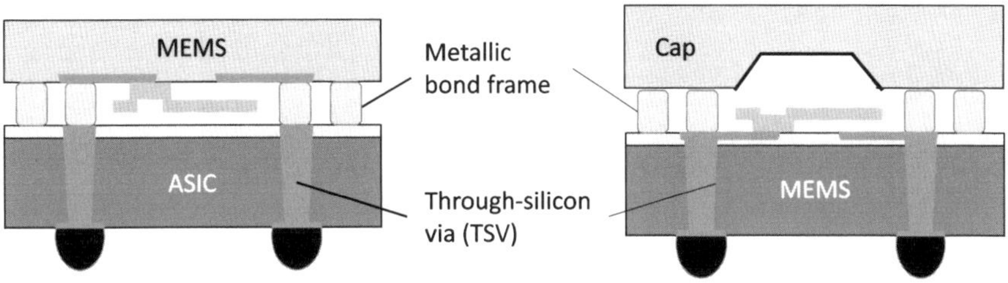

Fig. 1.21. Illustration of TSV in combination with wafer-level packaging (WLP) of MEMS to ASIC or more conventionally packaged MEMS device. The TSVs enable direct contact between the ASIC and MEMS, while also reducing the overall footprint of the packaged device. Also, no wire bonding is required to connect to the pads along the edge of the die.

same material, this is one reason why metal bonding is so appealing for MEMS devices. Initially the TSVs were tapered, but dry resist technology and sophisticated vertical via etching technology has enabled the vias to be more vertical. This allows a larger I/O count and smaller pitch between the vias. This is one reason why imaging sensors quickly adapted the TSV technology. In either case, electrical connections to substrates or other devices is obtained using bumps on the backside of the wafer.

STMicroelectronics was one of the first companies to use TSV in full volume production for their MEMS sensors. The use of TSV reduced the footprint and overall size of the package by more than 20%. The MEMS sensor is later mounted together with an ASIC die in a larger system-in-package that connects down to the PCB.

1.3.3.2 *Interposer technology*

The idea of an interposer is to create a substrate that interconnects many active and passive devices in one device. Together with TSV technology, the interposer will house both MEMS and active circuit elements, in a similar way to a system-in-package. The interposer technology is similar to a system-on-chip (SoC) or multi-chip module (MCM) in that it would attach directly to a PCB but differ in terms of including packaged and sealed MEMS devices, as illustrated in Fig. 1.22. There is no fundamental reason why the interposer may not be silicon technology, however CTE matching and a very mature technology platform reduces complication and overall cost. If miniaturization is important, silicon interposers are at an advantage since micro-machining techniques can create very small pitch TSV and interconnects.

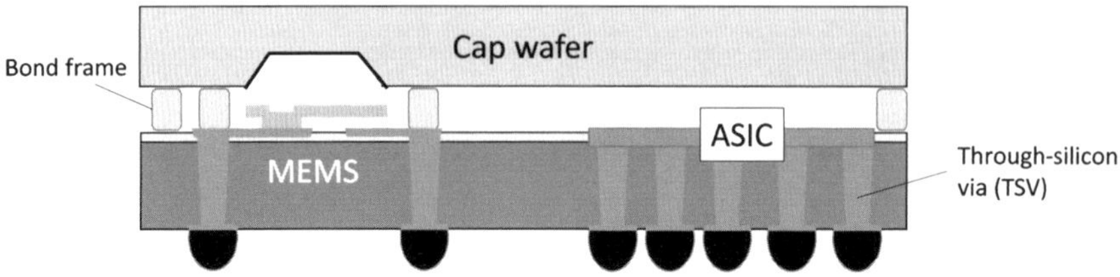

Fig. 1.22. Illustration of Si-based interposer where the MEMS devices is packaged and integrated directly with an ASIC circuit in one device. TSV creates electrical connection to the backside of the device.

A great example of a true 3D system is the e-CUBES project [27] where a heterogeneous stack of different technologies was combined into one cube, as shown in Fig. 1.23. Here two MEMS devices (pressure sensor and FBAR) are connected with two ASIC devices using TSV and Au-stud bumping and AuSn micro-bumps. The two ASICs were first bonded together using the AuSn micro-bumps. The two MEMS devices were flip-chip bonded to the ASIC stack using Au-stud bump bonding. The triple-stack assembled pressure sensor is connected to the ASIC using Si-based TSV in the bottom glass wafer.

1.4 Other Opportunities

MEMS devices are used in a wide spectrum of applications such as microelectronics, optoelectronics, RF, biomedical lab-on-a-chip, and micro-energy systems. Therefore, it is very important to review a few more examples to illustrate other opportunities, which are different from the configurations and processes reviewed as above. Specifically, we will present examples for the following: packaging technology to fabricate MEMS devices, MEMS devices to enhance packaging technology, and micro/nano systems.

1.4.1 *Packaging for MEMS*

Silicon processing is not the only means to fabricate MEMS devices. For example, we expect to see more and more MEMS devices to be fabricated using polymer materials such as flexible circuit boards. Ramadoss *et al.* [28, 29] demonstrated RF MEMS capacitive switches fabricated using PCB processes. Figure 1.24 presents the major components and processes with coplanar waveguide (CPW) lines for mounting switches and on-wafer multi-line TRL calibration patterned on the metallization layer

Fig. 1.23. Illustration and photo of assemble 3D system. Note the relative large size difference between the MEMS and electronic circuits. (Courtesy of SINTEF and e-CUBES partners)

of a Duroid substrate. Photosensitive benzocyclobutene (BCB) dielectric layer is spin-coated and patterned on CPW lines. All the layers are laminated. These switches can be integrated with other PCB-based RF circuits and antennas. There are other MEMS devices fabricated using the PCB technology for microfluidic, pressure sensor and other sensor and actuator applications.

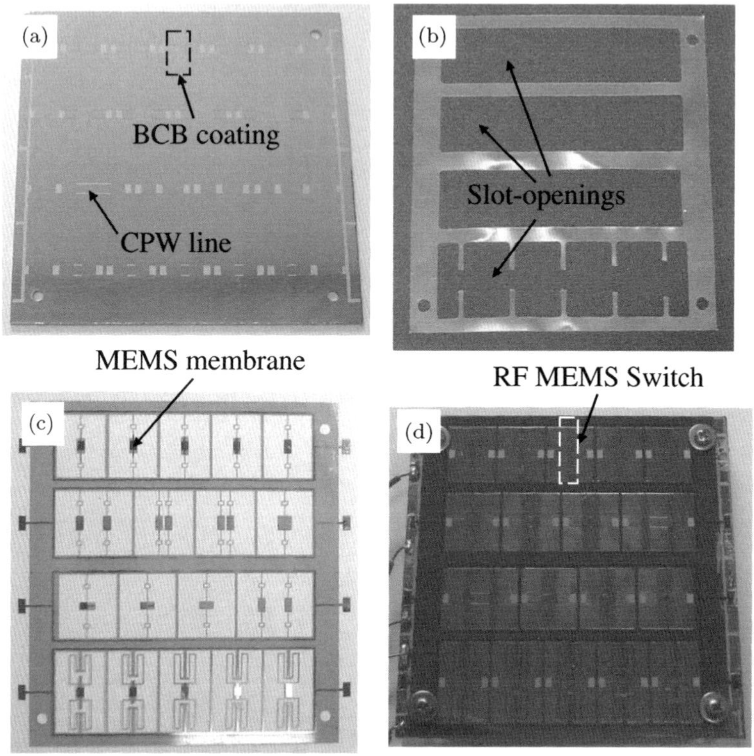

Fig. 1.24. RF MEMS capacitive switches fabricated using PCB processes: (a) CPW lines with patterned BCB dielectric layer on the Duroid substrate; (b) an adhesive spacer film with milled slot-openings; (c) Kapton-E polyimide film with patterned switch top electrode metallization and laser machined slot-openings; and, (d) an assembled prototype with multiple devices [28, 29].

1.4.2 *MEMS for Packaging*

TSV is reviewed in Section 1.3.3. 3D Integration is the most well-known MEMS technology applied to 3D stacking or packaging. In addition, we would like to present another novel approach that applies MEMS technology to fabricate co-axial cable-like RF circuits on PCBs (see Fig. 1.25). Recta-coax lines and components studied by Filipovic *et al.* [30] are fabricated using a new sequential micro-fabrication process. The recta-coax structures are built up layer-by-layer with uniform copper strata defined by photoresist. The copper thicknesses can be 10, 75, 75, or 50 μm. The inner conductor of the co-axial cable-like circuit is supported mechanically by using 15 μm thick polymer deposited and patterned. Once the structure is

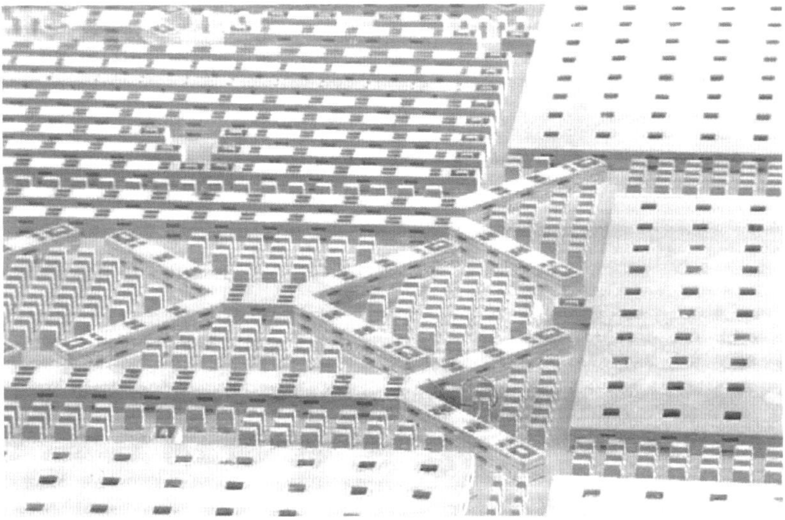

Fig. 1.25. SEM photograph of a recta-coax based cavity resonator (larger areas in the photo) and a number of coaxial lines with various separation distances ([30], © 2006 IEEE)

built, the resist is removed through release holes in the top and side walls. Figure 1.25 shows different recta-coax lines, calibration structures, test sets for isolation, and several different cavity resonators. With a proper design, the performance of a line built within the limits of the fabrication tolerances is almost the same as the performance of a perfectly fabricated structure. For example, Ehsan *et al.* [31] fabricate a 50 Ω line with outer conductor cross-section of $600\,\mu\text{m} \times 400\,\mu\text{m}$. The line handles 53W at 2:5 GHz with a 10% duty cycle. Integration with active components for a 1–10 GHz GaN MMIC amplifier shows minimal performance degradation relative to on-wafer measurements.

1.4.3 *Micro/Nano Systems*

Many novel micro/nano systems are demonstrated each year. Figure 1.26 illustrate an example for biomedical applications. This microfluidic device provides 300 parallel reverse transcription quantitative polymerase chain reaction (RT-qPCR) channels and executes all steps of single-cell capture, lysis, reverse transcription, and qPCR [32]. The device consists of a three-layer elastomeric structure with a blank bottom layer, a middle control layer with channels that act as valves by pushing up and pinching closed channels in the above flow layer. This and other similar devices/systems

Fig. 1.26. A microfluidic device with 300 parallel reverse transcription quantitative polymerase chain reaction (RT-qPCR) channels.

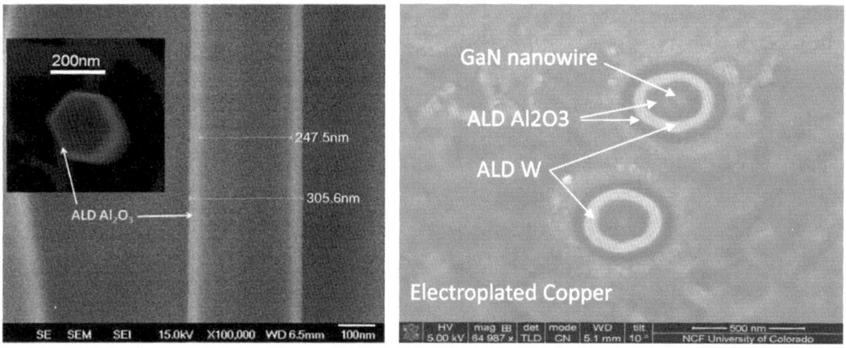

Fig. 1.27. GaN nanowires encapsulated by an ALD dielectric layer, e.g. Al_2O_3, and another ALD electrical conducting layer, e.g. W, for electrical and thermal interconnects. The electroplated copper is used to remove heat from the GaN nanowire devices with a negligible thermal interface resistance through the nanoscaled ALD Al_2O_3, ALD W and another ALD Al_2O_3 layers.

are fabricated and assembled by MEMS fabrication and hybrid system integration, i.e. packaging. The growth of microfluidic systems is increasing exponentially, and we expect to see many such systems to be integrated using MEMS and packaging technologies.

Figure 1.27 illustrates GaN nanowire-based nanoelectromechanical systems (NEMS) integrated by ALD layers [33]. One of the challenges for nanowire-based systems is how to interconnect these as-grown nanowires to form a device. With the ALD Al_2O_3 encapsulation, the semiconducting GaN nanowires are electrically insulated from the surrounding. Then, these nanowires are interconnected through an electrically conducting ALD W layer, which is further electrically insulated but thermally connected to electroplated copper by another ALD Al_2O_3 layer. The copper layer is an effective means to transfer heat to the next thermal management level. This packaging and interconnect technology illustrates a packaging concept for future MEMS/NEMS.

1.5 Summary

This chapter introduces MEMS packaging with an emphasis on its different configurations and processes developed to meet a unique challenge to package moving MEMS devices in a very compact size. The major configuration is a plastic package enabled by wafer-level capping. The major processes are monolithic packaging, wafer-level bonding and 3D integration. In addition, there are other solutions driven to meet requirements demanded by different applications in microelectronics, optoelectronics, RF, biomedical lab-on-a-chip, and micro-energy systems. These opportunities are represented by: a) packaging technology to fabricate MEMS devices, b) MEMS devices to enhance packaging technology, and c) micro/nano systems.

References

1. iPhone 7 plus teardown, https://www.ifixit.com/Teardown/iPhone+7+Plus+Teardown/67384, September 15, 2016.
2. iPhone 7 plus teardown (in Chinese), http://www.leiphone.com/news/201609/MQ7DuwAEqY6WmrGu.html, September 20, 2016.
3. Lee, Y. C., Kong, M., and Zhang, Y. (2017). Microelectromechanical Systems and Packaging, in Lu, D., and Wong, C. P. (eds.), *Materials for Advanced Packaging*, 2nd Edition, Springer, Switzerland, pp. 691–731.
4. Lee, Y. C. (2007). MEMS Packaging and Reliability, in Ephraim Suhir, Lee, Y. C., and Wong, C. P. (eds.), *Micro- and Opto-Electronic Materials and Structures: Physics, Mechanics, Design, Reliability, Packaging*, Vol. II, Springer, pp. 299–320.
5. Uzunlar E., Schwartz, J., Phillips, O., and Kohl, P.A. (2016). Decomposable and template polymers: fundamentals and applications. ASME. *Journal of Electronic Packaging* 138(2), pp. 020802–020802-15. doi:10.1115/1.4033000.
6. Fischer, A. C., Forsberg, F., Lapisa, M., Bleiker, S. J., Stemme, G., Roxhed, N., and Niklaus, F. (2015). Integrating MEMS and ICs, *Microsystems and Nanoengineering* 1, 15005.doi:10.1038/micronano.2015.5.

7. Morris, A. (2015). Monolithic integration of RF-MEMS within CMOS. *2015 International Symposium on VLSI Technology, Systems and Applications*, 27–29 April, Hsinchu, Taiwan.
8. Velten, T., Ruf, H. H, Barrow, D., Aspragathos, N., Lazarou P., Jung E., Malek, C. K., Richter, M., Kruckow, J., and Wackerle, M. (2005). Packaging of bio-MEMS: strategies, technologies, and applications, *IEEE Transactions on Advanced Packaging*, 28(4), pp. 533–546.
9. Tseng, A. A., Tang, W. C., Lee, Y. C., and Allen, J. (2001). NSF 2000 workshop on manufacturing of micro-electro-mechanical systems, *Journal of Materials Processing and Manufacturing Science*, 8(4), pp. 292–360.
10. Candler, R. N., Woo-Tae, P., Huimou, L., Yama, G., Partridge, A., Lutz, M., and Kenny, T. W. (2003). Single wafer encapsulation of MEMS devices, *Advanced Packaging, IEEE Transactions on*, 26, pp. 227–232.
11. Guo, B.,Wen, L., Helin, P., Claes, G., Verbist, A., Van Hoof, R., Du Bois, B., De Coster, J., De Wolf, I., Hadi Shahar, A., Li, Y., Cui, H., Lux, M., Vereecke, G., Tilmans, H. A. C., Haspeslagh, L., Decoutere, S., Osman, H., Puers, R., Severi, S., and Witvrouw, A. (2011). Above-IC generic poly-SiGe thin film wafer level packaging and MEM device technology: Application to accelerometers, in *Micro Electro Mechanical Systems (MEMS), 2011 IEEE 24th International Conference on*, pp. 352–355.
12. Kuan-Lin, C., Shasha, W., Salvia, J. C., Melamud, R., Howe, R. T., and Kenny, T. W. Wafer-level epitaxial silicon packaging for out-of-plane RF MEMS resonators with integrated actuation electrodes, *Components, Packaging and Manufacturing Technology, IEEE Transactions on*, 1, pp. 310–317.
13. Discera (2010). *MEMS Opportunities and Challenges*, Tom Willey, October 6. http://slideplayer.com/slide/7238152/.
14. Knechtel, R. (2012). Glass Frit Wafer Bonding, in Ramm, P., Lu, J. J.-Q., and Taklo, M. M. V. (eds.) *Handbook of Wafer Bonding*, Wiley-VCH Verlag GmbH & Co. KGaA, Weinheim, Germany, pp. 1–17.
15. Lapadatu, D. and Jakobsen, H. (1998). Building of silicon mechanical sensors by bulk micromachining and anodic bonding., In *CAS 1998 International Semiconductor Conference*, 21st edition, Vol. 1, Sinaia, Romania, pp. 33–44.
16. Hoivik, N. and Aasmundtveit, K. (2012). Wafer-level solid–liquid interdiffusion bonding, in Ramm, P., Lu, J. J.-Q., and Taklo, M. M. V. (eds.) *Handbook of Wafer Bonding*, Wiley-VCH Verlag GmbH & Co. KGaA, Weinheim, Germany, pp. 181–214.
17. Welch, W. C. and Najafi, K. (2008). Gold-indium Transient Liquid Phase (TLP) wafer bonding for MEMS vacuum packaging, in *Micro Electro Mechanical Systems(MEMS) 2008, IEEE 21st International Conference on*, Tucson, Arizona, pp. 806–809.
18. Farrens, S. and Sood, S. (2008). Wafer level packaging: balancing device requirements and material properties, in *IMAPS 41st International Symposium on Microelectronics*, Providence, RI - USA.
19. Yoon-Chul, S., Qian, W., Suk-Jin, H., Byung-Gil, J., Kyu-Dong, J., Min-Seog, C., Woon-Bae, K., and Chang-Youl, M. (2007). Wafer-level low temperature bonding with Au-In system, in *Electronic Components and Technology Conference*, Sparks, NV, USA, pp. 633–637.

20. Belov, N., Chou, T. K., Heck, J., Kornelsen, K., Spicer, D., Akhlaghi, S., Wang, M., and Zhu, T. (2009). Thin-layer Au-Sn solder bonding process for wafer-level packaging, electrical interconnections and MEMS applications, in *International Interconnect Technology Conference(IITC)*, Sapporo, Japan, pp. 128–130.
21. Hoivik, N., Kaiying, W., Aasmundtveit, K., Salomonsen, G., Lapadatu, A., Kittilsland, G., and Stark, B. (2010). Fluxless wafer-level Cu-Sn bonding for micro- and nanosystems packaging, in *Proceedings of the 3rd Electronic System-Integration Technology Conference (ESTC)*, Berlin, Germany, pp. 1–5.
22. Wang, K., Aasmundtveit, K., and Jakobsen, H. (2008). Surface evolution and bonding properties of electroplated Au/Sn/Au, in *Proceedings of the 2nd Electronics System-Integration Technology Conference (ESTC)*, Greenwich, UK, pp. 1131–1134.
23. Lapadatu, A., Simonsen, T. I., Kittilsland, G., Stark, B., Hoivik, N., Dalsrud, V., and Salomonsen, G. (2010). Cu-Sn wafer level bonding for vacuum encapsulation of microbolometer focal plane arrays, in *ECS Transactions*, pp. 73–82.
24. Lau, J. H. (2014). Overview and outlook of three-dimensional integrated circuit packaging, three-dimensional Si integration, and three-dimensional integrated circuit integration, ASME. *Journal of Electronic Packaging*, 136(4): 040801-040801-15. doi:10.1115/1.4028629.
25. Lau, J. H. (2016). Recent advances and new trends in flip chip technology, *ASME. Journal of Electronic Packaging* 138(3): 030802-030802-23. doi:10.1115/1.4034037.
26. Zhang, X., Han, X., Sarvey, T. E. *et al.* (2016). Three-dimensional integrated circuit with embedded microfluidic cooling: technology, thermal performance, and electrical implications, ASME. *Journal of Electronic Packaging* 138(1): 010910-010910-9. doi:10.1115/1.4032496.
27. Lietaer, N., Taklo, M., Schjolberg-Henriksen, K., and Ramm, P. (2010). 3D interconnect technologies for advanced MEMS/NEMS applications, *ECS Transactions*, 25, pp. 87–95.
28. Ramadoss, R., Lee, S., Lee, Y. C., Bright, V. M., and Gupta, K. C. (2006). RF MEMS capacitive switches fabricated using printed circuit processing techniques. *IEEE/ASME Journal of Microelectromechanical Systems*, pp. 1595–1604.
29. Ramadoss, R., Lee, S., Lee, Y. C., Bright, V. M., and Gupta, K. C. (2007). MEMS capacitive series switch fabricated using PCB technology. *International Journal of RF and Microwave Computer-Aided Engineering*, pp. 387–397.
30. Filipovic, D.S., Popovic, Z., Vanhille, K., Lukic, M., Rondineau, S., Buck, M., Potvin, G., Fontaine, D., Nichols, C., Sherrer, D., Zhou, S., Houck, W., Fleming, D., Daniel, E., Wilkins, W., Sokolov V., and Evans, T. (2006). Modeling, design, fabrication, and performance of rectangular μ-coaxial lines and components, *IEEE MTT-S International Microwave, Symposium Digest*, pp. 1393–1396.
31. Ehsan, N., Cullens, E., Vanhille, K., Frey, D., Rondineau, S., Actis, R., Jessup, S., Lender, J. R., Immorlica, A., Nair, D., Filipović, D., and Popović,

Z. (2009). Micro-coaxial lines for active hybrid-monolithic circuits, *Microwave Symposium Digest, MTT '09. IEEE MTT-S International*, pp. 465–468.
32. White, A. K., VanInsberghe, M., Petriv, O. I., Hamidi, M., Sikorski, D., Marra, M. A., Piret, J., Aparicio, S., and Hansen, C. L. (2011). High-throughput microfluidic single-cell RT-qPCR, *Proceedings of the National Academy of Sciences (PNAS)* 108(34), pp. 13999–14004.
33. Cheng, J.-H., Seghete, D., Lee, M., Schlager, J. B., Bertness, K. A., Sanford, N. A., Yang, R., George, S. M., and Lee, Y. C. (2011). Atomic layer deposition enabled interconnect technology for vertical nanowire arrays, *Sensors and Actuators: A. Physical*, 165 pp. 107–114.

Y. C. Lee is the S. J. Archuleta Professor at the Department of Mechanical Engineering, University of Colorado Boulder, USA. He is the Editor of *ASME Journal of Electronic Packaging.* He is also the President and CEO of Kelvin Thermal Technologies, Inc. for the commercialization of flexible thermal ground planes. He was the Director of DARPA Center on Nanoscale Science and Technology for Integrated Micro/Nano-Electromechanical Transducers (iMINT) from 2006 to 2012. Prior to joining the University in 1989, he was a Member of Technical Staff at AT&T Bell Laboratories, Murray Hill, New Jersey. Dr. Lee is recognized as a leader in packaging and interconnect technologies for microsystems integrating microelectronic, optoelectronic, microwave, microelectromechanical, and nanoelectromechanical devices. He received ASME InterPACK Achievement Award in 2013.

Ramesh Ramadoss received his Bachelor of Engineering degree from Thiagarajar College of Engineering, Madurai Kamaraj University, India in May 1998 and PhD degree in Electrical Engineering from the University of Colorado at Boulder in May 2003. From Jun 2003 to Dec 2007, he was employed as an Assistant Professor in the Department of Electrical and Computer Engineering at Auburn University, Auburn, AL. From Jan 2008 to Mar 2012, he was employed as a Manager at FormFactor Inc., Livermore, CA.

From Apr 2012 to May 2015, he was employed as a Senior Manager at Microprobe Inc. (Acquired by FormFactor Inc.), San Jose, CA. From Jun 2015 to Feb 2016, he was employed as a Technical Director at Applied Nanostructures Inc., Mountain View, CA. Currently, he is the President & CEO of IoT Technologies Inc., San Ramon, CA. He is the author or coauthor of 4 book chapters and 55 papers. He has conducted R&D projects for DARPA, NASA, US Army, US Air Force, Sandia National Labs, and Motorola Labs. In 2013, he was elevated to the grade of IEEE Senior Member. In 2014, he founded the IEEE San Francisco Bay Area MEMS & Sensors Chapter. In 2017, he was elected as the Chair of the IEEE Santa Clara Valley Section, the world's largest IEEE Section located in the heart of Silicon Valley.

Nils Hoivik received his PhD degree from the University of Colorado at Boulder in 2002 on flip-chip assembled RF-MEMS devices. He worked at IBM T.J. Watson Research Center in NY, USA as the technical team leader for the RF-MEMS project from 2002–2007. Dr. Hoivik is a Professor at the University College of Southeast Norway. He acted as Head of the Department (IMST) from 2011 to 2014. His research interests include design, fabrication, packaging, modeling, and testing of micro and nano devices for RF, optical, energy, and environmental applications. He is currently Director of Technology at EnergyNest AS, a Norwegian company focusing on thermal energy storage technology.

Chapter 2

Silex's TSV Technology: Overview of Processes and MEMS Applications

Tomas Bauer and Thorbjörn Ebefors

2.1 TSV Overview

The first TSV was patented by William Shockley in 1962, although most people in the electronics industry consider Merlin Smith and Emanuel Stern of IBM the inventors of TSV, based on their patent "Methods of Making Thru-Connections in Semiconductor Wafers" filed on December 28, 1964 and granted on September 26, 1967. The original idea was that TSVs would offer significantly better performance than wire bonds. TSVs in 3D package integration appear to be a cost-effective approach to manage increasing memory needs and the shrinking share of logic area. Added benefits include fewer packages, improved device yield from chip partitioning and lower power consumption. TSVs also enable integration of additional technologies for heterogeneous integration of multiple chips or components such as analog, radio frequency (RF), sensors, microelectromechanical sensors (MEMS), and passives. In 2.5D packaging, passive silicon interposers can support integration of multiple technologies including passive components (i.e. resistors, inductors, and capacitors). Currently, 2.5D integration is a less disruptive process while maintaining the benefits of eliminating keep out areas for bond pads, not impacting the IC performance and using layout techniques similar to printed wire board (PWB) design. The term "through-silicon vias" has in recent years become "through substrate via", due to the proliferation of substrate materials beyond silicon that are being used in 3D IC integration. However, the terms are often used interchangeably and "TSV" can mean both "through-silicon via" and "through-substrate via". This chapter will refer to the latter term when using the abbreviation "TSV".

TSVs are an important developing technology that utilizes short (vertical electrical connections or "vias" that pass through a substrate) to establish an electrical connection from one side of the substrate to the other side. This provides the shortest interconnect path in 3D integration. TSV technology offers greater space efficiencies and higher interconnect densities than traditional technologies, such as wire bonding and flip-chip stacking. When combined with micro-bump bonding and advanced flip-chip technologies, TSVs enable a higher level of functional integration and performance in a smaller form factor. By one step further, TSVs can provide single die or monolithic integration. An example of this could be a MEMS pressure sensor device, where the analog sensor is located on one side of the die, and the CMOS-based sensor interface is on the opposite side. The TSVs could carry analog sensor signals to the interface circuit where these signals are converted to the digital domain. Monolithic integration provides another step towards die size minimization and reduced costs.

The use of TSVs for 3D packaging and die integration has its fair share of challenges with a range of associated failure modes. For instance, there are defects related to the bonding process that result in short circuit and open circuit failures. Defects in the TSV manufacturing process can result in pinholes and microvoids, which can affect the electrical performance of the vias, such as shorting to the carrier. The use of materials, with different thermal coefficients for vias and the substrate, can result in cracks in the TSV insulation that can cause reliability issues. Testing and identifying these failure modes can be challenging, especially when trying to detect small microvias that could potentially lead to cracks in the via and associated reliability issues. This requires very sensitive and accurate measurements of structures that can be difficult to access. A further complication is the need for high test parallelism, which is usually required for cost sensitive, high density via substrates. There are also test challenges associated with the assembly and packaging processes. The complexity in stacking multiple die into the same package, with stringent test requirements, challenges existing economical test cost models for minimizing test cost. For example, some devices might be suitable for testing during assembly (i.e. partial stack). For other devices, this would be cost prohibitive. Table 2.1 summarizes common metal-based TSV defect mechanisms.

Table 2.1. Selected metal-based TSV defects and failure modes.

Process Parameter	Defect Mechanism	Failure Mode
Bond quality	Oxidation or contamination of bond surface	Opens
Misalignment	Bond misalignment	Shorts or opens
Microvoids	Holes	Impedance faults in high speed paths
Pinholes	Pinholes in TSV oxide	Shorts between TSV and substrate
Process stress	Micro-fracture caused on thinned wafer during process	Device failures over time
Thermal	Mismatch of thermal expansion coefficients	Cu "pumping" Cracked liner Reliability issues
IV degradation	IV curves degradation due to wafer thinning and TSV induced strain	Shift in performance
Pinched TSV	Not filled correctly	Opens Reliability issues
Micro-bump co-planarity	Open circuit connections after bonding	Opens Reliability Issues

2.2 TSV Process Options

TSVs can generally be developed at three stages during the manufacturing process; via first (e.g. polysilicon filled via), via middle (e.g. copper filled via), and via last (e.g. copper liner). Via first are created early in the device manufacturing process and must be compatible with subsequent CMOS and/or MEMS processing steps. Since the via first TSVs are created before the fabrication of transistors, they are not limited by a CMOS compatible thermal budget. The materials used for via first typically withstand high temperatures while complying with the approved materials used in subsequent processing steps by the processing facilities and fabs. Overall, it is more cost effective to create vias upstream in the processing flow rather than downstream. The main reason is that a scrapped wafer at the beginning of the manufacturing process is only at nominal value.

Via middle are fabricated before copper interconnects and via last are fabricated after bonding the wafer stack. Both these vias will typically not be exposed to extreme temperature excursions. This will minimize the focus on thermal stress. Via locations must be determined at the design stage and account for process variations to ensure that they do not interfere with other circuitry. Via middle and via last can create supply-chain issues and the possibility of scrapping wafers late in the manufacturing process deems them riskier than via first. A scrapped wafer late in the manufacturing process accounts for the nominal value of the wafer with the addition of all value-added process steps. This includes materials, labor, processing time, and overhead.

2.2.1 *Via Formation*

The development of a TSV starts with the formation of a via hole. The process choices include deep reactive ion etching (DRIE), laser drilling, and wet etching. DRIE, which here refers to the "Bosch DRIE process", is a cyclic process that alternates between etching and sidewall passivation. The isotropic plasma etch removes substrate material in the vertical direction, and the plasma deposition of polymer passivation adds sidewall protection, preventing lateral etching. The highly reactive fluoride part of the plasma-etch results in very high and selective etching rates. DRIE allows tight control for forming high-aspect ratio structures and features in a substrate. It has also been proven to define full wafer thickness vias (200 μm to 600 μm), particularly for MEMS applications. For rigid interposers that need a tight via pitch, DRIE is arguably the best option. BGA substrates used in flip-chip packaging require vias with a pitch between 130 μm and 200 μm, with 200 μm being the most common. If the wafer substrate can be thinned, laser drilling and wet etching can be used for these type of applications, though wet etching usually cannot get close to the DRIE via pitches. Laser drilling can create vias with diameters ranging from 10 μm to 80 μm, typically with diameters of less than 50 μm. There are tradeoffs between the technologies used to form the vias: cost (i.e. throughput) and quality (i.e. damage that can be tolerated by the substrate). For example, excessive heat from the laser drilling can damage the surrounding substrate areas, such as crystalline silicon in single crystal silicon wafers. This can be mitigated by shorter laser pulses that will increase the processing time and cost. The costs associated with laser drilling (versus DRIE) are higher when you get above a certain threshold; for example, 200,000 vias per wafer.

Silicon has traditionally been used in TSV applications. Glass is a widely known alternative, but the industry is constantly reviewing other materials, including polymers. Common glass is a low-cost material that has properties making it a very intriguing material for TSV applications. Glass is an insulator and has less capacitance than silicon. This makes it favorable for RF signal applications. The term "glass" refers to a broad set of materials with configurable properties based on the material composition. For example, a glass wafer composite can be tailored to specific mechanical and thermal property requirements. The process for forming vias and other microstructures in silicon do not directly translate to glass wafer micro-machining. For example, the most common glass composition, SiO_2, has a very strong silicon-to-oxide bonding. This bond makes it difficult to create fine structures and high aspect ratio features. There has been significant development in glass micro-machining that eliminates thinning and polishing of the wafer, and enables batch wafer processing. Micromachining methods for glass include powder blasting (i.e. sand blasting), laser machining, wet etching, ultrasonic, DRIE and magnetic neutral loop discharge (NLD) plasma etching.

2.2.2 *Through Silicon Insulator*

Through silicon insulator or TSI™ is Silex's proprietary technology. TSI™ refers to the embedding of vertical insulation features within a wafer, isolating a conductive via within the original, highly doped and conductive silicon starting material. An insulation material is deposited inside an etched trench that is defining the perimeter of the through wafer connection, thereby isolating the conductive part of the via from the rest of the conductive substrate. This patented TSI™ approach of isolating a conductive region through the wafer, is practically the inverse of the more intuitive approach of making a hole in the substrate and filling it with a conductive material. The TSI™ technology does not incorporate any metal and can withstand process temperatures up to 1100°C. This makes it ideal for MEMS designs where subsequent high temperature fusion wafer bonding allows creation of high vacuum devices (without getter), with electrical connections reaching into the cavities. TSVs that are created using the TSI™ technology are called Sil-Via® (see Fig. 2.1). These vias can be incorporated into the handle wafer of a silicon on insulator (SOI) wafer. This allows the combination of a highly doped handle wafer (for low TSV resistance) and a low doped device layer.

The TSI™ technology can be used to create isolating features for eliminating cross talk between electrical signals within a substrate, or as a divider between analog and digital signal lines in the same device. In many cases, the system will be sensitive to signal contamination between the analog and digital side of the system. To reduce signal contamination and crosstalk, the insulator technology can create vertical walls, separating certain regions of an IC chip or a MEMS die. Even pure ICs with mixed signal processing

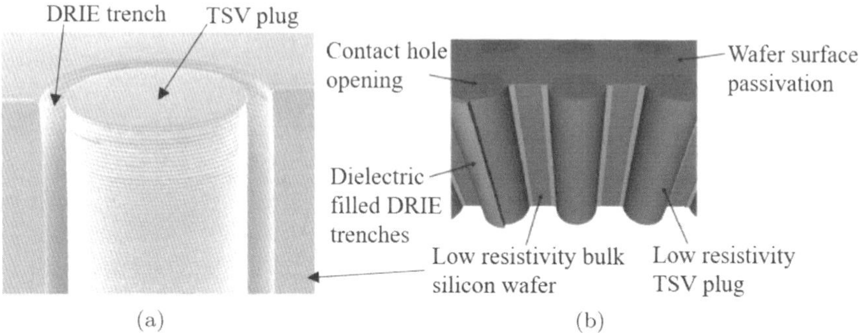

Fig. 2.1. Silex's TSI™ technology: (a) SEM picture of a TSV "plug", formed using DRIE in low resistivity bulk silicon wafer, and (b) cross section of Sil-Via® TSV wafer with dielectric filled DRIE trenches.

will benefit from this technology, as the isolating walls can be incorporated within the die layout. When the TSI™ technology is configured to eliminate signal contamination, it is called Zero-Crosstalk™.

The TSV isolation discussed in this section is primarily a via first process that requires via filling at high temperatures. The via processing is completed at the start of the wafer processing and before MEMS features are formed. This eliminates post-processing and associated wafer handling of delicate MEMS wafers to create via-last vias. The Sil-Via® substrate wafer resembles the SOI wafer in the sense that it is a wafer with buried isolator features inside the wafer and it allows more or less unlimited processing of the substrate. It also means that the via substrate wafer will enter the MEMS device manufacturing flow as part of the starting material.

2.2.3 *Via Fill*

There are a wide range of materials used as via fill material, including copper, tungsten, doped and un-doped polysilicon. Copper and gold are the most common metals used, for their ability to be seeded and plated, and their high current-carrying capabilities. The via material selection depends on the application, but also on where the vias are filled during the wafer processing. For example, via first must often have the ability to withstand high temperatures later in the manufacturing process. This can be done by etching a highly doped silicon substrate and filling isolating trenches around the conductive silicon feed-throughs (Sil-Via®). Another via first method uses doped polysilicon. While it has higher resistance than metals, doped poly or single crystal silicon vias offer better thermal matching to the

Table 2.2. Via fill materials versus applications.

Fill Material	Pros	Cons	"Best Fit" Applications
Single crystal via-first architecture (Sil-Via®)	• Single crystal, high repeatability. • Excellent thermal matching to substrate. • Supports rigid substrate for easier handling.	• Trench defined, so limits to via pitch at full wafer thickness. • Capacitance depends on trench width and via length.	• Sub-megahertz signals, power, ground. • Applications needing integrated blocking diodes.
Polysilicon Via	• Very good thermal matching and reliability. • Can support thick wafers.	• Poly fill at thick wafers is expensive. • Via matching dependent on highly accurate process controls. • Voids in via likely for high aspect ratio/narrow vias.	• Sub-megahertz signals, power, ground. • Thin wafer applications.
Metal Via	• Low resistance and very good thermal conductivity. • Via last technology for thin wafer and CMOS backside via preparation.	• Thermal mismatch between plating and surrounding substrate. • Metal fill for thick wafers very expensive.	• High frequency and RF applications, as well as power applications needing low via resistance.

surrounding substrate and processing compatibility with post-via processes. These are superior via technologies for non-RF, high speed input/output (IO), power and ground bussing, and control logic applications. Table 2.2 below lists fill materials and the associated pros, cons and "best-fit" applications.

Each via fill material and technology has its own set of challenges. Typical issues with metal based vias are the thermal mismatches with silicon that can cause micro-cracks and reliability issues. Thermally induced stresses due to thermal mismatch may lead to cracking of the silicon wafer. Furthermore, a metal via must be incorporated towards the end of a MEMS process, as the thermal budget does not allow high temperature post processing. Metal based via fabrication techniques, in particular, may induce tensile or compressive stresses in the substrate. This could lead to severe buckling, which in extreme cases can cause wafers to look like a "potato chip". Every device design must consider the mechanical stress induced by via structures. Other process challenges specific to metal-based TSVs include metal voiding during filling, uniform via wall material deposition and wafer handling.

2.2.4 *Via Fill Material*

Met-Via® full wafer thickness TSV is an all-metal via solution for interposers, MEMS substrates, and Through-Mold Vias (TMVs). This is robust and cost effective approach to 2.5D and 3D packaging. With substrate thicknesses up to 300 μm and 400 μm, the need for handling thin substrates and special bonding/de-bonding procedures are eliminated. The Met-Via® technology uses void free conformal metal plating to partially fill the via hole. Short plating times and partial via fill help minimizing the thermal mismatch between the via and the substrate. Both Sil-Via® TSV and Met-Via® TSV solutions support edge-to-edge via density at any via diameter. This is critical for high power, multi-via configurations, as well as shielded Met-Via® structures for RF applications. A 3D inductor based on the Met-Via® technology is displayed in Fig. 2.2. Table 2.3 below contains a comparison table between the Sil-Via® TSV and Met-Via® TSVs.

2.3 TSV MEMS Applications

In MEMS applications, TSVs are either part of the substrate on which the MEMS device is built, or part of the wafer cap. MEMS TSVs can be up to 600 μm thick with the typical standard thickness being around

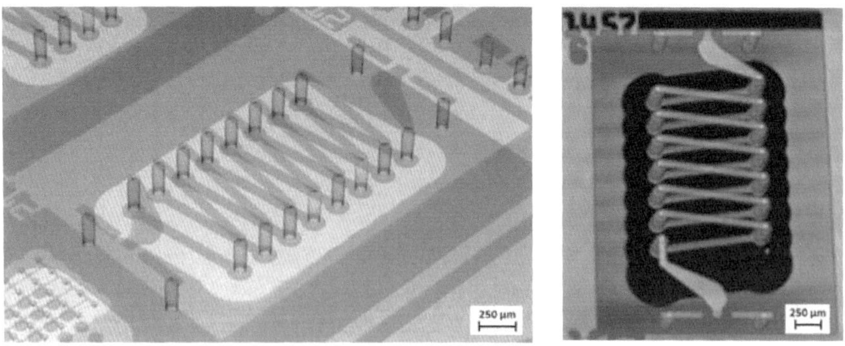

(a) (b)

Fig. 2.2. Met-Via® 3D inductors: (a) Oblique view with high magnification (OVHM) X-ray image, and (b) 3D computed tomography (CT) X-ray image.

Table 2.3. Sil-Via® vs Met-Via® comparison table.

	Sil-Via® (Highly doped Si substrate)	**Met-Via® (Cu plated via)**
Years in production	>8 years	>2 years, MEMS Capping
Via size	100 μm standard (lower limit is 50 μm diameter)	300 μm is standard 200 μm and 90 μm are available
Wafer thickness	430 μm is standard (lower limit is 350 μm)	300 μm to 400 μm
Reliability	Zero field failures in 8-year history	85°/85% RH and UHAST complete; will go through full mil/aero testing
Resistivity	Depends on geometry Typically <1 Ω per via	<20 mΩ per via
Thermal	Perfect matching	Hollow via construction buffers thermal coefficient (TC) differentials
Cost	Roughly equivalent to SOI wafer preparation	Dependent on via density, RDL requirements
Availability	6" and 8" wafers	6" and 8" wafers

400 μm. The substrate can in some applications be thinned down after wafer bonding to 100–150 μm thickness, as the bonded wafer or cap provides mechanical rigidity to the overall structure. The minimum required thickness for providing mechanical stability to an assembly stack is 300 μm. It does not matter if that thickness is made up of MEMS wafer or capping wafer. Full thickness wafer TSVs can be processed at sub 50 μm pitch for through wafer connections in up to 400 μm thick substrates. TSV technology enables integration of advanced interconnect functions in MEMS, sensors and microfluidic devices.

2.3.1 *MEMS Microphone*

The use of TSVs for MEMS microphones dates back to 2003 when Silex developed through full wafer thickness silicon via. The first implementation in MEMS microphones was in a 2.5D interposer substrate produced for a leading cellphone manufacturer. Volume production started in 2006, ramping up to 2,000 6" wafers per month. This patented TSV technology offered by Silex under the name Sil-Via® has been in volume production for more than 8 years. Figure 2.3 below displays a picture of one of the interposer vias along with a picture of the final microphone die assembly.

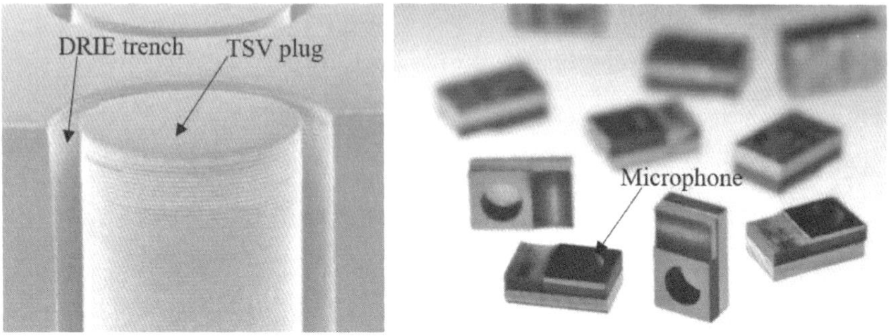

(a) (b)

Fig. 2.3. TSV applications in MEMS Microphones: (a) DRIE silicon vias, and (b) MEMS microphone die.

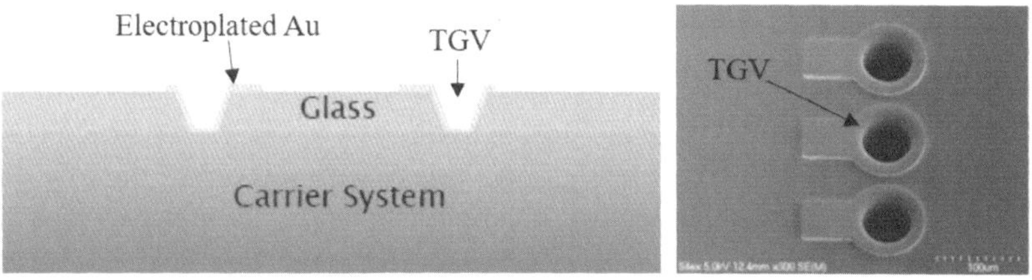

(a) (b)

Fig. 2.4. TSV applications in RF MEMS: (a) metal vias in glass, and (b) electroplated 50 μm TGVs.

2.3.2 *RF Device*

Through glass via (TGV) is a highly anticipated technology for RF devices. Glass has a low dielectric constant and provides great isolation for highly conductive via materials. This results in very low losses for high frequencies, which is desirable for RF applications. RF MEMS include a range of components such as resonators, oscillators, tunable inductors, switched capacitors, varactors, and a variety of switches. Through extensive process development and testing, Silex has developed a TGV structure for RF applications with very low resistance ($<20\,\text{m}\Omega$) and very low insertion loss (<0.1 dB @ 2.7 GHz). The test vehicle was a Borofloat glass wafer with a thickness of 75 μm. This wafer was temporarily bonded to a carrier. Vias were etched into the glass using dry and wet etch processes, followed by electroplating Au at $7 \pm 1\,\mu$m. (see Fig. 2.4).

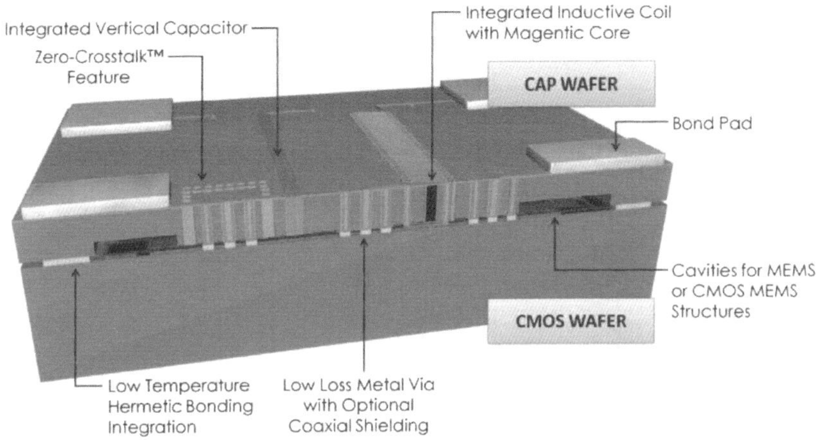

Fig. 2.5. Functional metal via capping process.

2.3.3 *Metal Via Capping*

Metal via capping is a method where metal vias are integrated into a cap wafer along with a cavity and potentially other functions, such as passive devices (resistors, capacitors, and inductors). The development of a standardized capping process platform minimizes cost and risk for new designs. The processing platform reduces the need for unique process development for specific device designs. There is a significant reduction in development cost, risk and time to market. Silex' partially filled Met-Via® approach solves the inherent reliability issues related to thermal mismatch between metals and standard wafer substrates. This enables the use of thick wafer metal vias. A specific Au based solution has proven very low resistance ($<25\,\mathrm{m}\Omega$ per via) in $300\,\mu$m thick wafer substrates. With wafer level capping becoming a major trend in today's market, the unique approach of hosting a robust 3D interconnect functionality in the required cap wafer, enables fast adoption into a wide range of applications. Figure 2.5 displays the concept of a metal via capping wafer that includes TSVs, a device cavity and passives (i.e. capacitor and inductive coil).

2.4 Summary

The future of TSV technologies looks bright for semiconductor, MEMS and sensor applications. TSVs are a high-performance alternative to printed interconnects and wire bonding in 2.5D and 3D packaging of integrated circuits. Silicon interposers with metal TSVs can achieve five to ten times the density of vias produced by conventional printed circuit board (PCB) techniques. TSV is an enabling technology for WLCSP and single die, monolithic integration. Monolithic solutions are especially attractive in MEMS solutions, where the signal conditioner (i.e. appli ation specific integrated circuit (ASIC)) can be located on one side of a die and the sensor on the other side. This minimizes the overall package size, which can be very attractive in high-volume markets that are highly cost and size sensitive. Since TSVs provide the shortest connection between two vertically stacked die, it is one of the leading technologies for 3D chip interconnections of memory, processor and CMOS image sensors. Shorter and direct signal lines help reduce latency and enables power optimization through discretely choosing fab process nodes. This chapter has presented a range of TSV technologies, from more traditional metal vias on silicon or glass to Silex's proprietary technologies such as Sil-Via® and Met-Via®. The field of TSV technologies is expanding and the next frontier is integrating TSVs directly into CMOS for true heterogeneous integration.

References

1. EE Times, 3D TSV Test: ATE challenges and potential solutions, 2011.
2. MEMS Journal (2012). Through Substrate Vias: Opportunities for CMOS MEMS and 3D Integration.
3. MEMS Journal (2015). Through Substrate Vias — Overview of Processes and Applications.

Tomas Bauer has for more than a decade been in charge of the Global Business Development and Foundry Strategy of Silex Microsystems; helping Silex to grow to a top position in its market space. With an eagerness for understanding markets and customer needs, he has deployed one of the more successful business development strategies in the MEMS Foundry space as well as contributed to several patents in the field of 3D integration. With Silex being one of the leading players in its field, his work at Silex has played a contributing role in shaping the global MEMS Foundry market evolution.

Thorbjörn "TOBY" (TOBY) Ebefors is a Co-founder of Silex Microsystems and served in various roles including Director of MEMS CAD & Mask Design (May 2005–Oct 2010), Director of Purchasing (Sep 2010–Jan 2011) and Chief Technologist (Jan 2011–Jan 2017). Since Feb 2017, he has been with Spinverse as the Managing Director. He is skilled in MEMS, Thin Films, Microfabrication, Semiconductors, and Materials Science. He is an experienced entrepreneur with a demonstrated history in the semiconductor industry. He is also a Ski Instructor focused on Alpine Skiing at Friluftsfrämjandet (Swedish Outdoor Association), Stockholm, Sweden.

Chapter 3

Vertical Interconnects for High-End MEMS

Maaike M. Visser Taklo and Sigurd Moe

3.1 Introduction

To interconnect microelectromechanical systems (MEMS) was previously not at all a concern for MEMS providers. The only aspect to consider was whether the customer preferred aluminum or gold sputtered contact pads, and if the backside of the MEMS should be metallized or not for any reason. Typically, a passivation layer was added and opened in regions of contact pads, the wafer was diced and shipped. Apart from some risk related to uneven quality of the sputtered metal layers or incomplete removal of passivation from small contact holes, the MEMS provider could trust on the system integrator to perform die-attachment and wire bonding and thereby ensure successful interconnection. However, lately, there is a demand for MEMS to be made smaller, stacked and 3D-integrated with electronics to reduce footprint. The mobile electronics industry has pushed the development to fast forward, but the benefits of reduced footprint have also become evident for high-end applications. At SINTEF, through silicon or glass interconnects have replaced wire bonding and thereby enabled the required miniaturization in products like pressure sensors, inertial sensors, and interposers.

Why do we see such a plethora of solutions for vertical interconnects in MEMS? The dry etching of silicon based on the Bosch process was adopted early by MEMS providers to form more advanced mechanical structures than what was possible using wet etching. Thus, it was a natural choice to use dry etching for vertical interconnects. Once the interconnection became an integral part of the process flow rather than a free-standing post process, the diversity of MEMS resulted in a corresponding variation of via solutions. The diversity of MEMS processing is still one of the keys for differentiation in this market although efforts towards standardization are in progress (e.g. the M&NEMS activities by Leti/Tronics [1]). A certain level of uniqueness is expected for MEMS in the years to come, although more appropriate for small volume niche products. When niche products are fabricated in labs without a dedicated process line, the selection of via solution will typically be a compromise between cost, material and tool availability, and risks of cross-contamination within the lab. For example, Cu, which is probably the most commonly used material for filling of through silicon vias (TSVs) in the world of 3D-IC, is not an acceptable material inside our SINTEF MiNaLab. The fast diffusion of Cu at room temperature and its ability to form charge-trapping centers are harmful to radiation sensors manufactured in our lab. This is just one example of how specific lab conditions can have an influence on the selection of via solution for MEMS fabricated in multipurpose labs. Another aspect that could be specific for a particular MEMS device is tight specification on mechanical stiffness and thus the thickness of the MEMS die/chip in order to ensure a certain performance, as we discussed in [2]. This will require etching of deep vias, where high-aspect ratio will be beneficial to reduce the pitch. Etching of TSVs with high-aspect ratio can also be an approach for bypassing lack of processes like chemical mechanical polishing (CMP) and thin-wafer handling in a lab. Typically, thin-wafer handling involves solutions for temporary bonding and de-bonding processes. In the following sections, examples of through silicon and glass vias for some high-end applications will be discussed.

3.2 Through Glass Vias (TGVs)

As discussed in [3], a vertical interconnect for MEMS can be integrated in the device wafer, or in the top or bottom cap wafer, depending on the specific design. In this section, two European project examples of TGVs for MEMS pressure sensor applications are discussed: the e-CUBES project (see Fig. 3.1), and the HISVESTA project (see Fig. 3.2). In the e-CUBES project [4], vertical interconnect vias were required in the top cap wafer for 3D-integration of transceiver and microcontroller dies. The obvious reason was the need for the pressure inlet in the bottom cap wafer to face the outside environment. Glass wafers with silicon vias were finally selected, in combination with anodic bonding for wafer-level encapsulation and Au stud bump bonding for chip-to-wafer 3D integration [5]. The compound glass-silicon wafer was designed by SINTEF and prepared by Planoptik [6].

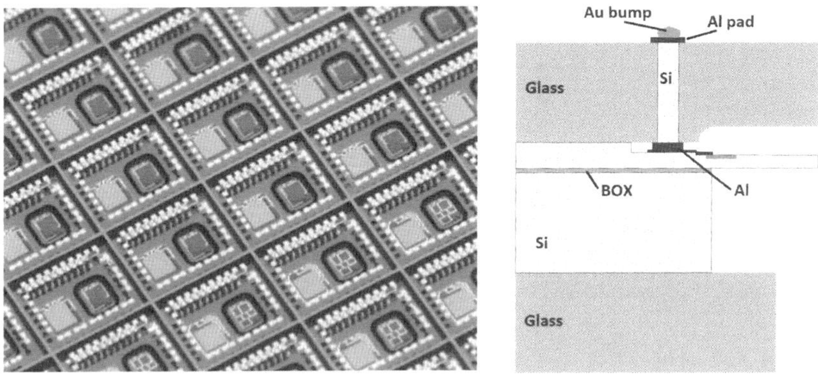

(a) (b)

Fig. 3.1. TGVs for MEMS pressure sensors from the e-CUBES project: (a) top view of the device, and (b) cross section illustration of TGV (BOX: Buried Oxide Layer).

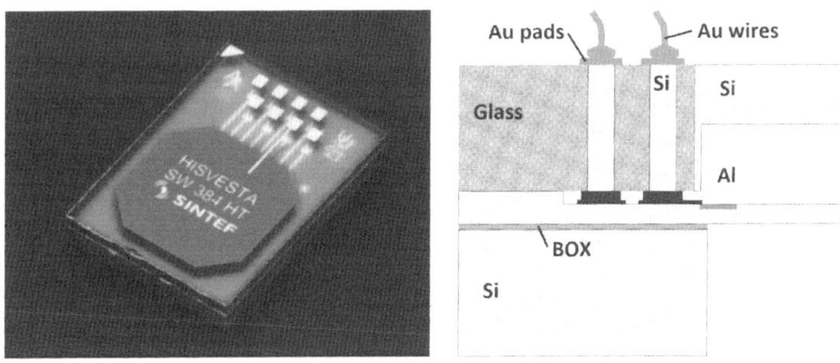

(a) (b)

Fig. 3.2. TGVs for MEMS pressure sensors from the HISVESTA project: (a) top view of the device, and (b) cross section illustration of TGV (BOX: Buried Oxide Layer).

Glass wafers with metal-filled vias were not commercially available at that time, but the resistance of the silicon vias (in the order of ohms) was acceptable for the application. The vias in the glass matrix were doped silicon pins originally formed by removing surrounding material by dry etching. The freedom in design with respect to silicon-to-glass area ratio for such compound wafers is relatively large. However, cost is a well-known limitation of dry etching processes. For example, patterns with a mix of finer and coarser structures require long process times and impose challenges on etch uniformity.

In TGV technology, the use of anodic bonding achieves hermetic sealing and electrical contacts simultaneously between the metal pads on the cap and device wafers. However, the TGVs would have electrically short-circuited the bonding process unless a process modification had been made; a CMP step was added after the anodic bonding and the TGVs were revealed. The final process steps involve the formation of backside electrical routing in aluminum, Au stud bumping, and wafer dicing. The same solution was applied to a high-temperature tolerant pressure sensor element in the European project called HISVESTA [7]. In this case, aluminum was replaced by gold in the top cap wafer for electrical routing to ensure high-temperature tolerance of the assembly.

3.3 Through Silicon Vias

3.3.1 *Hollow TSVs in Bulk Si Wafer*

In the e-CUBES project, SINTEF developed a process for TSVs for applications where a silicon cap wafer is preferred. Silicon is a preferred choice when there is a stringent demand to match coefficient of thermal expansion. A via solution with $50 \times 50\,\mu\text{m}^2$ hollow vias in $300\,\mu$m thick, 150 mm diameter wafers was presented in [8]. Holes of both $20 \times 20\,\mu\text{m}^2$ (shown in Fig. 3.3) and $50 \times 50\,\mu\text{m}^2$ were etched during process development, where the larger holes were finally preferred due to the high etch rate. An average etch rate of $16\,\mu$m/min was achieved for the larger holes on an Alcatel AMS 200 I-productivity etching tool. A layer of $1\,\mu$m thick thermal oxide was grown for insulation and $1\,\mu$m thick PolySi was deposited by low pressure chemical vapor deposition (LPCVD) and doped using phosphorous gas phase to achieve a conductive liner. For good electrical contact to the poly silicon layer, a $1.2\,\mu$m thick aluminum layer was sputtered on both sides of the wafer. A $10\,\mu$m thick Dupont MX-5010 dry-film resist was used to pattern

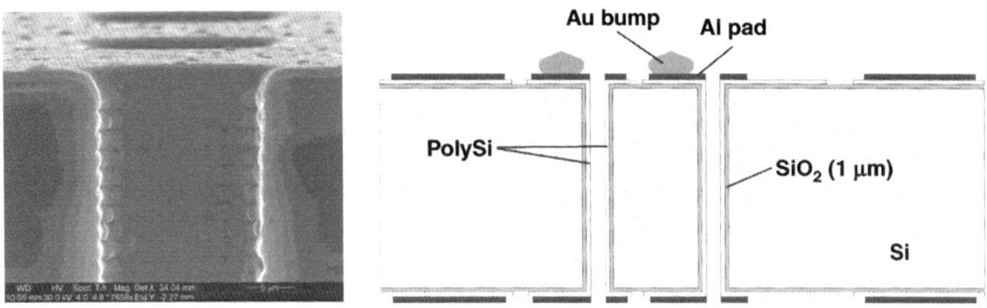

(a) (b)

Fig. 3.3. Hollow TSVs in bulk silicon wafers: (a) SEM image after growth of thermal oxide, deposition of PolySi and sputtering of aluminum, and (b) cross section illustration showing Au stud bumps added for further assembly.

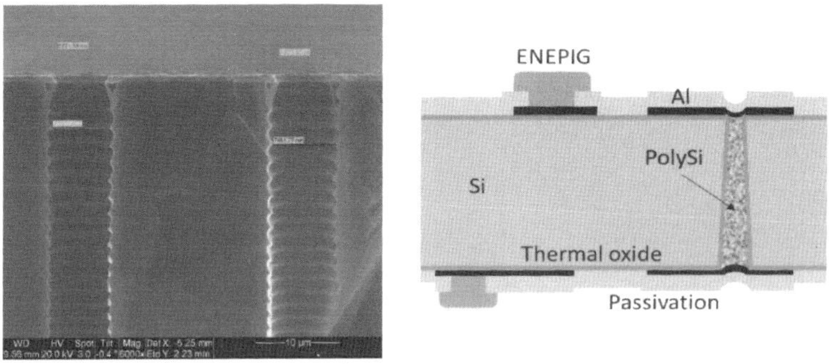

(a) (b)

Fig. 3.4. Filled TSVs in bulk silicon wafer: (a) SEM image of cross section made after dry etching, and (b) cross section illustration of the silicon interposer with PolySi filled TSV and ENEPIG surface finish on the contact pads.

polysilicon and aluminum on the perforated wafers. The wafers received Au stud bumps before dicing. For a daisy chain with 16 TSVs, the average resistance was 119.4 Ω with a standard deviation of 4.6 Ω. This means that the resistance of a single TSV interconnect was approximately 7.5 Ω.

3.3.2 *Filled TSVs in Bulk Si Wafer*

SINTEF has developed a process for filled TSVs with improved etching performance and high-aspect ratio. This process was applied to a 300 μm thick silicon interposer where components connected by TSVs were assembled on both sides [9]. The TSVs were formed in 150 mm diameter wafers using an aluminum hard mask on the front side and an aluminum etch stop layer on the backside. A layer of 1 μm thick thermal oxide was grown for insulation after stripping the aluminum. The TSVs were filled with doped PolySi using LPCVD and phosphorous gas phase processes. The vias were rectangular in shape and had a nominal width of 7 μm. The filling processes were repeated until the holes were filled. The inspection involves cross-sectioning and SEM inspection. Routing was fabricated by sputtering and patterning 1 μm thick aluminum on both sides of the interposer. Finally, a 1.5 μm thick passivation layer was deposited and opened at the contact pad locations. Then, the wafers were sent to PacTech [10] to obtain an 'Electroless Nickel Electroless Palladium Immersion Gold' (ENEPIG) surface finish before dicing. Figure 3.4 shows an SEM image of the etched vias and a schematic illustration of the final cross-section. The resistance of one TSV, as estimated from daisy chains with >80 vias, was around 4 Ω.

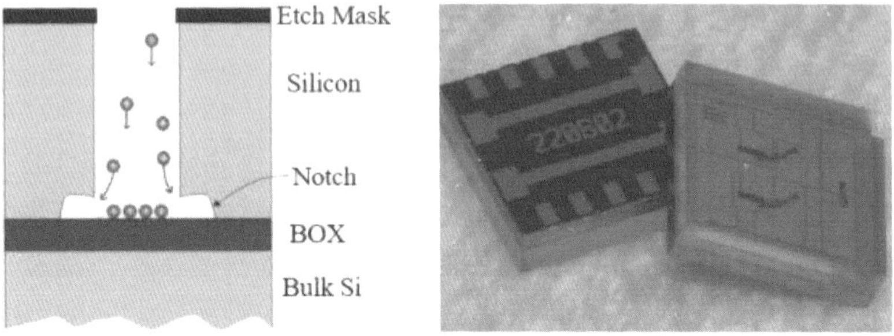

(a) (b)

Fig. 3.5. Filled TSVs in SOI wafer: (a) sketch illustrating the notching effect when reaching the BOX layer in an SOI wafer [12], and (b) MEMS switches with filled TSVs fabricated in SOI wafers [13].

3.3.3 *Filled TSVs in SOI Wafer*

In this section, filled TSVs in Si-on-insulator (SOI) wafer are discussed. One of the challenges in dry etching is the notching effect, illustrated in Fig. 3.5. Notching is lateral etching of sidewalls of the silicon structures. As part of the European project called JEMSiP_3D [11], SINTEF developed a solution for dry etching of vias in SOI wafers with relatively thick buried oxide (BOX) and device layers. Specifically, the SOI wafer had a 40 μm thick device layer and a 2 μm thick BOX on a 300 μm thick handle wafer [12]. The solution was based on a pulsed low-frequency bias. The TSVs were fabricated prior to the MEMS structures themselves. For the filled vias made in bulk Si wafers, aluminum was used on the front side as a hard mask and on the backside as a final etch stop layer. The vias were 7 μm wide. Using the same aluminum hard mask, the vias were etched through the device layer, BOX layer, and handle wafer. The high-aspect ratio of TSVs was possible by gradual adjustment of the etching conditions throughout the etching process. Final processing of the wafers is similar to that of the filled vias in bulk Si wafers. The resistance of a single TSV, estimated from daisy chains with >180 vias, was approximately 4.5 Ohms [13]. MEMS acceleration switches fabricated with these vias were exposed to environmental tests including high-temperature storage for 1000 h at 230°C and 1000 thermal shock cycles between −55°C and 125°C [14], but no degradation was observed. Specifically, no significant increase in resistance or visible crack formation was observed under SEM inspection of the cross sections.

3.4 Summary

In this chapter, SINTEF's TSV activities for MEMS and interposers applications have been discussed. Specifically, design and processing of TGVs in cap wafers for pressure sensors, hollow vias for interposers, and filled vias for bulk silicon and SOI wafers have been discussed. All these interconnect solutions are based on vertical electrical paths through regions of highly doped silicon or poly silicon. Several design features and processes can be easily applied to other applications. However, the application specific need for either a glass or a silicon cap wafer, at the top or the bottom of the device, will still dictate certain limitations in design freedom.

References

1. Leti Activity Report 2013, www.leti.fr/en/content/download/2177/27134/file/Leti-activity-report-2013.pdf
2. Taklo, M., Mielnik, M., Schjølberg-Henriksen, K., Storås, P., Rosquist, H., Lietaer, N., and Johannessen, R. (2007). Technologies enabling 3D stacking of MEMS, *IEEE workshop on 3D System Integration*, Munich, Germany, Oct 1–2.
3. Taklo, M., Lietaer, N., Tofteberg Rosquist, H., Seppännen, T., Prainsach, J., Weber, J., and Ramm, P. (2009). 3D MEMS and IC integration, *MRS Fall Meeting Symposium E*, p. E04-04.
4. www.e-CUBES.org
5. Taklo, M., Schjølberg-Henriksen, K., Lietaer, N., Elfving, A., Prainsack, J., Weber, J., Klein, M., Schneider, P., and Reitz, S. (2009). 3D Integration of MEMS and IC: Design, technology and simulations, *Advanced materials and technologies for micro/nano-devices, sensors and actuators*, St. Petersburg, Russia, Jun 29–Jul 02, 2009.
6. http://www.planoptik.com/produkte/produkte_en.html
7. http://www.sintef.no/projectweb/hisvesta/
8. Lietaer, N., Taklo, Maaike., Klumpp, A., Weber, J., and Ramm, P. (2009). 3D Integration technologies for miniaturized tire pressure monitor system (TPMS), *International Conference and Exhibition on Device Packaging*, pp. 106–111.
9. Nilsen Wright, D., Belle, B., Nguyen, V., Gakkestad, J., Kristiansen, H., Schjoelberg-Henriksen, K., and Taklo, M. (2016). Anisotropic conductive film for fine-pitch interconnects, *49th International Symposium on Microelectronics*, Oct 11–13, 2016, Pasadena, CA, USA.
10. www.pactech.com
11. JEMSiP_3D: http://cordis.europa.eu/project/rcn/201936_en.html
12. Summanwar, A. and Lietaer, N. (2011). Etching burried oxide at the bottom of high aspect ratio structures, *MME 2011: Proceedings of the 22nd*

Micromechanics and microsystems technology Europe workshop, pp. 326–329, Vestfold University College Tønsberg, Norway.
13. Lietaer, N., Bakke, T., Summanwar, A., Dalsjø, P., and Gakkestad, J. (2011). Wafer-level packaged MEMS switch with TSV, *International Wafer Level Packaging Conference*, Oct 3–6, Santa Clara, CA, USA.
14. Tollefsen, A., Taklo, M., Bakke, T., Lietaer, N., Dalsjø, P., and Gakkestad, J. (2012). Reliability of TSV and wafer-level bonding for a 3D integrable SOI based MEMS application, *International Wafer-Level Packaging Conference*, Nov 05–08, Santa Clara, CA, USA.

Dr. Maaike M. Visser Taklo is employed as a Chief Scientist at SINTEF Digital in Norway at the Department of Smart Sensor Systems which she joined in 2010. She is Research Manager for the group "Advanced Packaging and Interconnects" within this department. From 1998 until 2010 she was employed at the Department of Microsystems within SINTEF Digital where she worked in the SINTEF Minalab on MEMS processing and was responsible for their wafer level bonding activities. She received her PhD degree in Physical Electronics from the University of Oslo in 2002 for her thesis entitled "Wafer bonding for MEMS".

Sigurd Moe took his MSc at the University of Oslo, Norway in 1996. The topic of the thesis was "Technology for silicon tunnelling sensors". He has worked at SINTEF Digital, Department of Microsystems 1996–2000 and 2002–present. From 2000 to 2002 he was Manager for Semiconductor Design in Optiswitch (later Polight) in Horten, Norway, concentrating on optical components for telecom. He is currently Research Manager for the silicon sensor technology group at SINTEF Minalab. His work involves process development, process integration and production of MEMS components, with a main focus on pressure sensors.

Chapter 4

Using Wafer-Level Packaging to Improve Sensor Manufacturability and Cost

Paul Pickering, Collin Twanow and Dean Spicer

4.1 Introduction

Sensor technology has been a force in advancing markets like industrial automation for over 150 years. One of the earliest examples of a sensor grew out of the observation of the predictive sensitivity of various metals to temperature which occurred in the early 1800s. In 1860, Wilhelm von Siemens observed these principles and later developed a temperature sensor based on a copper resistor. These observations set the stage for a flurry of subsequent inventions and patents that created an industry of temperature sensors used in harsh environments where conventional thermometers could not function. By the turn of the century in 1900, temperature sensors could be found in thousands of industrial applications and played a significant role in the Industrial Revolution that drove the global economy at the start of the 20th century. This is merely one example of the sensor technology evolution that has taken place in the last century.

This evolution is far from being over and, in fact, is fueling waves of change in industrial automation as well as medical, automotive, consumer, aerospace, and nearly every other major industrial sector. In most cases, these markets are driving the evolving needs for feature scalability, lower cost, and size reduction. The widespread use of sensors is also fueling the engine of change. We do not have to look far to realize that sensor technology touches nearly every person on the planet on a daily basis and in most cases, the average person in an industrialized nation is in constant connection with sensor technology through their smartphone, tablet, keyboard, and dozens of other IoT systems. There are numerous publications and organizations predicting that global consumption of sensors will eventually eclipse a trillion sensors annually. This lofty goal won't be achievable without significant change to the current materials, packaging, and overall manufacturing capacities. For the time being, each participant in the cause of advancing sensor packaging technologies will play a vital role in getting the industry to these levels. The goal of this chapter is to look deeply at wafer-level packaging (WLP) as a technology that can provide numerous advancements to the various applications of sensor technologies.

The broad adoption of sensor technology has created significant opportunities for sensor manufacturers to continue to evolve the technology to improve functionality, size, cost, and manufacturability. With sensor component volumes now exceeding tens of billions of units annually, the need for lower cost manufacturing processes and materials is more pressing than ever. However, the process of advancing the various sensor technologies presents several hurdles to overcome. Sensor applications are very diverse and the manufacturing materials cover a wide array of elements, alloys, chemicals, and gases. Adding to the diversity is the fact that material choice can be driven by the device performance, maturity of established processes in foundries, process compatibility, and even foundational technology compatibility. The challenge of choosing a set of areas to focus on for improvement can be daunting. A common approach to managing this type of challenge is to look for areas of the technology that can be improved by exploiting other adjacent technologies and quickly, or not so quickly, provide a new advantage not previous realized in the existing technology. Such is the case for the Micralyne MicraSilQ^{TM} platform development which provides a novel approach to standardize a process for integrating silicon WLP with a through silicon via (TSV) technology to create a low-cost packaging solution for inertial sensors.

The primary goal of the MicraSilQ^{TM} developers was to provide an alternate packaging technology to the standard ceramic and metal hermetic packages that dominate the packaging approach for precision inertial sensors. The key advantages of a new packaging approach would be threefold; cost reduction, improved manufacturability, and decreased size. Key attributes of the former packaging technologies that would need to be supported in the new packaging approach are hermeticity and the ability to support a vacuum cavity that could retain a sub 1mTorr vacuum if necessary. The developers of MicraSilQ^{TM} decided to focus on a platform approach to WLP that is technically robust and cost effective. The following sections of this chapter cover the various aspects of the MicraSilQ™ approach to WLP along with the test data, supporting documentation and process limits of the technology.

4.2 WLP Description

The term WLP can have a couple of different meanings, so it is best to clarify the specific description. For the purpose of this chapter, WLP will mean a process that leverages the bonding of multiple silicon or glass wafers to create a fully encapsulated device that has an internal cavity space for the device layer or proof mass to move freely within. In most applications, WLP has either bond pads or solder balls so the signals from the sensor stimulus can be connected to a driver, controller, or sensor hub device. The final defining feature is the ability to use the WLP device with no additional packaging. The silicon provides the robust structure and acts as the device package. No additional overmolding or ceramic carrier is required.

In Fig. 4.1, various components of Micralyne's MicraSilQ™ WLP approach are shown. The package starts with a base wafer. In this case,

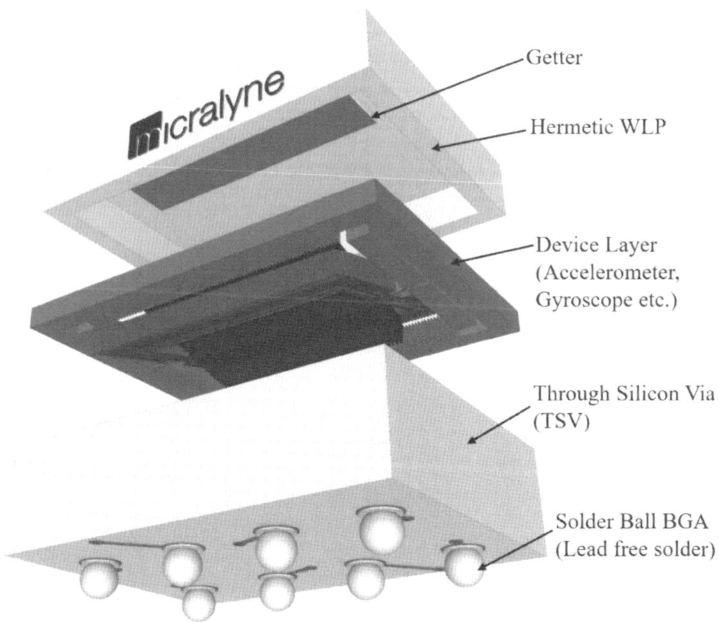

Fig. 4.1. Illustration of various components in Micralyne's MicraSilQ™ WLP approach.

the base wafer uses TSVs to bring the signals out of the package. The TSV wafer is manufactured first and serves as the starting platform for the development of the subsequent layers. The device layers implement the functional sensing mechanism and can be one or multiple layers of silicon depending on the design. The device layers can implement elements like comb drives, springs, proof mass, and cavity structures. One or more layers of silicon are bonded to the base layer to support these added features. The additional device layers are typically bonded using a fusion bond method. The lid or capping layer is made out of silicon or glass and usually has a cavity structure etched into it. The fabrication of the lid wafer is done as a separate process in preparation for later bonding to the device structure in one of the final steps. There is a good bit of flexibility in the lid structure in terms of thickness and cavity size. The standard lid wafer thickness is roughly 500 microns. The cavity in the lid needs to support the space required for the movement of the device layer or sensing structure. The lid wafer often has an embedded getter layer for improved vacuum performance.

The bonding of the lid or capping wafer needs to support the ability to create a robust, hermetic seal. In the case of MicraSilQ^{TM}, a gold silicon eutectic alloy (Au/Si) bond is used. The gold deposition is done on the device wafer during fabrication and serves as the hermetic seal between the lid and the device/base wafer. During the bonding process, a eutectic alloy is created between the two wafers and thus sealing the wafer in specific locations. The lid wafer is aligned and bonded under vacuum in a conventional wafer bonding tool. After bonding, the resulting wafer stack is completely sealed and the device layer is protected. Figure 4.2 shows a typical wafer stack configuration for MicraSilQ^{TM}.

The final steps in the MicraSilQ^{TM} WLP manufacturing process are Under Bump Metallization (UBM), Solder Ball attach and dicing. Once the wafer stack has completed the bonding process, a two-patterned metal layer is deposited on the backside of the wafer to create electrical routing from the vias to the solder ball landing pads and then a compatible metal that can withstand reflow of the solder. This involves two photolithography steps, subsequent etching of the aluminum trace metallization, to mask the area for UBM deposition. Various metals are used for the UBM depending on the application but the most common metal stack is an adhesion layer such as Ti, TiW, or Cr, then a barrier layer, such as Ni, followed by a solder wettability layer that is typically gold. The solder ball attach is a well-known process across the industry and can be

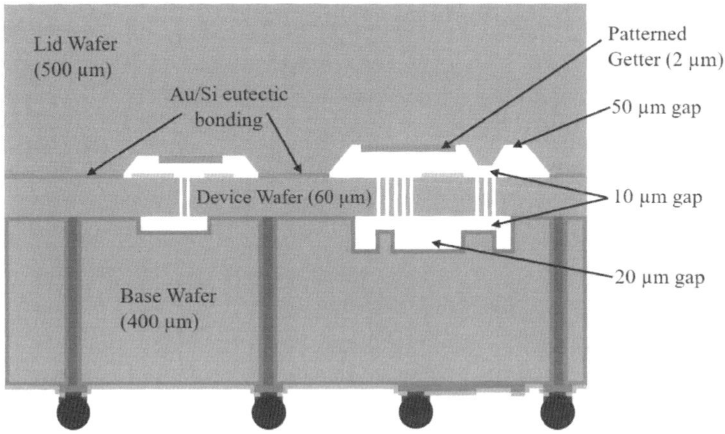

Fig. 4.2. A typical wafer stack configuration for Micralyne's MicraSilQ^{TM}.

done in various ways depending on the size of the balls, the pitch, and the materials used.

Lastly, once the wafer processing, wafer bonding, and ball attach have been completed, the wafer is ready to be diced into individual components. One of the primary benefits of using WLP is the scalability of the manufacturing process. All the processes used are common to the semiconductor/microchip manufacturing processes on silicon wafers. In the case of MicraSilQ^{TM}, a single wafer will usually have nearly 1,000 components on it. All the manufacturing processing steps are done at the "wafer level" so the cost of the processing is spread across a thousand or so components. These "wafer level" economics have fueled the semiconductor industry for the last 50 years and directly apply to the manufacturing process for WLP. The only steps in the process that are done at the component level are the dicing or segmentation of the wafer, and in some cases, final die test. If possible, electrical test should be completed at the wafer level with automated wafer probers, once the backside metal is in place. In most cases, the dicing of the WLP wafer is done in a conventional wafer dicing tool which uses a diamond saw. This process is a wet process due to the use of water to cool the saw blade. Fortunately, the wafer stack is fully hermetic and therefore the delicate device inside is completely protected. After dicing, the resulting components can be picked directly from the mounting tape and soldered directly to the printed circuit board (PCB) or other substrates used in the end application.

4.3 WLP for Inertial Sensors

Inertial sensors are a class of devices that provide a navigational aid through a dead reckoning mechanism that continually calculates the relative position, orientation, and velocity of a device. These devices can be applied to simple applications like game controllers that simply sense the motion, direction, and velocity. More complex inertial sensors are used to supplement or even replace GPS and serve applications for high precise control like autonomous vehicles. The signal output of an inertial sensor is analyzed to create various useful pieces of information necessary to determine position, speed, attitude, etc. Inertial sensors are some of the most commonly used sensors and are found in smartphones, cars, robots, planes and nearly every moving vehicle. The initial target application for the MicraSilQ^{TM} platform was inertial sensors for several reasons. First, the end application requirements were well known to the developers so the benefits offered by WLP were also understood and believed to be significant. Secondly, the market for inertial sensors is quite broad with many applications that are currently using precision sensors in large volumes. As with all new technologies, it is helpful to have a large and diverse market to target. Lastly, there are new applications that would consider using a precision inertial sensor except for the fact that the current cost is prohibitive. These new markets would benefit greatly from a low-cost device.

A latent benefit of using WLP comes from the standard platform approach to designing the sensor device. By standardizing the process into a manufacturing flow and creating a Process Design Kit (PDK) approach, the designers of sensors can focus on the device design and leverage the previous engineering for the rest of the solution from the standard platform. The various elements that surround the device, mainly the TSV wafers and the encapsulation process, are already completed and qualified. As much as 90% of the process can be reused from one application to another. This provides for lower cost and greatly reduces the development time.

Since the inertial sensor market was the primary focus for the MicraSilQ^{TM} platform, much of the initial platform definition was driven by the technical requirements for inertial sensors. Features like the vacuum sealed device cavity and the TSV wafer provide for a compelling solution for most inertial sensor requirements. It is important to note that the same features also make this platform very useful for many other applications like micro-mirrors, thermopiles, bolometers, and gas sensors. So, the ease of design and manufacturing economics of a platform approach to WLP can be applied to a variety of microelectromechanical systems (MEMS) and sensor applications.

For the rest of this chapter, we are going to focus on the specific requirements of inertial sensors and how the MicraSilQ^{TM} platform provides a competent solution to achieve even the most demanding specifications. With regard to the specific technical requirements of an inertial sensor, the key attributes are usually measured around the sensitivity of the device. The largest factor in sensitivity degradation is usually various types of noise. Therefore, the Signal to Noise Ratio (SNR) is critical to create a high performance inertial sensor. The primary goal of most MEMS-based inertial sensors is to remove the potential sources of noise contribution within the proof-mass cavity of the sensor. The cavity requires significant vacuum pressure for this reason. The vacuum minimizes the "Brownian-motion noise" which results from the impact of air molecules against the mass [1]. The cavity within the sensor should be free of all elements which could impede the movement of the proof-mass in any way. Getter is also used to maintain the elimination of gases and other organic elements resulting from out-gassing over time. From a packaging standpoint, sustaining the cavity vacuum over the lifetime of the device is one of the most critical factors in achieving stable inertial sensor performance. For this reason, the vacuum cavity aspect of the solution is placed as one of the highest priorities in the process definition for MicraSilQ^{TM}. Fortunately, the concept of wafer bonding under vacuum pressure is not a new process and various equipment manufacturers offer bonding tools with integrated vacuum chambers. For more than 20 years, vacuum wafer bonding techniques have been practiced by many universities and companies, including Micralyne. The various methodologies have also been well documented in books and various journal articles.

4.4 Hermetic Wafer Bonding

The cap wafer or lid wafer is etched with a cavity over the region of the wafer that has active elements. Optionally, a getter may be deposited in this cavity. The presence of a getter inside this cavity makes it possible to achieve very low cavity vacuums, in the order of 0.5 Pa, and maintain these specifications over the course of five or more years. At these cavity pressures, resonant modes of the active elements of the device are unaffected by viscous or squeeze-film damping and can exhibit extremely high-Q factors.

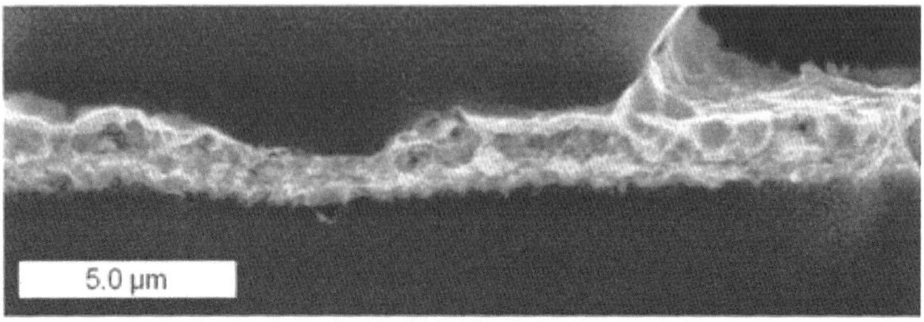

Fig. 4.3. Gold–Silicon eutectic forms a high-performance bond between the active layer of the device and the cap wafer which may contain a getter.

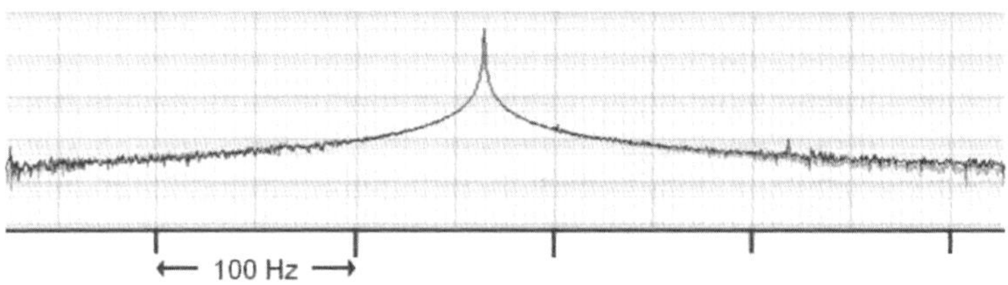

Fig. 4.4. The resonant signal of a MicraSilQ^{TM} device (log y-axis of signal amplitude).

The top wafer of the stack (see Fig. 4.3) is bonded to the second layer of the device on a bond ring that is structurally independent of the moving electrodes. Micralyne has developed a high performance hermetic bond that relies on the formation of Gold–Silicon Eutectic above 363°C to perform this sealing. Leak-rates less than 10^{-17} Pa.m^3.s^{-1} of non-getterable species have been repeatedly achieved for each die on wafers bonded with Micralyne's eutectic bonding process. The cavity of MicraSilQ^{TM} device may be evacuated during eutectic bonding. Resonant structures inside a cavity may be designed to exhibit extremely high Q factors (see Fig. 4.4). The result of this vacuum encapsulation is a very low noise and extremely sensitive device.

The well-established method for testing a package's hermeticity uses helium as the tracer gas and a mass spectrometer for measuring the rate at which the helium flows through a seal. This well-established technique for measurement is based on following a specific protocol to fill up a package with helium and then measuring the rate at which it leaks back out. The package to be tested is subjected to helium ambient environment at or above ambient pressure. This exposure extends over a period of hours or days to fill the package with a quantity of helium dependent on the rate at which helium can leak into the package. The package is then removed from the helium ambient and placed in a tool to measure the rate of helium outflow. This mass spectrometer provides a measured rate of helium outflow from the device which can be used in conjunction with the Howell–Mann equation (Eq. (4.1)) relating the measured outflow with the characteristic parameter describing how permeable the package is to helium [2–3]. The Howell–Mann equation relates the measured leak rate R over time t, to the characteristic leak rate of a package L with volume V, as a result of exposure to helium at pressure P_{B} for time t_{B}. The ratio $M_{\mathrm{A}}/M_{\mathrm{He}}$ corrects for the expected variation in leak rate between helium and air.

$$R = L\left(\frac{M_{\mathrm{A}}}{M_{\mathrm{He}}}\right)^{\frac{1}{2}} P_{\mathrm{B}}\left\{1 - e^{\left[-\frac{Lt_B}{V}\left(\frac{M_{\mathrm{A}}}{M_{\mathrm{He}}}\right)^{\frac{1}{2}}\right]}\right\} e^{\left[-\frac{Lt}{V}\left(\frac{M_{\mathrm{A}}}{M_{\mathrm{He}}}\right)^{\frac{1}{2}}\right]} \tag{4.1}$$

Testing a MEMS chip using established helium bombing protocols and calculations with the Howell–Mann equation are limited in their detection capabilities. The limitations are straightforward, the cavity volume of a typical MEMS chip could be in the order of 10^{-4}–10^{-3}cc, and the detection limits of many common helium mass spectrometers are limited to leaks in the order of $5 \cdot 10^{-12}$atm $\cdot$ cc $\cdot$ s^{-1} [4]. The device's small volume itself limits to how much helium can be filled into the chamber and together with the characteristic leak rate of the device being tested, it becomes impossible to measure test protocols such as very high exposed helium pressures for impractically long durations. The complication of extremely small device volumes severely inhibits the testing of even poorly sealed MEMS devices shown in Fig. 4.5. For each measured leak rate, there are two physical solutions to the Howell–Mann equation (Eq. (4.1)), one representing a relatively quickly filling and quickly discharging cavity and another representing a slowly filling device. Due to the low measured

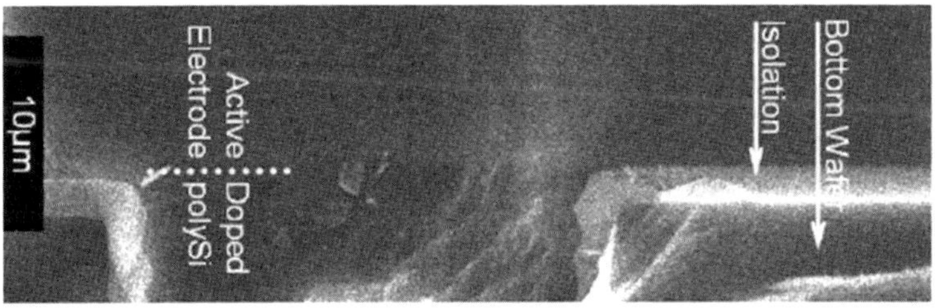

Fig. 4.5. The electrical connection of a MicraSilQ^{TM} formed with doped polysilicon TSV.

leak rates and the devices' physical parameters, these solutions fail to distinguish themselves in a statistically significant manner from the noise floor of the detector. While Fig. 4.5 shows a die suspected of having a characteristic leak rate of approx. $4 \cdot 10^{-11}\text{Pa} \cdot \text{m}^3 \cdot \text{s}^{-1}$, the alternative theoretical solution of $9 \cdot 10^{-8}\text{Pa} \cdot \text{m}^3 \cdot \text{s}^{-1}$ is also completely within the window of the experimental repeatability of the tool.

In summary, the historic helium bombing protocol and the equipment that has been designed to perform the testing cannot be used to accurately delineate between device performances at even one million times the specification limit imposed by a 5-year device lifetime. As the measurement is incapable of assigning an accurate numerical value to the hermetic performance of such devices, it is very much incapable of distinguishing between variations of hermetic performance at the leak levels required for high-performance MEMS accelerometers. An alternative method must be developed to meet this need and provide a means by which to analyze and validate the optimization of a eutectic bonding process.

As an alternative to measuring leak rates using a helium bombing test, the leak rate of a device can be calculated directly based on the pressure change within a device cavity. This method of calculating leak rates requires an accurate and repeatable cavity vacuum measurement. The stipulation requiring cavity vacuum measurements limits hermeticity testing to MEMS packages that are constructed in a way that allow such measurements to be taken. In the case of a mass-on-springs geo-sensor, it is possible to resonate the device and measure the damping provided by the residual gas in the cavity. Measurement of Q-factors for various devices can be taken (see Fig. 4.4) and the cavity vacuums plotted over time. Leak parameters equal to $1 \cdot 10^{-17}\text{Pa} \cdot \text{m}^3 \cdot \text{s}^{-1}$ for a device with a 10^{-4}cc volume will result in a cavity pressure rising by 8.6 mPa per day under exposure to air at standard pressure. The presence of a getter capable of absorbing water vapor, nitrogen, and oxygen will effectively mask many leaks in the short term as these species will be trapped and eliminated from the gas phase within the cavity until the getter is saturated. Any species that cannot be absorbed, either because the getter has exhausted its capacity to sink reagents or because the species is non-reactive with the getter, will result in a fraction of the leak that is un-masked. Under atmospheric conditions, a presence of approximately 1% argon will contribute to the pressure rise within the MEMS cavity even while the getter is active as it is noble gas and non-getterable species. To test the true leak rate, independent of the getter, MEMS dies are exposed to a 100% argon ambient at standard pressure. This

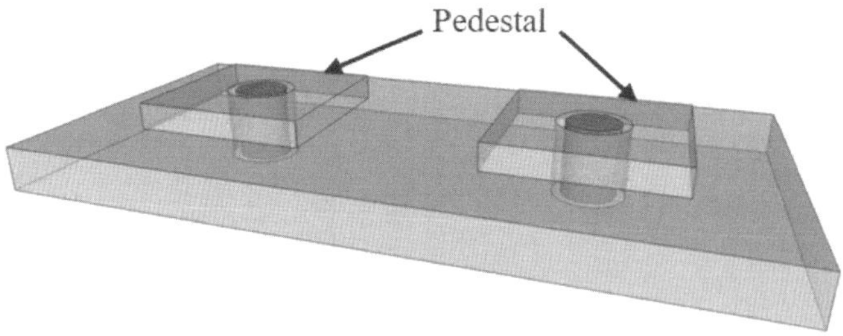

Fig. 4.6. The bottom wafer of the MicraSilQ^{TM} stack is composed of a cavity with pedestals. A through wafer via electrically connects the active layer that is fusion bonded to these pedestals to the outside of the die.

accelerates the leak of non-getterable species to approximate the complete leak of all species as if no getter were inside.

Testing of MEMS die is completed with a protocol based on this theory. MEMS dies were exposed for approximately one week at a time to an argon ambient at room temperature and 14.7 psi (1atm) pressure. Q-factors were measured before and after testing and the test was repeated twice on each population of die, and the cavity pressure is calculated using the devices' resonant Q-factors. Such an exposure protocol simulates 75% of the 5-year lifetime of the device, and further repetitions could be executed if a notable evolution of behavior was observed between tests. As an example, a pressure increase of less than 0.06 Pa $= 8.6\,\text{mPa/day} \times$ 1 week exposure for a 10^{-4}cc cavity volume device would be considered a permissible change assuming $1 \cdot 10^{-17}\text{Pa} \cdot \text{m}^3 \cdot \text{s}^{-1}$ as a specification limit. The specific leak parameters for the configuration shown in Fig. 4.6 exhibit average leak parameter of $4.2 \cdot 10^{-18}\text{Pa} \cdot \text{m}^3 \cdot \text{s}^{-1}$ based on the die's specific geometries and argon exposure protocol.

4.5 Through Silicon Vias

For the benefit of wafer real estate and ease of automated pick and place for board assembly, TSV technology is used in the MicraSilQ^{TM} platform. The vias are used for routing electrical signals from the internal device to a control chip outside. Maintaining low signal noise in this electrical connection is paramount to the performance of all high-performance sensor devices. This high performing TSV is built with $>100\,\text{M}\Omega$ isolation from the substrate, $<5\,\text{pF}$ parasitic capacitance and $\sim 10\,\Omega$ series resistance (through a $400\,\mu\text{m}$ substrate) using doped polysilicon.

A patent protected (US No. 8,445,304) process flow and device architecture is used to route these electrical signals in an optimal configuration for performance and device reliability. The MicraSilQ^{TM} process flow for fabricating TSVs is both inherently hermetic and inherently electrically isolated. The typical via has a resistance of 10–15 Ω/via which fits the requirements of most sensor devices. The via process eliminates the need for wire bonding which also limits the space needed for bond pads on the top surface. The device manufactured with TSV technology typically results to a die size savings of 25% to 40%. This translates to many more dies per wafer and a significant cost savings to the component. The MicraSilQ^{TM} TSV technology use the doped polysilicon approach because it is relatively inexpensive as compared to other via technologies. The via technology used in this case would be considered in the industry as "via-first", which means the via processing is done at the front end of the fabrication process. Conversely, a "via-last" process incorporates the via at the end of most of the processing steps. The "via-first" was selected for MicraSilQ^{TM} because it provides the greatest amount of flexibility and configurability in the rest of the device. Since MicraSilQ^{TM} is a platform that is intended for use in many different, customer defined configurations, "via-first" was the obvious choice.

4.6 Device Layer Etching

The device layer on the MicraSilQ^{TM} platform is where the customization and differentiation take place. The designer is given flexibility to create a sensor that meets the needs of the application. The designer has the ability to modify the size of the proof mass within a broad range in the x and y dimensions. The target device size in the design rules is roughly 5.6 mm × 3.4 mm so the guidelines encourage the designer to create a device that fits within those dimensions given the other design considerations such as springs, torsion beams, and comb drives. A larger package size can be accommodated, if necessary.

The target thickness of the device layer after is $60\,\mu$m+/$-1\,\mu$m. This thickness can be achieved with a grind/CMP (chemical mechanical polish) process after bonding to the base wafer. It can also be accomplished through a SOI (silicon-on-insulator) buried oxide (BOX) which is used as an etch stop mechanism. Again, there is some flexibility to modify the device layer thickness based on the requirements of the device. After the grind/CMP, the device layer wafer is ready for the device etch process which defines the specific device structure.

The process of etching the device structure is used for both the definition of the structure and releasing of the device layer structures using deep reactive ion etching (DRIE). The active electrodes of a device built on the MicraSilQ^{TM} platform are patterned using non-contact stepper lithography with excellent repeatability, consistency, alignment accuracy, throughput, and quality.

Established DRIE technology provides a flexible method of etching deep cavities with nearly vertical walls (see Fig. 4.7). The depth of DRIE etching capabilities varies from under $10\,\mu$m to over $700\,\mu$m in depth. One of the challenges in this process is the heat buildup associated with the deep etch process. Standard DRIE recipes are incapable of releasing heat sensitive membrane structures which are often used in sensor applications. The problem is due to excessive heat buildup that results in a loss of sidewall passivation and ultimately destruction of the device. For the purposes of the MicraSilQ^{TM} platform, specific low heat generating recipes have been developed and successfully implemented into multiple designs and process flows. These DRIE recipes have been successfully used in over 50 product designs. Many designs require very specific etch specifications such as sidewall angle, sidewall roughness, etch rate, uniformity, selectivity, max temperatures, etch stop, low polymer production, and others. The deep etch process is the most critical process step in terms of the actual formation of the MicraSilQ^{TM} device layer. Figure 4.8 shows the active layer of the device patterned with stepper lithography and etched with DRIE to release

(a) (b)

Fig. 4.7. A few examples of features etched using Micralyne's DRIE process. (a) Comb drive and (b) hinge.

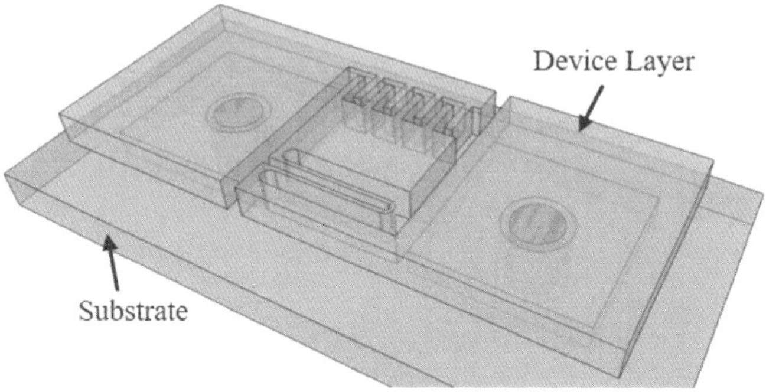

Fig. 4.8. The active layer of the device is patterned with stepper lithography and etched with DRIE to release precision features in the device layer.

precision features in the device layer. The features range from sensing and driving capacitive elements to springs and resonant cantilevers.

4.7 Device Integration (Board Attach)

The final stage of manufacturing involves the mechanism to attach the device to the printed circuit board (PCB) or other substrate for interfacing with the rest of the system. As stated previously, the MicraSilQ^{TM} platform uses a TSV wafer to bring the signals out of the package. The TSVs connect to a redistribution layer at the bottom of the device. This provides good electrical connection to the silicon as well as means to redistribute the solder ball placement. Uniform solder ball placement on the die is important in setting up robust downstream die attach processes (uniform pressure distribution). This results in a ball grid array (BGA) (see Fig. 4.9) that can be easily handled by automated pick and place tools. BGA can be soldered to the board using standard solder method with normal industry defined temperature profiles. In the solder ball attach process, the solder spheres are placed on metal traces at the bottom surface of the MicraSilQ^{TM} platform. Placement of the solder balls is done at the "wafer level" to improve the efficiency and provide significant cost savings as compared to doing this process on individual components post wafer dicing or wafer singulation. Additionally, the placement of the metal traces used for the BGA allows for low cost wafer-level device testing to ensure product quality. The UBM layer was optimized such that the die can be attached reliably to low cost FR4 boards. This contrasts with the reduced reliability requirements when considering the more expensive ceramic boards.

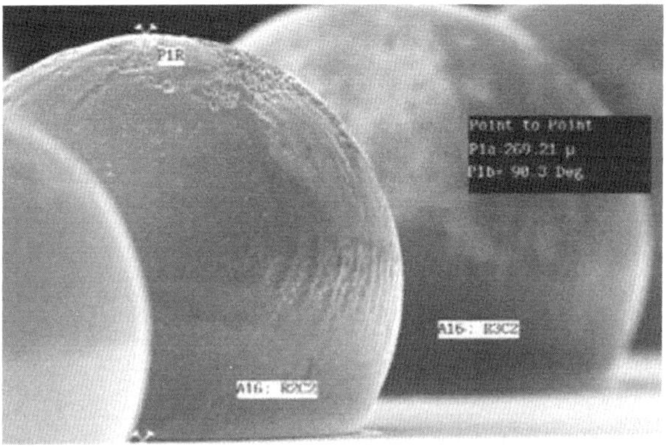

Fig. 4.9. BGA of lead-free solder spheres placed directly onto the wafer prior to die singulation.

Other integration techniques are possible including direct bonding of the MicraSilQ™ stack to a complementary ASIC wafer. In addition to BGA which is preferred as a low-cost solution, it is also possible to wire bond MicraSilQ™ devices into ceramic packages to facilitate the prototype testing of MEMS devices under development.

4.8 Additional Considerations

The premise of this chapter is to present a standardized approach to using WLP in MEMS sensors to reduce the cost and improve the manufacturability. The authors believe that the production-level WLP is still in its infancy. A large variety of sensors will take advantage of the cost benefits as the technology advances and becomes even less expensive. A point of clarification that should be noted as part of the closing comments is the relationship between low cost and manufacturability. These two qualities are often found together and are closely linked to one another. However, manufacturability is very critical for many industries for reasons other than cost. A primary concern for many industries is reliability. This must be considered as one of the primary aspects of the sensor packaging solution. The automotive, industrial and aerospace industries, for instance, have incredibly high standards for reliability due to the tremendous costs and inconvenience of dealing with failed parts after components have been installed in end applications. Therefore, creating a packaging technology that is itself reliable and can be assembled into other systems in a reliable way is critically important. This requirement for reliability often presents a set of challenges for sensor devices and the companies that use them in systems. Many conventional component packaging technologies were adopted from the semiconductor industry. This is also the case for many PCB assembly techniques. However, sensor products have a different set of requirements and therefore mainstream methods of encapsulating semiconductor devices do not work for sensors. A good example is overmolded plastic packaging that is a mainstay in the semiconductor industry and is almost never used for sensors. There is a list of reasons why molded plastic packaging does not work well on most sensors but the most common one is the added stress of the molded plastic connected to the sensor distorts the performance. The hot plastic tends to create unpredictable stress as it cools. Numerous attempts to use plastic over-molding have resulted in sensor device performance that is not predictable and often completely non-functional. The point of this discussion is to illustrate the challenge of creating packaging technologies that perform well with sensor requirements, can be manufactured using similar board assembly processes as semiconductors, and meet the reliability requirements for the end systems.

The quest to overcome this challenge has increased the focus of researchers on the use of WLP as a solution. WLP offers a reasonable cost effective way of encapsulating a sensor that conforms with most of the industrywide board assembly techniques. The resulting device produced using the MicraSilQ^{TM} platform is a BGA that can be picked, placed, and soldered in the same way that most of the semiconductor products are assembled on PCB. Additionally, the silicon WLP device has a similar TCE (temperature coefficient of expansion) as PCB materials. These similarities provide for a reliable solder bond to the PCB that is also able to leverage the cost benefits of standardization across the industry. For these reasons, the authors believe that WLP provides an exceptionally good solution to the challenge of cost effective sensor packaging.

4.9 Summary

The concepts expressed in this chapter are intended to provide a handful of elements that can have a material impact on the cost of packaging complex sensors. The strength of WLP for sensors has many dimensions to be considered. It takes advantage of existing processes that are well established and known to be reliable. Wafer bonding offers the ability to encapsulate thousands of devices in a single process step. The WLP encapsulation process provides a robust bond that is hermetic, a requirement for many sensor devices. The fabrication process used in assembling the sensor wafers leverages the same economics, which the semiconductor industry has used for more than 50 years.

The addition of a cost effective TSV process to the bonded wafer solution adds a compelling advantage because it establishes the mechanism to use the existing, high volume board assembly processes. The TSV enables the packaged sensor to adopt the same BGA format that has been widely used for several decades. The key benefits are the reduced size of the device and the ease of manufacturing.

The development of new sensor technology is always a significant undertaking. The various elements necessary to design, fabricate, package, test, and manufacture all need to be well understood. The development process can easily get derailed before it even starts by uncertain challenges faced in any of these areas. By using a standardized WLP flow that uses industry proven capabilities, designers can simplify the development process and accelerate the time to market.

References

1. Laine, J. and Mougenot, D. (2014). A high-sensitivity MEMS-based accelerometer, *The Leading Edge*, pp. 1234–1242.
2. Greenhouse, H. (2000). Understanding Helium fine leak testing in accordance with method 1014, MIL-STD-883, in Chapter 6, *Hermeticity of Electronic Packages*, New York: Noyes Publications, pp. 210–258.
3. Test Method Standards Microcircuits — Method 1014, MIL-STD-883H, 2010.
4. VS Series Helium Mass Spectrometer Leak Detectors, No. 699909942 rev. F, Varian, 2010, pp. C-1.

Paul Pickering has worked for more than 35 years in semiconductor and micro-fabrication technology companies with a specific focus on Medical and Bio-MEMS applications for the last 5 years. He is currently overseeing the Product Marketing and Business Development teams for Micralyne Inc based in Edmonton Alberta, Canada. During his career, he has held various product line executive roles in large semiconductor manufacturers as well as several start-up companies. He co-founded two venture-backed companies and has consulted with numerous other start-up companies. Prior to joining Micralyne, he served as the Executive VP, Sales and Marketing at Exar Corporation, an analog and mixed signal semiconductor manufacturer. He is a graduate of West Chester University of Pennsylvania and also attended Widener University in Chester Pennsylvania.

Collin Twanow has worked at Micralyne for more than 15 years and is currently the Vice President of Technology. Previous roles have included Vice President of Engineering, Vice President of Sales and Business Development, Director of R&D, Product Line Manager for Optical MEMS, R&D Project Manager, and Development Engineer. He is primarily responsible for Micralyne's technology development roadmap. He is also responsible for technical problem solving and communicating Micralyne's capabilities to the MEMS market. Collin has managed development programs for many devices including optical telecom switches, silicon optical bench, implantable medical devices, MEMS microphones, pressure sensors, and components for military applications. Throughout these projects, he has gained extensive experience in wafer processing, design layout, device test, and failure analysis. Collin is a Professional Engineer and holds a BSc in Engineering Physics (1997) as well as a Master's of Electrical Engineering degree with a specialization in Microfabrication (2002). Collin has also lectured on microfabrication topics at the University of Alberta.

Dean Spicer joined Micralyne in 2006 as a Research and Development Engineer, specializing in Advanced Wafer Bonding processes (eutectic, fusion and anodic). During his tenure at Micralyne, he has served in various leadership roles in technology development and engineering. Dean currently serves as Director of Engineering. Prior to working at Mycralyne, he worked for JDS Uniphase in Ottawa for over three years. Dean is passionate about MEMS and works to bring customers' products to market efficiently through the entire development cycle, from concept to finished product. Both detail-oriented and forward-thinking, Dean is Micralyne's internal champion for developing process platforms that enable multiple projects to be fabricated from standard processes. Dean earned a Bachelor of Science degree in Engineering Physics and a Master's of Science degree in Electrical and Computer Engineering from the University of Alberta.

Chapter 5

Nasiri Fabrication Process for Low-Cost Motion Sensors in the Consumer Market

Steven Nasiri, Sandra Winkler and Ramesh Ramadoss

5.1 Introduction

While the military was once the prevailing market driver for semiconductor products, the consumer market has been dominating the demand for electronic goods in recent years. The consumer market is very price-sensitive and changes happen quickly and frequently. The semiconductor market is aptly suited to these changing needs because it relies on many standardized parts that are created in a standardized CMOS process, which can be mixed and matched as per the consumer needs. Semiconductors, which include both integrated circuits (ICs) and discrete devices, have been shrunk to amazingly small size and placed in small packages. This is very desirable in smaller consumer products which are often handheld. The package provides the mechanical and electrical interface to the printed circuit/wiring board (PCB or PWB) and allows the device to be tested before it is attached to a PCB. The addition of the package basically converts a semiconductor "die" into a "chip" [1].

Microelectromechanical systems (MEMS) devices are inherently small and thus well suited to the consumer market. However, the lack of standardized products and production processes have hampered the ability of these small devices to have a larger impact in the fast-paced consumer market. A MEMS device by itself is just a sculptured (micro-machined) piece of silicon. Electronics and signal conditioning circuitry must be integrated with the MEMS device for "reading" the real-world parameters. Traditionally, packaging micro-machined silicon with electronics, subsequent signal conditioning and calibrations, have generally accounted for over 70% of the overall cost of all MEMS devices. The reason for this is that while creating the MEMS device can be somewhat straightforward, packaging and testing of these devices is a lot trickier, as well as non-standardized. MEMS devices use a release mechanism, which is activated when a tipping point is reached. The release mechanism can neither be used during the packaging process nor during the testing of the final device. Finding ways to package and test MEMS devices without activating the release mechanism have resulted in multiple packaging approaches as the MEMS devices are all different. Further, all these packages deviate from the standards in the world of semiconductor packaging. Designing and assembling unique packages, and figuring out how to test them without activating them, results in higher costs due to the loss of mass manufacturing economies of scale. In addition, MEMS devices must be packaged so that they are hermetically sealed and protected from the environment. Another key obstacle in the growth of traditional MEMS is the fact that every MEMS product historically had its specialized process, package, and test equipment. Thus, the time to market in developing any new MEMS product typically ranged from 5 to 7 years with excessive investment over $50M [2].

5.2 MEMS Motion Sensor Products

MEMS motion sensors (MMS) comprise of MEMS accelerometers and gyroscopes. Motion sensors can be used to detect accident in a vehicle by detecting a loss of speed and tilt. This application has been hailed as life savers in deploying air bags in cars. Motion sensors are now in every vehicle sold in the USA and have ignited a whole new wave of applications: eliminating the hand jitter in digital still cameras, dead reckoning and heading for assisted GPS, advance motion sense-enabled game consoles, gesture recognition and intuitive user interfaces, GPS-assist for personal navigation devices (PND), and the location-based services (LBS) for the cellphones. Critical requirements for all these motion-sensing applications include low cost, multi-axis, and integrated gyroscopes [2, 3].

5.2.1 *Difficulties Associated with Packaging Gyroscopes*

Gyroscopes are exceptionally challenging sensor products to design and fabricate than other high-volume MEMS products such as accelerometers or pressure sensors. MEMS gyros operate on the principle of vibrating masses and hence are equivalent to manufacturing two high-performing MEMS devices into one single unit: a self-tuned resonator for creating the sensing force (Coriolis), and a micro-g accelerometer for sensing this force, which is proportional to the rate of rotation. Both must work in tandem to produce precise results. The Coriolis force arises in a rotating reference frame and is proportional to the angular rate of rotation. The absolute magnitude of the Coriolis force sensed is orders of magnitude lower than any high-volume production MEMS accelerometer. Generally, capacitive sensors are used for measuring these minute changes in capacitance. Gyroscope performance is very sensitive to any potential manufacturing variations, packaging stresses, linear acceleration, and changes in ambient temperature. To achieve high performance and low cost, great care must be taken during the initial design to achieve a solution that is inherently less sensitive to such potential variations.

5.3 The Nasiri Fabrication Solution

The Nasiri Fabrication (NF) process (named after its inventor, Steven Nasiri, Founder, InvenSense, Inc.) is the first standardized process that leverages off-the-shelf CMOS process and allows for the fabrication of fully packaged MEMS devices using the same standard process [2–6]. The NF process was developed to provide a high-performance motion sensing solution, which meets the challenging size and cost requirements of the mobile consumer electronics market. Initially, InvenSense developed a 2-axis gyroscope, then a 3-axis gyroscope, and subsequently the world's first 6-axis motion sensor that integrates a 3-axis gyroscope with a 3-axis accelerometer on a single chip. The patented NF process overcomes the major challenges for adoption into handheld consumer products, allowing for the smallest footprint, lowest cost, and highest functionality of any motion sensor in the market.

5.3.1 *Motion Sensing Products*

InvenSense pioneered the world's first integrated motion sensor (3-axis and 6-axis) products that meet the high volume, low cost, low power, and small size requirements of mobile consumer products. Thus, addressed the key challenges when incorporating motion sensors in these products. The company's product couples a MEMS device with companion CMOS silicon electronics in a single hermetically sealed wafer-level package (WLP) to combine the mechanical portion of MEMS with the "intelligence" portion of electronics simultaneously. The MEMS devices were fully integrated with the electronics at wafer level and could be functionally tested prior to the final packaging. This approach provides an enormous efficiency in yield and overall cost advantage. Another key advantage of this approach is that it allows the integration of MEMS with standard off-the-shelf CMOS process, drastically cutting the cost and time to market. By integrating the MEMS and CMOS at wafer level, the resulting known-good-dies (KGDs) were packaged just like any other semiconductor product and thus dropping the backend cost to around 15% of the overall cost.

5.3.2 *NF Process*

NF process is a simple five-mask, MEMS-specific fabrication process that is compatible with the standard CMOS process (see Fig. 5.1) [6]. First the MEMS wafer is created and then stacked on top of an off-the-shelf CMOS wafer. Its patented wafer-to-wafer bonding process allows for direct integration of a fabricated MEMS wafer with any off-the-shelf CMOS wafer. Although the wafer-bonding process uses off-the-shelf wafer-bonding equipment, the bonding process itself is InvenSense's proprietary process. The process involves eutectic bonding of the MEMS wafers directly to the aluminum layer on the CMOS wafer without addition of any other material layer on top of the aluminum.

The handle wafer is a conventional double-side polished wafer with alignment marks patterned on one side using the first mask. Using these alignment marks, a cavity is defined on the opposite side of the wafer using the second mask. This cavity accommodates the moving structures of the MEMS device. On the cavity side of the handle wafer, an oxide layer is deposited by PECVD. The handle wafer is then fusion-bonded to a standard wafer, which is called the MEMS device wafer. The oxide layer on the handle wafer provides electrical isolation between the device wafer and the handle wafer. The device wafer is then thinned back to the desired MEMS membrane thickness by conventional grinding and polishing techniques. The final membrane thickness can range from tens to hundreds of microns. The result of the thinning process is an "Engineered" silicon on insulator (SOI) wafer.

The third mask is used to define standoffs on the device wafer. These standoffs have two functions: one to define the hermetic seal ring area for the MEMS device, and the second to provide the electrical interconnections to the CMOS wafer. The height of the standoff etch determines the vertical

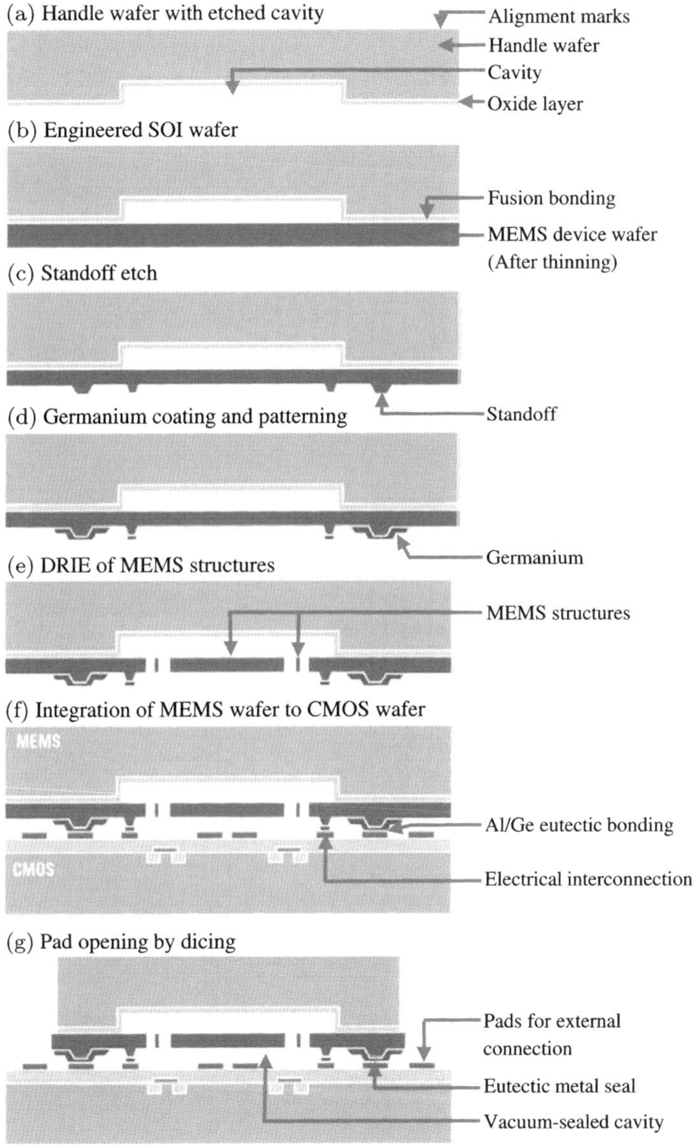

Fig. 5.1. NF five-mask process flow: (a) handle wafer with etched cavity, (b) engineered SOI wafer, (c) standoff etch, (d) germanium coating and pattering, (e) DRIE of MEMS structures, (f) integration of MEMS wafer to CMOS wafer, and (g) pad opening by dicing [6].

gap between the MEMS and CMOS wafer and it can range from few to several hundred microns. On the device wafer, a germanium-based material is deposited. Then, the fourth mask is used to pattern the germanium layer over the seal ring and standoffs. The deposition and etching of this material is readily achievable.

The fifth and most critical mask defines the actual micromechanical structures of the MEMS device, where masses, beams, and comb fingers are patterned into the membrane of the device wafer. Deep Reactive Ion Etching (DRIE) process is used to create these high-aspect ratio MEMS structures. When etching is completed through the thickness of the membrane, the movable MEMS structures are released. It is important to note that for device performance or fabrication issues, any fine tuning of the MEMS design can be achieved by making changes to only this fifth mask.

Wafer integration involves aligned eutectic bonding in a sub-atmospheric pressure environment. Specifically, the integration involves eutectic bonding between the patterned germanium (Ge) layer on the MEMS wafer and the top aluminum (Al) layer on the CMOS wafer. The surfaces of the MEMS and CMOS wafers are prepared for bonding with commonly available wet and dry processing steps. The bond is a thermocompression eutectic metal bond, which creates a hermetic seal encompassing the MEMS structures as well as the hundreds of electrical interconnections for each device.

The finished wafer then goes through yet another patented and proprietary pad opening process that uses a standard sawing technique to remove any unnecessary silicon from the MEMS wafer. This pad opening step exposes all the electrical pads necessary for external electrical connections. Finished wafers are then tested on standard automated wafer probers. Due to the novel wafer-scale packaging, the subsequent packaging can be completed cost effectively in plastic packages. Thus, packaging can be performed at any industry-standard contract assembly house, using a QFN package for instance. Figure 5.2 shows InvenSense's 6-axis motion sensor that integrates a 3-axis accelerometer with a 3-axis gyroscope on a single die. The die was packaged inside a $3 \times 3 \times 1$ mm QFN plastic package [7].

The NF process is compatible with already fabricated CMOS wafers. Further, single-crystal silicon (SCS) is a more consistent, reliable, and high-yield material than polysilicon films. It permits the design of much thicker (10× to 20×) proof masses, compared to surface micro-machining process. Thus, SCS enables fabrication of high sensitivity and high-performance gyroscopes per square footprint.

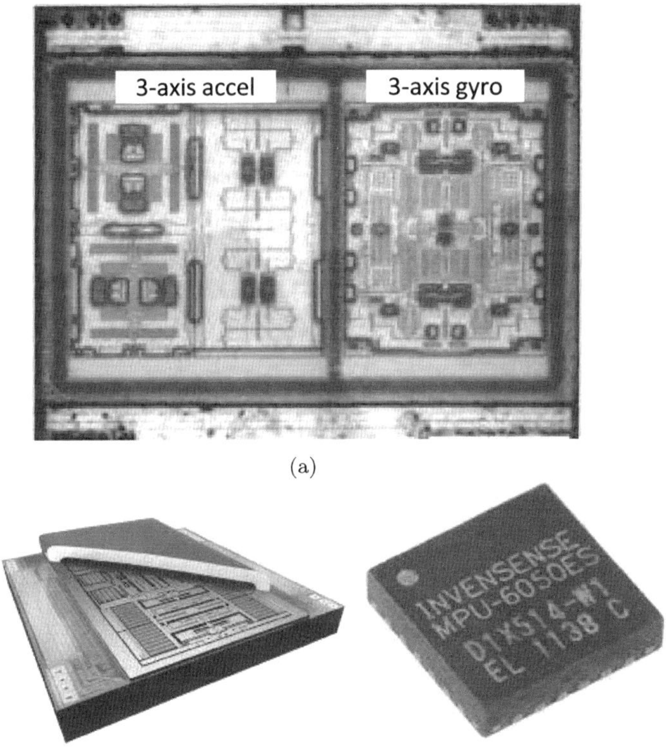

(a)

(b) (c)

Fig. 5.2. InvenSense's single-chip 6-axis motion sensor: (a) an infrared image of the die with a 3-axis accelerometer and a 3-axis gyroscope, (b) a schematic view of the die, and (c) the die packaged inside a $3 \times 3 \times 1$ mm QFN plastic package [7].

InvenSense's motion sensors have been integrated into many consumer electronic devices with initial market entry into digital still cameras and later into mobile phones, GPS devices for dead reckoning, and 3D peripherals (e.g., 3D mice, TV remote controllers, game controllers, etc.). InvenSense has been the market leader in the motion sensors industry and can be credited with the creation of this market. Further, InvenSense's motion sensors have been the key enabler of products such as drones, hover boards, virtual reality, wearables, IoT devices, and many other new consumer products that incorporate motion as a main differentiating innovation.

The NF process can support a wide range of MEMS devices, including gyroscopes, accelerometers, rotation sensors, pressure sensors, resonators, ultrasound transducers, finger print sensors, microphones, actuators, and potentially many other MEMS devices. This process is particularly suitable for MEMS capacitive-type sensing devices and products requiring actuators and/or requiring very low-level signal detection.

5.3.3 *Benefits of NF Process*

The advantages of NF process originate from its simplicity and efficiency in integration of bulk silicon MEMS with CMOS electronics at the wafer level. Highly integrated electronics enable the tuning and trimming capabilities necessary for high-performance sensors such as gyroscopes.

The five-mask process provides versatility and supports a multitude of MEMS devices including actuators and sensors. Again, the simplicity and integration of the electronics with MEMS makes the process valuable. Low noise, low capacitance detecting circuits can be integrated with high-voltage drivers delivered from the CMOS electronics.

The process flow can be slightly tuned to optimize the device thickness and etch depths to support different electrostatic gaps. Additionally, the cavities can be shaped to tune damping characteristics. Standard design tools and verification platforms can be provided for the electromechanical design and other parameters needed by the MEMS designer to develop the final product. This not only includes generic fabrication design rules but extends to electronic integration rules such as routing and shielding parameters to ensure CMOS-to-MEMS integration performance.

A successful high-volume MEMS product demands not only a versatile process but also a robust design to counteract inherent variations within the process. Complex high-performance motion sensors must be modeled to reliably predict performance over process corners. The simplicity of the NF process makes feasible the accurate modeling of complete parametric variations including critical line widths, gaps, thicknesses, cavity pressures, etc. Custom system simulation tools have been developed and validated by finite-element analysis and verified through experimental characterization.

The operation of the InvenSense's motion sensor products requires careful design of the structural resonant modes. Also, a full-system electromechanical model to predict performance attributes such as noise and sensitivity that could impact the physical structure of the device. The simulation platform has been leveraged to improve robustness and yield of both existing products and new designs. The modeling too can predict device yield based on known manufacturing variations.

In 2012, InvenSense announced the availability of the NF process to external users through its NF-Shuttle service [8]. The NF-Shuttle apportions costs amongst multiple users as they can purchase "seats" on the same mask set. This service reduces upfront development costs for participants to a fraction of the total costs and provides an opportunity to verify advanced designs and prototypes in silicon. This service effectively launched the standardization of a fabrication process to the MEMS industry. InvenSense has used the same exact NF platform to develop MEMS microphones and pressure sensors.

5.4 Concluding Remarks

NF process is the first standardized process that leverages off-the-shelf CMOS fabrication and allows for the development of various MEMS devices using the same standard process. The strength of the NF platform is that by adding additional five-mask process steps, it allows for electronics integration and full wafer level packaging of MEMS devices. It shortens time to market from 5 to less than 2 years. It enables the cumulative learning and yield enhancement from previous products to be passed down to subsequent products as the process and packaging stays the same. Yields of more than 95% can be achieved with the NF process.

The NF process offers a cost-effective solution to several integration and packaging challenges faced by MEMS motion sensors products. Specifically, the NF process enables direct integration of the MEMS wafer to the CMOS electronics wafer through eutectic bonding between a germanium layer deposited on the MEMS wafer and the top aluminum layer on the CMOS wafer. This bonding step simultaneously provides electrical interconnections between the MEMS and CMOS wafers, and hermetic sealing of the sensitive MEMS structures. This novel packaging approach enabled a new price-performance benchmark that is more than 10× lower than its closest MEMS counterpart.

In 2006, InvenSense shipped its first generation of 2-axis MEMS gyroscopes for the digital still camera market. In 2009, InvenSense announced the world's first 3-axis gyroscope with embedded Digital Motion Processor™ (DMP) for the smartphone market. In 2010, InvenSense announced the world's first 6-axis MotionTracking™ device that integrated a 3-axis gyroscope and 3-axis accelerometer on a single chip. In 2012, InvenSense announced the world's first 9-axis motion sensor that combined a 3-axis gyroscope, 3-axis accelerometer, and 3-axis magnetic compass with an onboard DMP. In 2016, InvenSense announced the world's lowest power 9-axis MEMS motion sensor at only 2.65 mW and packaged in a $3\times3\times1$ mm QFN plastic package. To date, InvenSense has shipped over 1 billion devices and its products are prevalent in consumer electronics.

References

1. Winkler, S. (2008). *MEMS Packaging Markets and Trends*, Electronic Trend Publications.
2. Nasiri, S. and Lim, M. (2007). New innovations in MEMS fabrications are responsible for meeting the demand for low-cost inertial sensors in consumer market, *40th International Symposium on Microelectronics*, IMAPS, San Jose, CA, USA.
3. Nasiri, S. (2007). Wafer-scale packaging and integration are credited for new generation of low-cost mems motion sensor products, *International Wafer Level Packaging Conference*, IWLPC, San Jose, CA, USA.
4. Nasiri, S. *et al.* (2006). Vertically integrated MEMS structure with electronics in a hermetically sealed cavity, *US Patent No. 7,104,129.*
5. Seeger, J., Lim, M., and Nasiri, S. (2010). Development of high-performance, high-volume consumer MEMS gyroscopes, in *Solid-State Sensors, Actuators, and Microsystems Workshop*, pp. 61–64.
6. Daneman, M. (2012). Applying the CMOS test flow to MEMS manufacturing, *4th Annual MEPTEC MEMS Testing and Reliability Symposium*, Santa Clara, CA, USA.
7. http://www.chipworks.com
8. Assaderaghi, F., Daneman, M., and Lim, M. (2012). *NF-Shuttle: Standardizing the Fabless MEMS Industry*, InvenSense Inc., Tech. Rep., San Jose, CA, USA.

Steven Nasiri earned an MBA degree from Santa Clara University, a MS degree in Mechanical Engineering from San Jose State University and a BS degree in Mechanical Engineering from the University of California, Berkeley. Currently, he is an active angel investor at Nasiri Ventures, Chairman of Nasiri Foundations, Co-Chair of Wearables and Consumer group at Band of Angles. Mr. Nasiri has been a MEMS veteran and a serial entrepreneur with experience in launching six startup companies. His most recent and successful venture was InvenSense, which he founded in 2003 and served as the President, Chief Executive Officer, and Chairman until he stepped down in October 2012. Under his leadership, the company became the pioneer and global market leader in motion processing solutions for motion-based user interfaces in consumer electronic products, such as smartphones, tablets, game consoles, wearables, and drones. In November 2011, Mr. Nasiri took the company through the initial public offering (IPO) and got listed on the New York Stock Exchange (NYSE) under the symbol INVN. Prior to founding InvenSense, Mr. Nasiri held various key positions as a co-founder and/or executive of several pioneering startup companies in Silicon Valley, including SenSym (acquired by Honeywell), NovaSensor (acquired by General Electric), Integrated Sensor Solutions (acquired by Texas Instruments), and Intelligent Sensing Solutions (acquired by Maxim Integrated). Mr. Nasiri has been the inventor or co-inventor in nearly 100 patents and patent applications, and has authored many papers and articles on MEMS and its applications in consumer electronic products. In 2010, he was selected by Ernst & Young as Entrepreneur of the year for Northern California. In 2013 he was given Alumni Awards of Distinction by San Jose State University.

Sandra Winkler has an MBA degree from Santa Clara University. Ms. Winkler is currently the Webmaster and Editor of the IEEE San Francisco Bay Area e-Grid, having full control over content and design, selling and creating ads, and managing the production schedule and accounts. Before being Editor of the E-Grid, Ms. Winkler was an Industry Analyst producing market research reports on IC (integrated circuits) packaging. From conceptualization through editing, Ms. Winkler expanded the breadth of reports on this topic, increasing revenue from IC packaging reports by a 500% in four years. Ms. Winkler has spent the last ten years as a member of the executive committee for the IEEE CPMT (components, packaging, and manufacturing technologies) Chapter in Santa Clara Valley Section.

Ramesh Ramadoss received his Bachelor of Engineering degree from Thiagarajar College of Engineering, Madurai Kamaraj University, India in May 1998 and PhD degree in Electrical Engineering from the University of Colorado at Boulder in May 2003. From Jun 2003 to Dec 2007, he was employed as an Assistant Professor in the Department of Electrical and Computer Engineering at Auburn University, Auburn, AL. From Jan 2008 to Mar 2012, he was employed as a Manager at FormFactor Inc., Livermore, CA. From Apr 2012 to May 2015, he was employed as a Senior Manager at Microprobe Inc. (Acquired by FormFactor Inc.), San Jose, CA. From Jun 2015 to Feb 2016, he was employed as a Technical Director at Applied Nanostructures Inc., Mountain View, CA. Currently, he is the President & CEO of IoT Technologies Inc., San Ramon, CA. He is the author or coauthor of 4 book chapters and 55 papers. He has conducted R&D projects for DARPA, NASA, US Army, US Air Force, Sandia National Labs, and Motorola Labs. In 2013, he was elevated to the grade of IEEE Senior Member. In 2014, he founded the IEEE San Francisco Bay Area MEMS & Sensors Chapter. In 2017, he was elected as the Chair of the IEEE Santa Clara Valley Section, the world's largest IEEE Section located in the heart of Silicon Valley.

Chapter 6

PCB Based MEMS and Microfluidics

Ramesh Ramadoss, Antonio Luque and Carmen Aracil

6.1 Introduction

Microelectromechanical systems (MEMS) has proved to be a revolutionary technology in many applications, including inkjet nozzles, pressure sensors, accelerometers, gyroscopes, microphones, micromirrors, and microfluidics. Conventional MEMS devices consist of deflecting or vibrating mechanical microstructures (levers, springs, membranes, and hinges) fabricated primarily in silicon. MEMS processes involve thin-film processes and other specialized micro-machining techniques that have evolved primarily from the semiconductor industry [1].

In electronics packaging and assembly industry, Printed Circuit Board/Printed Wiring Board (PCB/PWB) is used as a substrate to provide mechanical support and electrical interconnections among various electronic components. PCB technology has been standardized and reliability has been continuously improved over the past several years. The manufacturing infrastructure for PCB is not expensive as it involves low-cost processes and equipment. Typical manufacturing processes for PCBs include metallization (plating, lamination or evaporation), photolithography, wet etching, multilayer lamination, and via interconnection. The substrate is typically a dielectric polymer with laminated copper sheets. FR4 (woven glass and epoxy) and PTFE (Teflon) are the most commonly used substrate materials. The conductive traces, pads, and other circuit features are etched into the copper sheets. Various electronic components are attached to the copper traces through solder reflow process to form functional electronic circuits. PCBs can be single-sided (one copper layer), double-sided (two copper layers) or multilayer (outer and inner layers). Vias are used to connect conductors on different layers [2].

Flexible PCB (FPCB), also known as Flexible circuits, is a technology for assembling electronic circuits by mounting electronic components on flexible plastic substrates, such as polyimide, Liquid crystal polymer (LCP), Polyethylene terephthalate (PET), and Polyethylene naphthalate (PEN). The circuits are formed by etching copper metallization (laminated or evaporated), plating metal or printing of conductive inks. Types of flexible circuits, include one metal layer, double-sided, multilayer, and rigid flex circuits [2].

In the past two decades, several research groups have explored the use of PCB and/or FPCB processes for fabrication of MEMS devices, known as PCB MEMS [3–18]. The advantages of PCB MEMS include low cost, ease of integration with electronics, and suitability for large area applications. Table 6.1 compares some common parameters and processes of conventional silicon MEMS and PCB MEMS technologies. Table 6.2 shows

Table 6.1. Comparison of common parameters and processes between silicon and PCB MEMS.

Parameters	**Silicon MEMS**	**PCB MEMS**
Resolution	5 μm	100 μm
Clean room class	100–1000	1000–10000
Metal layers	2–4	2–12
Metallization process	Additive	Subtractive/Additive
Layer build up	Anodic bonding, Direct bonding, Adhesive bonding	Eutectic bonding, Adhesive bonding (Lamination)

Table 6.2. Comparison of mechanical and electrical properties of silicon with commonly used PCB MEMS materials [1, 2].

Parameters	**Si < 100 >**	**FR4**	**PTFE**	**Kapton**	**LCP**
Density ($g\,cm^{-3}$)	2.3	1.85	2.16	1.42–1.53	1.4
Tensile strength (MPa)	7000	70–75	31	200–234	180
Tensile modulus (GPa)	130	21–24	0.5	2.5–4	7–22
Glass Transition Temperature (°C)	—	115–125	116	360–410	145
CTE (ppm/°C)	2.5	12–17	100	20	0–30
Dielectric constant	11.7	4.4	2.1	3.5	2.8
Moisture absorption	—	<0.1%	<0.01%	2.8%	<0.02%

Note: CTE is an acronym for coefficient of thermal expansion.

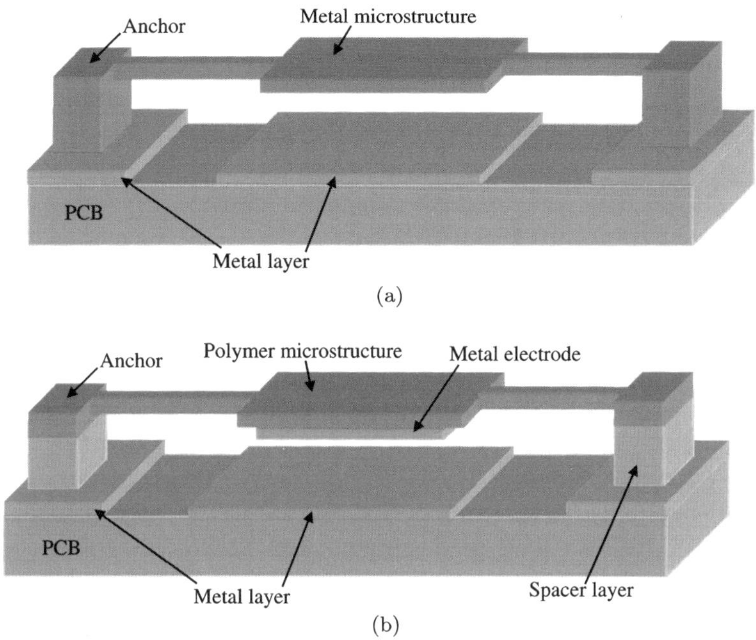

Fig. 6.1. Two types of PCB MEMS devices: (a) metal microstructure and (b) polymer microstructure.

a comparison of mechanical and electrical properties of silicon and some materials commonly used for PCB MEMS [1, 2].

As shown in Fig. 6.1, there are two types of PCB MEMS devices depending on how the active mechanical microstructures are fabricated on the PCB: metal microstructure (*Type A*) and polymer microstructure with patterned metal layer on one side or both sides (*Type B*). Since these devices are built using packaging processes and materials, it is easier to design the package into the device and the package is manufactured at the same time as the device. Thus, this approach often enables fabrication of self-packaged MEMS devices. The goal of this chapter is to provide a state-of-the-art review of various actuators, sensors, and microfluidic devices fabricated using PCB MEMS technology.

6.2 Actuators

In this section, electrostatic and thermal actuators fabricated using PCB MEMS technology are discussed.

6.2.1 *Electrostatic Actuator Based RF MEMS Switches*

Electrostatic actuator based RF MEMS series switch on micro-strip line [3] and shunt switch on coplanar waveguide (CPW) [4] have been implemented using metal structure (*Type A*). In this section, the RF MEMS switch from Motorola Labs [3] designed using a cantilever beam metal structure is discussed. Figure 6.2 shows the single-pole, single-throw (SPST) series MEMS switch that consists of a two-part metal cantilever beam suspended over a micro-strip line on a PCB substrate. One part forms the top electrode of the electrostatic actuator and the other part serves as the RF signal contact pad for the micro-strip line. In the cantilever beam up-position, the switch contact pad is in the "Open" state (OFF state) and no signal passes through the micro-strip line. When a dc voltage is applied on the actuator electrodes, the top electrode is pulled down and causes the contact pad to collapse onto the micro-strip line to "Close" the RF circuit by providing an ohmic contact in the beam down-position (ON state). When the applied dc voltage is removed, the mechanical stiffness of the cantilever beam brings it back to its "Open" state in the up-position. The MEMS switches were

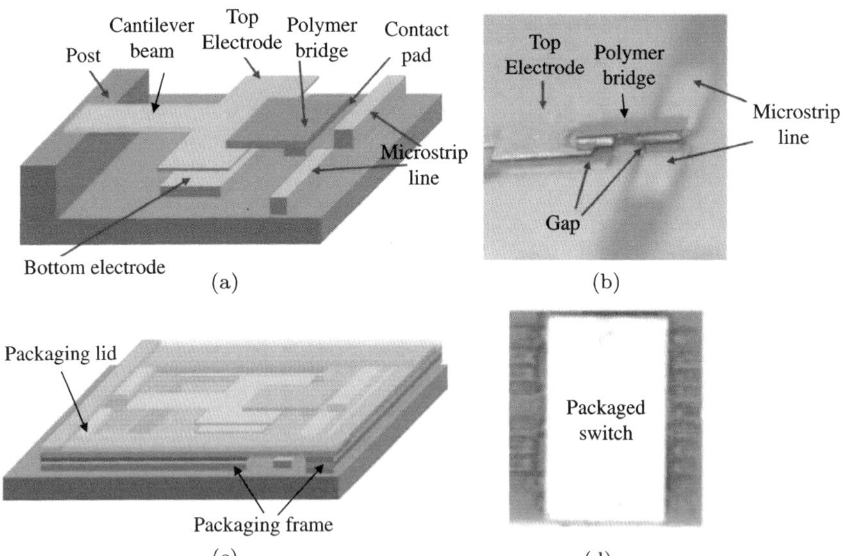

Fig. 6.2. Metal cantilever beam based RF MEMS series contact switch on PCB: (a) a schematic configuration, (b) a photograph of the switch, (c) a schematic configuration of the package, and (d) a photograph of the packaged switch [3].

fabricated using standard PCB processes on Rogers EXP4350B substrate. A dielectric packaging frame was co-fabricated during the switch fabrication process. The package lid was prepared on a 2 oz copper clad epoxy laminate and then attached to the packaging frame and cured. This non-hermetic package provides protection from air borne particles as well as mechanical protection from shock and impact loading. The pull-down voltage is >30 V. For a nominal air gap of 12 μm between the microstrip line and contact pad, the switch provides an isolation >30 dB at 2 GHz in the OFF state (up-position). In the ON state (down-position), the switch provides an insertion loss <0.3 dB in the frequency range of 0.01–10 GHz.

In [5], Kapton polyimide membrane (*Type B*) based RF MEMS capacitive switches fabricated using PCB MEMS technology was reported. As shown in Fig. 6.3(a), the switch consists of three layers: RT/Duroid 6002 substrate (30 mils), polyflon adhesive spacer film (2 mils), and Kapton-E polyimide film (2 mils). CPW lines are patterned on the copper layer of the substrate. A 1 μm thick photosensitive benzocyclobutene (BCB) layer is spin-coated and patterned over the CPW lines. MEMS top electrode is patterned on the copper metallization on the Kapton film. To reduce warpage due to thermal mismatch, identical copper electrodes were patterned on the top and bottom side of the Kapton switch membrane. The Kapton film was machined using an excimer laser to create the required mechanical suspension flexure design. The polyflon adhesive spacer film present between the substrate and Kapton film provides the required up-position gap height. The top electrode is biased using resistive dc bias lines. When a dc bias voltage is applied between the top electrode (through the resistive bias line) and the bottom CPW metallization on the substrate, the Kapton membrane deflects due to electrostatic force of attraction. When the applied voltage exceeds the pull-down voltage, the top electrode snaps down onto the BCB dielectric layer. Figure 6.3(b) shows a photograph of a typical capacitive shunt switch with various parametric dimensions: G/S/G = 0.1 mm/1.6 mm/0.1 mm, d = 0.2 mm, l = 2 mm, w = 1 mm, L_A = 1.375 mm, L_B = 0.75 mm, L_C = 1.375 mm, L_D = 1.0 mm, z_A = 47 μm, and z_D = 1 μm. The pull-down voltage is in the range of 100–120 V. The switch provides an insertion loss <0.4 dB in the ON state (up-position) and isolation >10 dB in the OFF state (down-position) in the frequency range of 8–30 GHz.

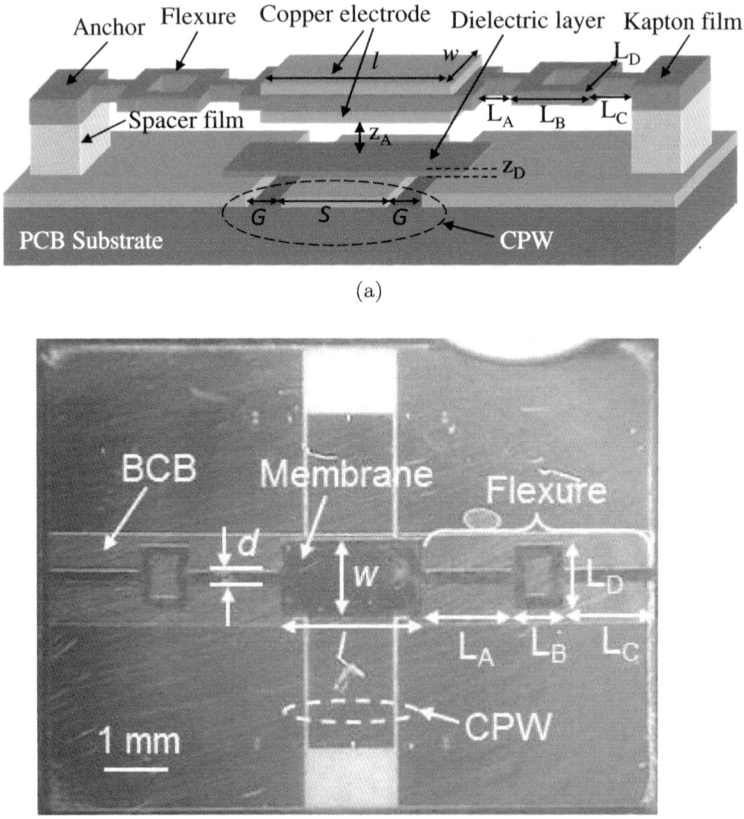

(a)

(b)

Fig. 6.3. Kapton polyimide membrane based RF MEMS capacitive switch over a CPW line on the PCB: (a) a schematic configuration, and (b) a photograph of an RF MEMS switch [5].

6.2.2 *Thermal Actuators*

In comparison with electrostatic actuators, thermal actuators hold promise for larger displacements and low actuation voltages. A process for fabricating MEMS metallic thermal micro-actuators (*Type A*) directly on a PCB has been described in [6]. Figure 6.4 shows nickel-based micro-actuators with two- and four-arm thermal actuator configurations for RF MEMS switch applications. The process starts with microwave-grade PCBs, RO-3000 series from Rogers Corporation, with approximately 15 μm thick laminated copper. A photoresist mold for the electroplating of Ni was formed on the copper layer by spin-coating a photoresist material followed by

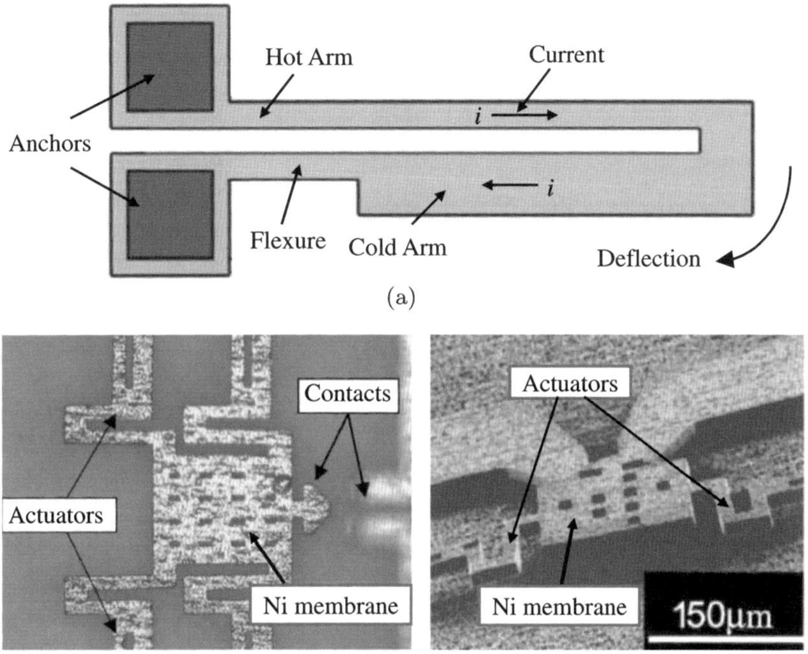

(a)

(b) (c)

Fig. 6.4. PCB MEMS metallic thermal micro-actuator: (a) a schematic configuration of the thermal actuator, (b) a photograph of a micro-actuator with four-arm thermal actuators, and (c) an SEM image of a micro-actuator with two-arm thermal actuators [6].

photolithography. A commercially available nickel sulfamate plating bath (Microfab Ni100, Enthone, RI) was used to deposit 16.5 μm of nickel, using a 290 mA (1.1 ASD) deposition current over 65 mins of plating time. The micro-actuator was released through timed wet etching of the underlying copper in ammonium persulfate ($(NH_4)_2S_2O_8$) solution. The minimum lateral feature size of these micro-actuators is in the range of 10–15 μm. The thermal micro-actuator provided a maximum measured displacement of 13 μm with a driving voltage of 0.21 V and a driving current of 0.21 A per actuator.

6.3 Sensors

Several sensors such as pressure sensor, accelerometer, tactile sensor, flow sensor, and microphone have been demonstrated using PCB MEMS technology.

6.3.1 *Pressure Sensor*

Capacitive-type pressure sensors manufactured using PCB/FPCB techniques have been reported in [7–11]. Three representative examples are discussed in this section. In [8], PCB MEMS capacitive pressure sensors (*Type B*) fabricated using LCP materials is reported. In Fig. 6.5, the schematic cross section and a photograph of the fabricated prototype are shown. The pressure sensor consists of a deflectable circular membrane formed by a sandwich of a 2-mil thick top LCP layer (with the top electrode), a 2-mil thick LCP spacer layer (with the circular hole), and a 30-mil thick LCP substrate (with the bottom electrode). The bottom and top electrodes form a parallel-plate capacitor configuration with a sealed cylindrical cavity at atmospheric pressure. When the external pressure (outside the sensor cavity) is increased above the atmospheric level, the circular top LCP membrane deflects down towards the bottom electrode and thus increases the capacitance of the sensor from its initial value. An IC chip was used to convert the change in capacitance to a corresponding change in the output voltage. A typical pressure sensor with a diaphragm diameter of 3.2 mm provides a total capacitance change of 0.277 pF for an applied pressure in the range of 0–100 kPa.

In [9], a flexible wireless capacitive pressure sensor (*Type B*) was developed using LCP by Cardio MEMS, Inc. In 2014, this spinoff startup company from Georgia Tech was acquired by St. Jude Medical, Inc. As shown in Fig. 6.6, the pressure sensor consists of a cavity formed by two parallel capacitor plates interconnected with an inductance. The sensors were

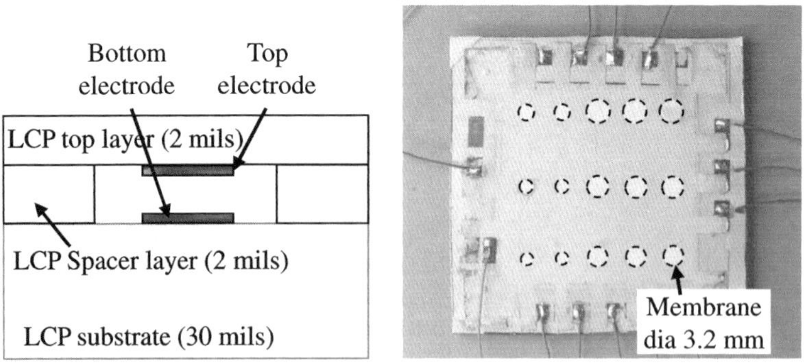

(a) (b)

Fig. 6.5. PCB MEMS capacitive pressure sensor fabricated using LCP: (a) a schematic cross section and (b) a photograph of the fabricated prototype [8].

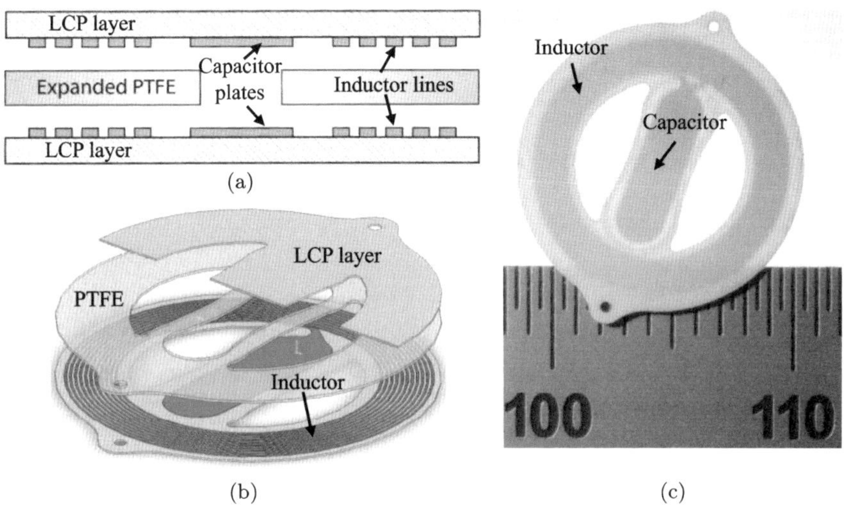

Fig. 6.6. Flexible LCP based MEMS capacitive pressure sensor: (a) a schematic cross section, (b) layer stack up configuration, and (c) a photograph of the fabricated sensor shown next to a ruler (scale in cm) [9].

fabricated using standard FPC processing techniques from sheets of copper-clad LCP and expanded PTFE-based inner bonding layers. The copper cladding on LCP was patterned to form the capacitor plates and inductor by photolithography and wet-chemical etching. Then, the three layers were laminated at atmospheric pressure to form the self-packaged pressure sensor. The distributed inductance and capacitance of the pressure sensor form an LC circuit, whose resonant frequency can be wirelessly measured using an external magnetic loop. When the external pressure increases above the internal pressure of the sensor, the LCP layers deflect inward and thus increases the capacitance of the sensor. This increase in capacitance of the sensor decreases the resonant frequency of the LC circuit. Thus, the sensor pressure can be wirelessly measured by calibrating the change in resonant frequency to a corresponding change in external pressure. The inductor consists of 12 spiral turns with line width of $60\,\mu$m, line spacing of $80\,\mu$m, line thickness of $18\,\mu$m, and starting radius of 3.8 mm. The capacitor plate area is $13.1\,\text{mm}^2$. The sensor design layout range from 1.7 to 2.5 mm and gaps from 25 to $50\,\mu$m. The measured resonant frequency range from 30 to 50 MHz with a linear pressure sensitivity characterized between -1 and $-20\,\text{kHz/mmHg}$.

In [10], a capacitive pressure sensor (*Type B*) was fabricated using two PCBs consisting of a rigid FR4 PCB substrate and a flexible polyimide substrate. The circular electrodes (bottom and top) of radius 3.5 mm were defined on the two PCBs. Using tin/lead solder alloy, these two PCBs were bonded together to form a parallel-plate capacitor with an initial gap of 16.5 μm. The capacitance increased from 20.6 pF to 39.2 pF as the pressure was increased from 0 to 3000 mbar.

6.3.2 *Accelerometers*

Capacitive-type accelerometers manufactured using PCB MEMS techniques have been reported in [12–14]. In [12], a capacitive-type accelerometer (*Type B*) fabricated using the Kapton-E polymide film as the structural layer was reported. As shown in Fig. 6.7(a), the accelerometer proof-mass with four suspension beams is defined in the 2-mil thick Kapton polyimide film. The proof-mass is suspended above a 30-mil thick PTFE substrate using a 2-mil thick spacer. The deflection of the proof-mass is detected using a pair of parallel-plate capacitive sensing electrodes. The top electrode of the accelerometer is defined on the top surface of the Kapton film. The bottom electrode is defined in the metallization layer on the PTFE substrate. The initial air-gap height is equal to the distance between the top surface of the bottom electrode and the bottom surface of the Kapton film. For an applied external acceleration or deceleration (normal to the proof-mass), the proof-mass deflects toward or away from the fixed bottom electrode due to inertial force. This deflection causes either a decrease or increase in the air-gap height thereby either increasing or decreasing the capacitance between the top and bottom electrodes. An example PCB

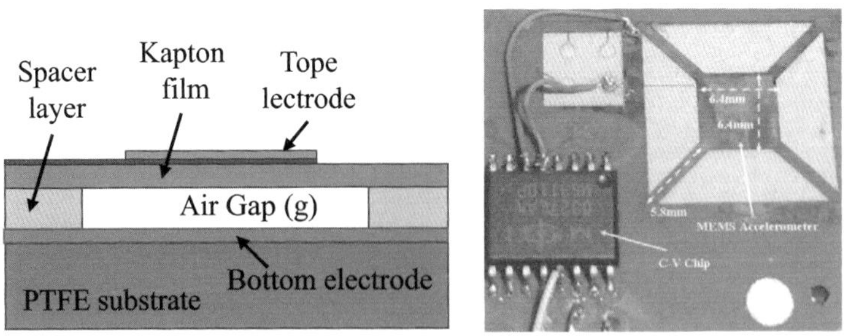

(a) (b)

Fig. 6.7. PCB MEMS accelerometer fabricated using Kapton polyimide membrane: (a) a schematic cross section and (b) a photograph of the fabcricated device [12].

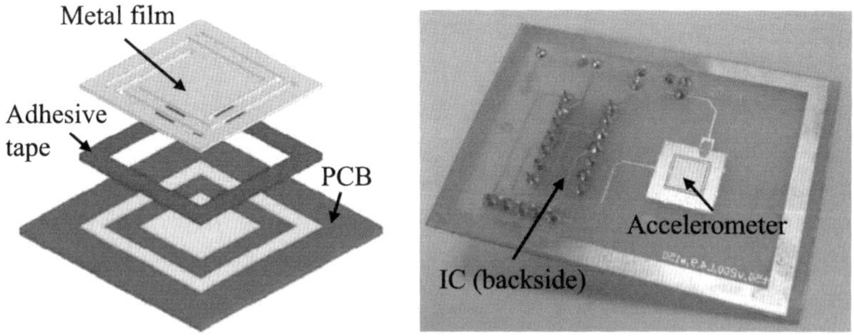

(a) (b)

Fig. 6.8. PCB MEMS accelerometer fabricated using metal film membrane: (a) a schematic configuration and (b) a photograph of the fabricated accelerometer [13].

MEMS accelerometer (see Fig. 6.7(a)) with a square membrane area of 6.4 mm^2 exhibited a resonant frequency of 375 Hz with a Q-factor of 0.52 in air.

In [13] a single-axis PCB MEMS accelerometer (*Type A*) was reported for human motion sensing applications. As shown in Fig. 6.8, the accelerometer consists of a metal film membrane (proof-mass), an adhesive tape, and a PCB substrate. The metal membrane with a suspension structure was etched into the center of the copper film. The bottom electrode was patterned on the PCB substrate. The metal membrane was attached to the PCB using an adhesive tape to form a parallel-plate capacitor configuration with an air gap. When the device is subjected to acceleration, the metal membrane moves and changes the air gap and thus causes a change in the capacitance. The change in capacitance is converted to a voltage output using an IC chip mounted on the backside of the PCB substrate. A typical accelerometer with an outline dimension of 10 mm $\times$ 10 mm exhibited a sensitivity of 0.103 pf/g. The devices were designed to have a dynamic range of ± 5 g and a bandwidth from 0 to 20 Hz, which covers the range of acceleration for human limbs. These results indicate that these devices can meet the requirements for human motion sensing applications.

6.3.3 *Tactile Sensors*

Typically, MEMS tactile sensors are fabricated on silicon and these sensors suffer from the lack of mechanical robustness. In [15], a tactile sensor array (*Type B*) fabricated using LCP material was reported. As shown in Fig. 6.9, the sensor consists of a 20 μm thick LCP membrane with nickel–chromium (NiCr) strain gauges connected at the fixed edge. The membrane

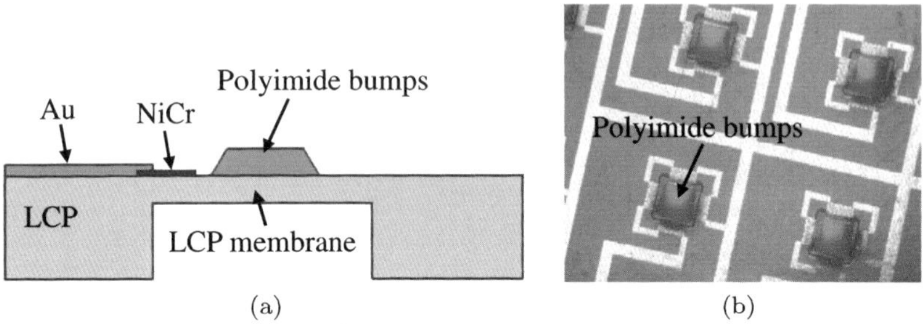

Fig. 6.9. LCP-based MEMS tactile sensor: (a) a schematic cross section and (b) a 2×2 fabricated sensor array [15].

was formed by selective (oxygen-based) plasma etching of a 2-mil thick LCP layer. On top of the membrane, $20\,\mu$m thick polyimide bumps were fabricated using spin coating and photolithography. These polyimide bumps serve as contact surfaces for the tactile sensors. Figure 6.9(b) shows the photograph of a 2×2 array of individual tactile sensor nodes with each sensor node built on a $500\,\mu$m square LCP membrane. The tactile membrane displacement was measured using a precision linearly variable differential transformer (LVDT). For displacements of $5\,\mu$m and $15\,\mu$m, the measured resistance changes were 200 PPM and 800 PPM, respectively.

6.3.4 *Flow Sensors*

In [15], a flow sensor (*Type B*) fabricated using LCP was reported. As shown in Fig. 6.10(a), a fluid flow imparts momentum on the cantilever beam and causes it to deflect and thus induces strain at the fixed base of the cantilever. The strain is sensed using a piezoresistive sensor made of a thin metal film. Figure 6.10(b) shows a $3000\,\mu$m long by $1000\,\mu$m wide LCP cantilever with NiCr strain gauges fabricated at the base. The sensors were subjected to wind tunnel testing with flow rates from 0 to $20\,\text{ms}^{-1}$ and showed a velocity-squared relationship as expected. For a flow-rate of $20\,\text{ms}^{-1}$, the resistance change was 120 PPM.

6.3.5 *Microphone*

In [16], PCB MEMS based fully packaged microphones (*Type A*) has been reported. These microphones consist of thin gold membrane diaphragms fabricated on a copper foil by electroforming. The copper foil is integrated into a standard multilayer PCB laminate stack during fabrication. The copper foil was etched to produce free-standing gold metal membranes. Then,

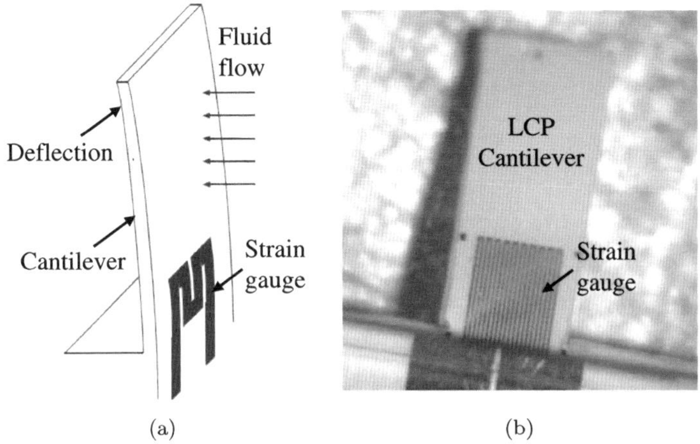

Fig. 6.10. LCP based cantilever flow sensor: (a) schematic illustration of the sensor operation and (b) a photograph of the fabicated flow sensor.

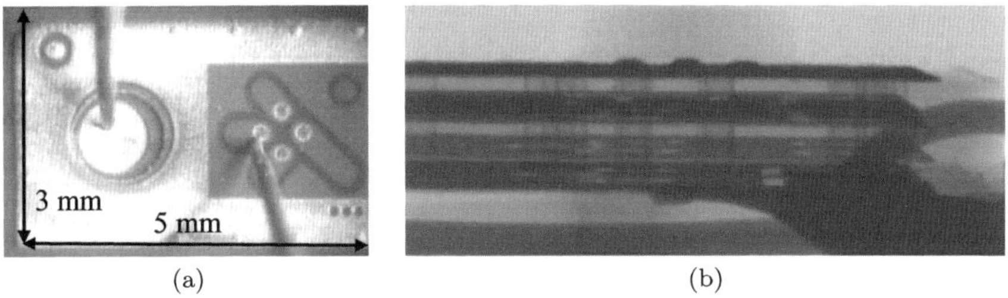

Fig. 6.11. PCB MEMS microphone: (a) a photograph of the self-packaged microphone and (b) an X-ray image of the 12-layer PCB stack up [16].

the gold metal membranes are suspended $25\,\mu$m above a flat metal condenser plate to serve as acoustic transducers. During operation, a suitable high voltage (approx. 60 V) is applied to the membrane and the acoustic vibrations impinging on the membrane produces a corresponding ac voltage on the condenser plate. This induced ac voltage is amplified to generate an acoustic signal from the microphone. Figure 6.11(a) shows a microphone of overall size 3 mm $\times$ 5 mm with an opening acoustic cavity and a vibrating gold membrane ($1\,\mu$m thick, 1 mm diameter). On the back side, the microphone consists of an acoustic cavity and acoustic holes. This microphone was fabricated using a multilayer PCB process consisting of a 12-layer stack up shown in the X-ray cross section image in Fig. 6.11(b). Several ground planes were built into the PCB laminate stack up to provide electromagnetic shielding. The laminate structure serves as a structural material as well as the final packaging material for the microphone. Several microphones were manufactured on a panel and singulated into individual units without the need for additional processing steps for packaging. Thus, this approach enables manufacturing of self-packaged MEMS devices.

6.4 Integration and Applications

6.4.1 *PCB MEMS Module*

In [17], PCB MEMS module (shown in Fig. 6.12) by monolithic integration of mechanical switches, and sensors for pressure, acceleration, temperature, tilt, and humidity has been demonstrated. For operation, the switches use electrostatic actuation and all sensors (except the temperature sensors) use capacitive sensing. The temperature sensor is a resistive device. The processing techniques were similar to those of RF MEMS switches described in Section 6.2.1 [5]. The modules were fabricated using three layers: a PCB substrate, Kapton-E film, and spacer. As shown in Fig. 6.12(a), the copper

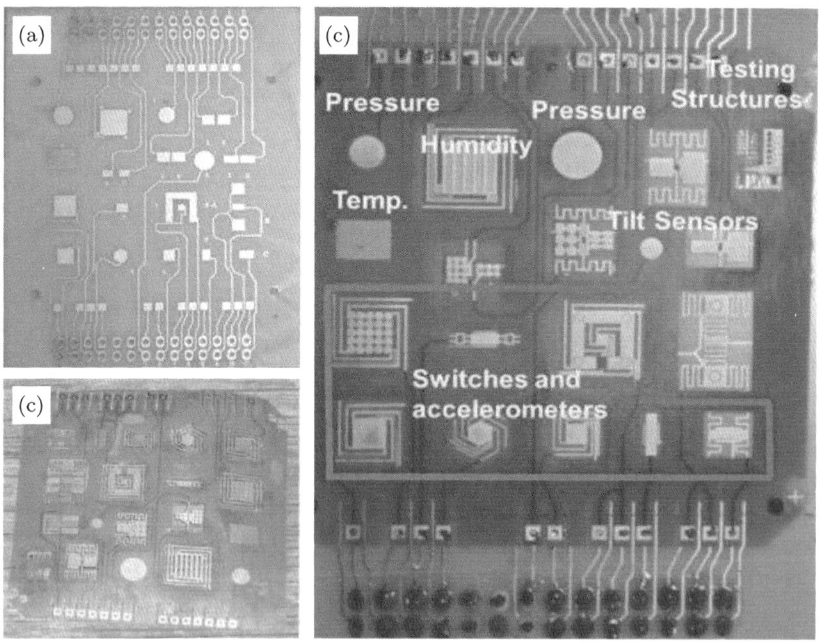

Fig. 6.12. PCB MEMS module with integrated switches and sensors: (a) PCB substrate with patterned ground electrodes, (b) MEMS structures defined in Kapton film using RIE, and (c) an assembled module containing switches and various sensors [17].

layer on the PCB substrate is patterned to create the ground electrodes of capacitive devices, electrical traces, and solder pads. The thickness of the copper layer is 15–20 μm and traces are 100 μm wide. A 1–2 μm thick benzocyclobutene (BCB) dielectric layer is coated over the copper layer.

The Kapton film is 50 μm thick with 5–10 μm thick copper cladding on both sides. The copper cladding on the bottom side of the Kapton film is patterned to define the sensing/actuation electrodes for the capacitive devices. The copper cladding on the top side is patterned to reduce film stress and prevent warping of the Kapton film. The film is micro-machined by reactive ion etching (RIE) process to create MEMS structures shown in Fig. 6.12(b). Finally, the Kapton film is bonded to the PCB substrate by thermocompression bonding using a 1-mil thick Polyflon bonding film. The thickness of this bonding film defines the gap between the ground electrode on the PCB and its sensing/actuation electrodes on the bottom side of each device in the Kapton film. The BCB dielectric layer prevents shorting between the ground and movable electrodes. Finally, solder is used to electrically connect the sensing/actuation electrodes on the bottom side of the Kapton film to their corresponding traces on the PCB. Figure 6.12(c) shows a typical PCB MEMS module after assembly. For some modules, Aluminum Oxide (Al_2O_3) deposited on top of the BCB dielectric layer by atomic layer deposition (ALD) was used as an additional protection layer. The characterization results for various devices were presented in [17].

6.4.2 *Environmental Sensor Module*

Oceanographic biologgers are miniature animal-attached instruments for logging and relaying data about a marine animal's movement, behavior, physiology, and environment. Technologically advanced biologgers can measure conductivity, temperature, and depth (CTD), which allows researchers to calculate salinity. Salinity information can define the importance of physical water mass features (such as frontal systems, currents, eddies, or ice edges) and the ecology of marine animals. Commercially available biologgers are bulky and expensive. In [18], a low-cost, compact CTD biologger for medium-sized marine animals has been reported. PCB processing techniques have been used to fabricate a polymer sensor subassembly, which could be integrated with the electronics of a biologger. A photograph of the completed rigid/flex CTD board (90 mm $\times$ 28 mm) is shown in Fig. 6.13. The board was fabricated using PCB processes and consists of two distinct regions on a continuous single LCP substrate: a rigid sensor section (18 mm $\times$ 28 mm) and a flexible interconnect section (72 mm

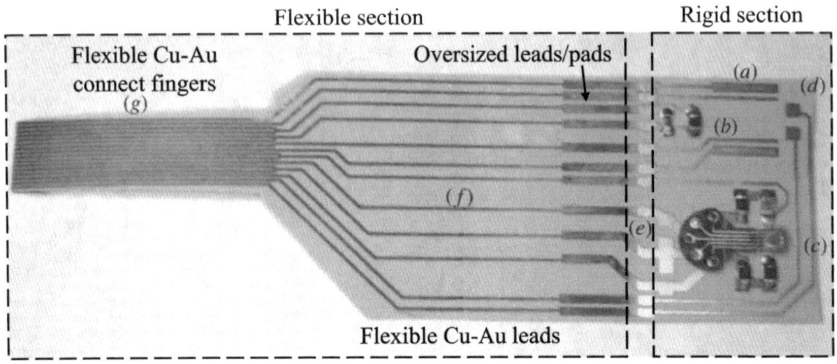

Fig. 6.13. A photograph of the completed rigid/flex CTD board (90 mm × 28 mm): (a) a conductivity cell, (b) a thermistor, (c) a pressure sensor module, (d) a Pt wet/dry sensor, (e) oversized leads/pads, (f) flexible Cu–Au leads, and (g) flexible Cu–Au connector fingers [18].

long). The rigid multi-metal sensor section consists of a conductivity cell, thermistor, pressure sensor module, and Pt wet/dry sensor. The second flexible interconnection section consists of oversized leads/pads, flexible Cu–Au leads, and flexible Cu–Au connector fingers (0.5 mm pitch). The 95% confidence intervals for the conductivity, temperature, and pressure sensors were demonstrated to be $\pm 0.083\,\text{mScm}^{-1}$, 0.01°C, and ±0.135 dbar, respectively. It should be noted that the PCB MEMS pressure sensors discussed in Sec. 6.3 are suitable for integration with this module.

6.5 Microfluidics

A considerable amount of research work has been carried out by several research groups on the development of microfluidic systems based on PCB, known as PCB microfluidics [19–37]. In this section, continuous flow microfluidic systems [19–33] and discrete flow digital microfluidic systems [34–37] based on PCB technology are discussed.

6.5.1 *Continuous Flow Microfluidics*

A low-cost PCB microfluidic system that enables integration of microfluidics and electronics on a single board, at a price that could make the whole device or part of it cheap enough to be disposable, makes the technology quite promising. However, the integration of microfluidic components and systems into PCB presents new challenges in addition to those typical of sensor and actuator systems. The challenges include the need for a good sealing of the parts to avoid fluid leakage, the ability to create a stable and predictable flow rate inside the board, and a way to provide energy to impulse the fluid and make it flow through various components.

6.5.1.1 *Fabrication methods*

Monolithic Integration: One of the initial approaches to build a continuous flow PCB microfluidic system is shown in Fig. 6.14 [19]. This technology has been successfully used to create pumps, valves, pressure sensors, flow sensors, bubble detectors, pH sensors, and flow-injection analysis (FIA) systems [20]. In this fabrication method, control electronics and microfluidic elements are monolithically integrated on FR4 substrates, which is commonly used in the PCB industry. Specifically, FR4 substrates are used for wiring of electronic components as well as a carrier for microfluidic elements. The basic microfluidic element is a micro-channel, which is created

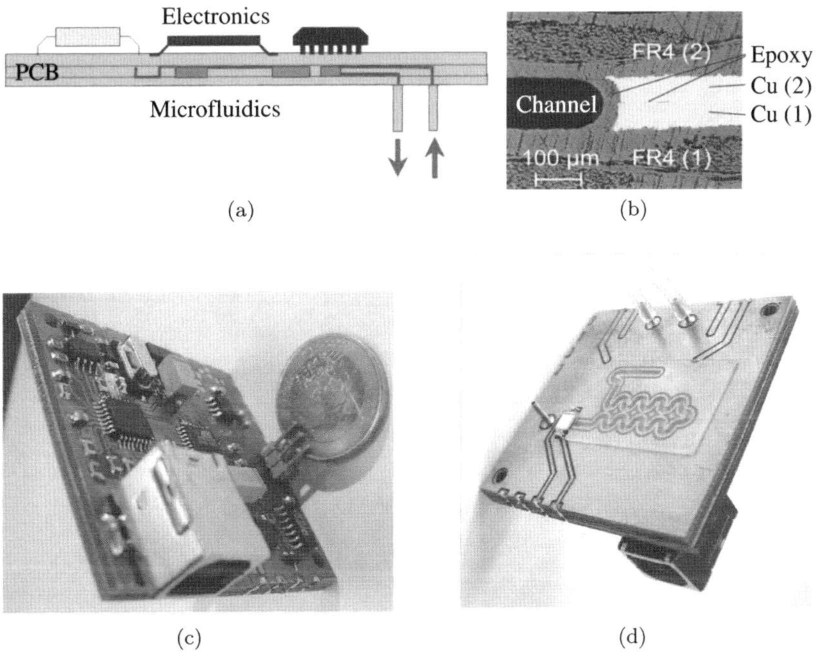

Fig. 6.14. PCB microfluidic system based on monolithic integration of microfluidic elements and discrete electronic components: (a) an illustration of monolithic integration, (b) microfluidic channels etched in copper layers of FR4 substrates, (c) a photograph of the electronics side of the FIA system, and (d) a photograph of the microfluidic side of the FIA system [19, 20].

by etching the copper layer on an FR4 substrate. Then, the microfluidic channels and cavities are formed by stacking two FR4 substrates patterned with the same microfluidic layout, one on the bottom and other on the top side. The FR4 stack is glued together using an adhesive epoxy material. The thickness of the copper layer defines the final depth of the channels. The channels and cavities on various layers of the stack can be connected using through hole via-ducts. The FIA system was used for the detection of Fe^{3+}-ions in the range of 0.16–10 mM with an accuracy of 10%.

In [21], Motorola Labs reported a PCB-based microfluidic system for DNA amplification and detection, known as DNA eSensor®, shown in Fig. 6.15. Micro-channels were patterned by photolithography in a photosensitive dielectric layer. Typical micro-channels have heights in the range of 50–60 μm and widths from 100 μm to 800 μm. It should be noted that the shape of the channels varies from "dome" to "flat channel" with increasing channel width. The high-density interconnect (HDI) based embedded

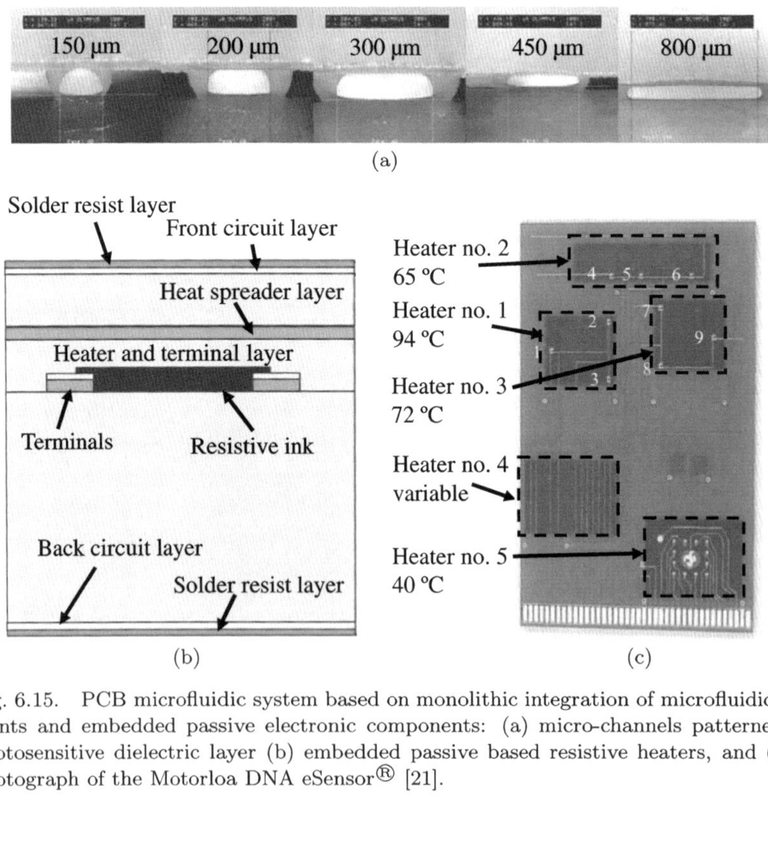

Fig. 6.15. PCB microfluidic system based on monolithic integration of microfluidic elements and embedded passive electronic components: (a) micro-channels patterned in photosensitive dielectric layer (b) embedded passive based resistive heaters, and (c) a photograph of the Motorloa DNA eSensor® [21].

passive (EP) resistor ink technology was used to fabricate the heaters in the PCB. These resistive heaters are used to perform continuous flow polymerase chain reaction (PCR) using three thermally isolated heating zones of 94°C for denaturing (Heater No. 1), 65°C for annealing (Heater No. 2), and 72°C for extension (Heater No. 3). An additional heater (Heater No. 4) is used as a thermal pump to maneuver the workings fluids through the device. Genotyping was performed on an HFE-H assay (1.4 μl genomic DNA, 2% rabbit blood) using a 50/50/50 second amplification protocol for 40 cycles. For comparison, the PCR assay was run on the microfluidic DNA eSensor® and a traditional thermocycler from MJ Research, Inc. Both tests showed similar trends and reached saturation after roughly 90 mins, which indicated detection of the assay.

Hybrid Integration: Another fabrication method for creating microfluidic structures on PCBs is by adding a new material on top, which is then molded, milled, or formed, to create the microfluidic channels. The patterned PCB can be used as a substrate in the fabrication of microfluidic systems, incorporating even other materials on top of it. Thus, a wide variety of designs can be implemented. In this way, hybrid integration of electronics and microfluidics can be easily achieved. Various materials that are compatible with this method include Poly-dimethyl-siloxane (PDMS), Poly-methyl-methacrylate (PMMA), SU-8, etc.

PDMS is a popular elastomer for biomedical devices due to its biocompatibility and excellent optical properties. PDMS is also a typical choice for microfluidic systems [22, 23]. PDMS is available in liquid form that needs to be cured. Thus, it is easy to embed electronic components inside PDMS [24]. In this case, the PCB is used to house electronic components and interconnect them. In some cases, PCB copper traces are used to provide heating to the microfluidic part by Joule heating effect. Microfluidic platforms can be easily manufactured in PDMS by soft lithography. However, the integration of metallic electrical connections in PDMS is challenging. So, the possibility of manufacturing the microfluidic platform on a PCB presents a wide range of design options and functionality. The microfluidic circuit is fabricated using an SU-8 or aluminum mold. If a closed structure is required, a thin layer of liquid PDMS is spun over the PCB, and then, both parts are cured together. PDMS can also be used for insulation of the electrical connections from the fluid.

Other option is to fabricate a multilayer system with PDMS and PCB, as shown in Fig. 6.16. The microfluidic circuit is manufactured in PDMS.

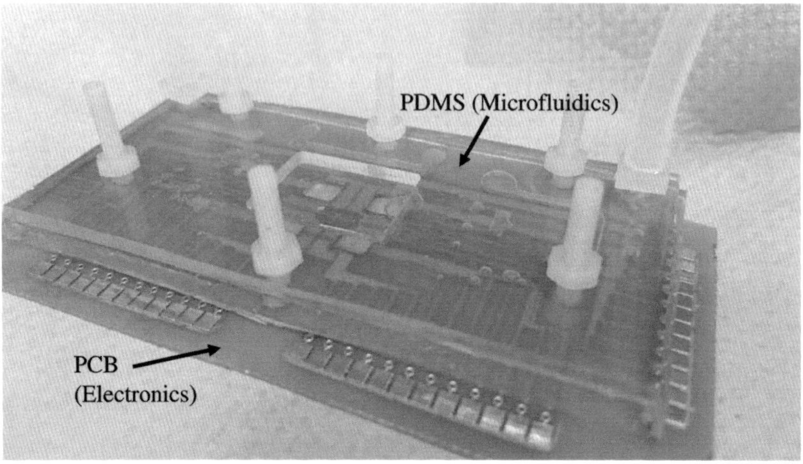

Fig. 6.16. PCB microfluidics based on hybrid integration of PDMS microfluidic circuit on top of a patterned PCB substrate.

The PCB and PDMS microfluidic circuit are fixed with polymer screws, and a PMMA layer is included in the stack to increase the stiffness of the system. Thus, various parts are detachable and the whole system can be disassembled. FPCBs, in combination with PDMS, which is also flexible, have been used to realize microfluidic systems. One such system reported in [25] allows the direct integration of chips inside the microfluidic walls, permitting them to be in close contact with the fluids. This is very convenient when the chips are used to sense the fluid properties.

SU-8 is one of the most commonly used MEMS materials known for its high-aspect ratio structures and can be patterned using photolithography. It is a negative photoresist and has good transparency. It can be deposited onto the patterned PCB and then crosslinked to create the microfluidic channels. The fabrication process associated with SU-8 photolithography is fully compatible with PCB processes. It is thus possible to construct a microfluidic circuit in a few hours, a task very difficult to achieve with silicon-based techniques. The thickness of the deposited SU-8 can be varied by adjusting the spin-speed during the coating process. The final thickness of SU-8 after processing will determine the height of the micro-channels. However, SU-8 is not electrically conductive and therefore requires additional metallic layers. The integration of SU-8 on PCB allows electrical connections to be created using the copper layers on PCB. This feature enables fabrication of smart microfluidic platforms.

PMMA is another alternative for the fabrication of microfluidic circuits. Hot embossing techniques are commonly used for patterning microfluidic circuits in PMMA. Once the microfluidic circuit is patterned on PMMA, a thin layer of UV glue is deposited over the PCB, and the parts are bonded together. In summary, microfluidic technology based on PCB processes is a powerful enabling technology, highly compatible with many commonly used fabrication processes and materials.

6.5.1.2 *Lab-on-PCB*

As mentioned above, most of the microfluidic research aims to develop micro-total analysis systems (μTAS), or point-of-care (POC) diagnostic systems, cheap enough to be easily deployed and disposed [26, 27]. These systems, usually called Lab-on-chip or LOCs, can be considered a Lab-on-PCB system when a PCB is used as a substrate. A generic architecture of a Lab-on-PCB systems is shown in Fig. 6.17. The fluidic samples come from outside sources or from reservoirs, and are subjected to a pretreatment (mixing, heating, separation, etc.). Then, the main reaction for the analysis is processed and the results are measured using integrated or external sensors through optical, thermal, or chemical means. Finally, the remnant waste is disposed. For the system to work, power needs to be transferred to the fluid samples so that they can be moved through the system, and a control mechanism (typically using valves) needs to be in place to direct these samples to the appropriate parts at any given moment.

Each of these components will be described in the following sections.

6.5.1.3 *Impulsion system*

One of the most critical elements in a Lab-on-PCB is transportation of a certain amount of fluid. This is the function of the impulsion system.

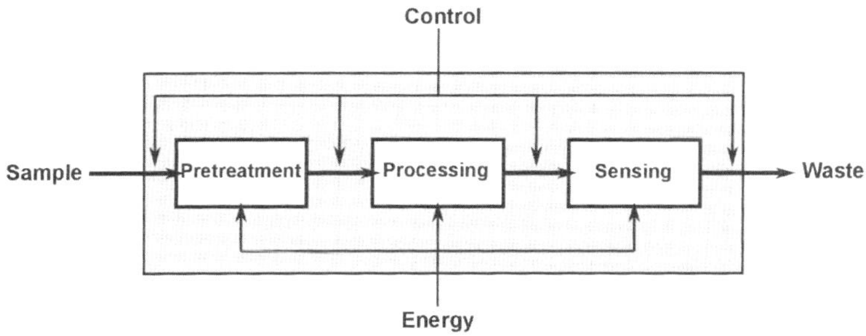

Fig. 6.17. A generic architecture of a Lab-on-PCB system.

The impulsion is necessary in different moments of the operation and thus needs to be controlled. Various impulsion mechanisms can be implemented on PCB:

Peristaltic pump: In this case, the impulsion is based on the volumetric expansion of air due to an increase in temperature. It can be implemented in a two-level platform that is separated by a deformable membrane. For example, PDMS can be used as the material for the walls and micro-channels. The basic level is composed of three compartments that are connected by micro-channels. The part of the pump that is located under the micro-channel, called the working micro-channel, can be impulsed. The whole structure is manufactured on a PCB substrate, where the micro-heaters are patterned in the copper layer. Initially, the system must be purged to start the operation. When the peristaltic pump is in operation, the first micro-heater is activated and this expands the air in the first chamber, which subsequently pushes the top PDMS membrane into the working micro-channel. Thus, through a certain sequence of activation of the micro-heaters, the fluid impulsion is accomplished [22].

Pressure pump: In this case, the mechanism that impulses the fluid is a differential pressure created between two parts of the system. The compatibility of PCB with SU-8 processes enables the implementation of this kind of design. The microfluidic circuit is manufactured with SU-8 on top of the PCB, and the pressure is stored in part of the SU-8 microfluidic platform, like a chamber, and then released in a controlled manner. The activation of the pump is carried out by a microvalve that separates the stored chamber from the rest of the system. The microvalve is designed as a fuse, which is implemented by a narrow electrical trace that is patterned in the copper layer. When a current is passed through the fuse, it is blown and both parts of the system get connected, and thus releases the pressure [28].

Electroosmotic: The electroosmotic effect appears when a fluid composed of polar molecules (like water) is in contact with a solid wall. A static charge appears on the surface of the solid that is in contact with the liquid. Simultaneously, opposite charges are induced in the thin layer of the water close to the wall to maintain electrical neutrality. If an electrical field is created inside the channel, electric forces will move the positive charges toward the negative potential and thus dragging the liquid. This type of micro-pump can be easily fabricated with the PCB technology. The electrodes are manufactured in the copper layer in the PCB and SU-8 is processed on top of the PCB to create the microfluidic channels [29].

6.5.1.4 *Control*

Flow control is usually achieved by placing valves between the microfluidic modules. The valves need to be actuated externally using a smart control method that typically makes use of feedback sensors [30]. Actuation of these valves is very often performed by electrical means. For example, by passing a current through a copper line to heat a chamber filled with air that expands and deflects a flexible membrane, which in turn closes a microfluidic channel. Occasionally, the deformation of such a membrane is directly performed by using pressurized air coming from an external source, making the valve actuation purely pneumatic instead of electrothermal.

6.5.1.5 *Pretreatment*

Pretreatment includes various processes that need to be performed on a fluid sample before it can be used for the intended reaction. This typically involves filtering, heating, mixing, and separation. As mentioned above, the ability to heat the samples using Joule heating effect on copper lines is one of the main advantages of PCB microfluidics. Mixing [26] and separation [31] are also accomplished using the techniques described here.

6.5.1.6 *Processing*

Typically, the analysis happens in a reaction chamber where the required reagents are combined in specific conditions of temperature, pressure, and time. The fabrication processes of these chambers are similar to that of micro-channels described above. PCB microfluidics have been used for cell detection and counting [31], cell manipulation [23], PCR [21, 32], electro-luminescence analysis [28], and nucleic acid and protein analysis [30].

6.5.1.7 *Sensing and measurement*

The last step associated with a Lab-on-PCB is that the sensors must be integrated in the microfluidic platform. The sensors can be integrated through soldering onto the PCB copper layer [33] or by a multilayer structure [21]. In the first case, the height of sensors can be an issue in certain configurations. A possible solution is to drill the PCB substrate to obtain some workspace for housing the sensors. The pads of the sensors can be connected to external pads by wire-bonding. Also, the pads can be isolated using a thin SU-8 or PDMS layer coated as a seed layer on the PCB. Then, the microfluidic structure is fabricated as described earlier.

6.5.2 *Digital Microfluidics*

Digital microfluidics (DMF) uses the principle of Electrowetting-on-dielectric (EWOD) for the manipulation of microliter or nanoliter sized discrete droplets. An EWOD device consists of a planar bottom substrate and top plate. The bottom substrate contains a two-dimensional metal electrode array with individually controllable electrodes. The top plate contains a metal layer coating that serves as a continuous ground electrode. A unit cell in the array includes a pair of electrodes that acts as two parallel plates. The electrodes are approximately the characteristic size of a droplet. The electrodes are coated with a dielectric layer. To increase the electrowetting effect on the droplet, a hydrophobic layer is coated on top of the dielectric layer. A droplet resting on an electrode is moved by applying a control voltage to an electrode adjacent to the droplet and simultaneously deactivating the electrode just under the droplet. This electronic method of wettability control creates interfacial tension gradients that move the droplets to the charged electrode. Using this electrowetting phenomenon, droplets can be moved to any location on the two-dimensional array. By varying the sequence of control voltage activation, many fluid-handling operations, such as droplet dispensing, merging, splitting, mixing, and diluting of droplets can be accomplished. This platform is reconfigurable and thus fluidic operations can be performed anywhere on the array. These systems do not require external pressure sources and thus eliminate the need for pumps and valves used in continuous flow based microfluidics [34].

A multilayer PCB has been used as an EWOD substrate by adapting the surface-mount solder pads to serve as electrodes for EWOD actuation in [35]. Figure 6.18 shows an EWOD device fabricated using a four-layer PCB (FR4) substrate with an 8×8 electrode array consisting of a liquid reservoir and driving electrodes (5×4). The size of the electrodes is $1.5\,\text{mm}^2$, which are pattered in $25\,\mu\text{m}$ thick copper layer. The gap between the electrodes is $75\,\mu\text{m}$ and the diameter of connecting via at the center of each electrode is $200\,\mu\text{m}$. The electrodes were coated with Parylene-C dielectric layer and 2000-Å thick Teflon hydrophobic layer. A series of tests has been conducted to study the droplet actuation voltage, in air and oil, as a function of the thickness of the Parylene-C dielectric layer. For example, a 500 V driving voltage was needed for a successful droplet actuation in air

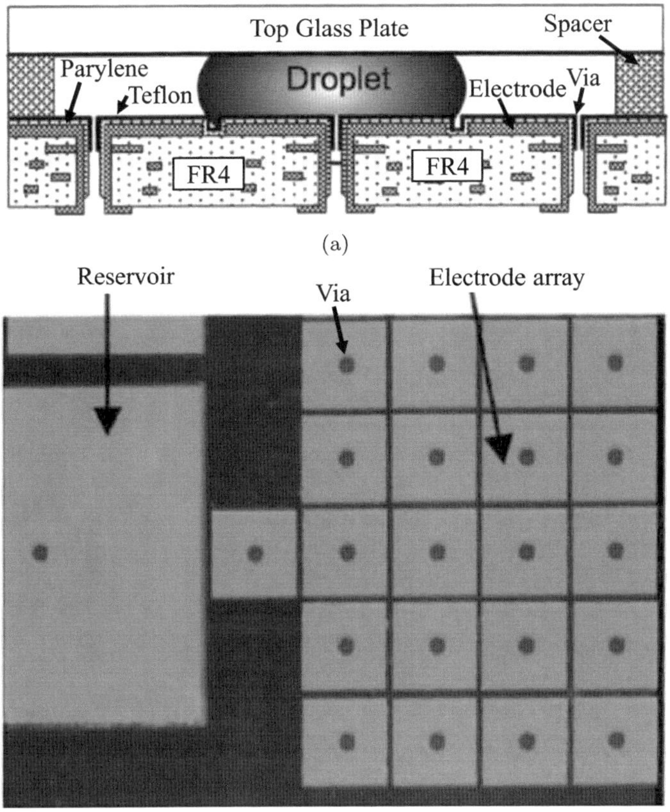

(a) (b)

Fig. 6.18. Digital microfluidic based EWOD device fabricated on a four-layer PCB: (a) a cross-sectional view of the device configuration and (b) part of the fabricated array showing the liquid reservoir and driving electrodes [35].

with a 7 μm thick Parylene-C layer. The driving voltage was reduced to around 200 V when the device was immersed in oil.

In [36], DMF-based EWOD device has been fabricated on PCB by Duke University group for adaptive cooling of IC. Figure 6.19 shows the schematic configuration and a photograph of a fabricated EWOD device. The system consists of reservoirs of liquid sandwiched between a substrate and a top plate. The substrate consists of an array of electrodes connected to the reservoir with a cooling liquid. The top plate can either be the IC chip itself, or a thermally conductive material on top of which rests the IC chip to be cooled. Through activation of the electrodes, the cooling droplets are

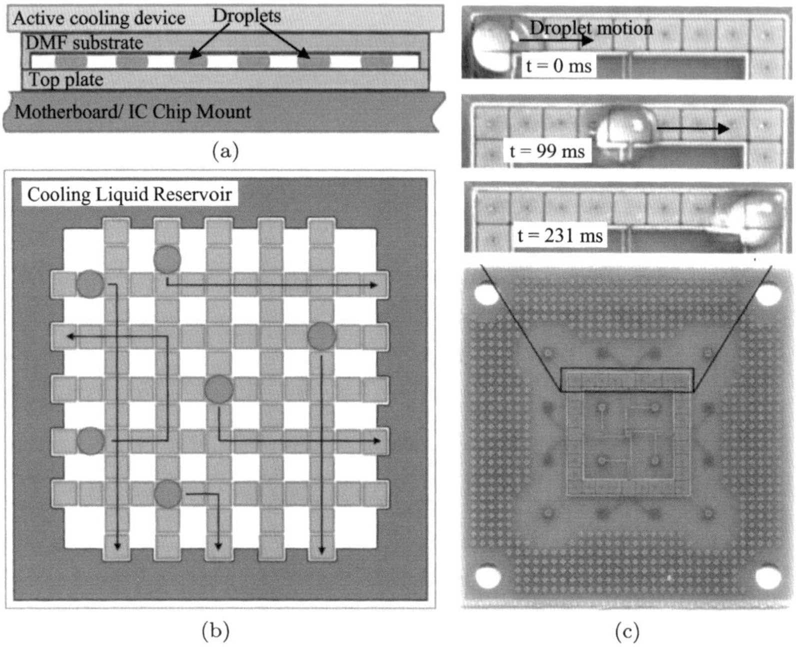

Fig. 6.19. Digital microfluidic-based EWOD device fabricated on PCB for adaptive cooling of IC: (a) a schematic side view of the device, (b) a schematic top view of the device, and (c) a photograph of the device showing movement of a droplet via actuation [36].

transported across the surface of the IC. Figure 6.19(c) illustrates the movement of a droplet from the top left pad to the top right pad in 231 ms by electrowetting actuation. The droplets transfer, store, and carry heat away from the IC back to the cooling reservoir. In this system, many droplets can be transported in a user-defined pattern over a two-dimensional electrode array. Thus, a variety of flow paths can be programmed for optimal cooling of the IC. This system enables location of thermal hot spots on the IC chip and reconfiguration of the flow of droplets for optimal adaptive cooling. This type of devices has been developed and commercialized for biomedical diagnostics by Advanced Liquid Logic, Inc. [37]. In 2013, this spinoff startup company from Duke University was acquired by Illumina, Inc.

6.6 Concluding Remarks

Several MEMS and microfluidic devices fabricated using low-cost PCB processing techniques, known as PCB MEMS and PCB microfluidics, respectively, have been discussed. A few companies have explored this approach for product development and commercialization, including Cardio MEMS, Inc. [9] (Acquired by St. Jude Medical Inc.), Motorola Labs [21], Advanced Liquid Logic, Inc. [37] (Acquired by Illumina Inc.), and Integra Devices, Inc. [38]. The manufacturing infrastructure needed for this approach is not expensive as it involves low-cost processes commonly used in the PCB industry. In some cases, the PCB MEMS approach enables co-fabrication of the package during the device fabrication and thus can yield self-packaged devices. Currently, there is a significant interest in the field of printed sensors fabricated using conventional printing methods (Screen, Gravure, Offset, Inkjet) and conductive inks (Silver, Carbon, Copper). There is a great commercial potential for implementation and/or integration of various devices and technologies discussed here with the emerging field of flexible printed sensors.

References

1. Liu, C. (2012). *Foundations of MEMS.* 2nd edition, Prentice Hall, NJ.
2. Coombs, C. and Holden, H. (2016). *Printed Circuits Handbook.* 7th edition, McGraw-Hill Education, NY.
3. Lian, K., Eliacin, M., Lempkowski, R., Chason, M., O'Keefe, M., and Drewniak, J. (2010). RF-MEMS switches on a printed circuit board platform. *Circuit World*, 36(4), pp. 12–17.
4. Chang, H. P., Qian, J., Cetiner, B. A., De Flaviis, F., Bachman, M., and Li, G. P. (2003). RF MEMS switches fabricated on microwave-laminate printed circuit boards, *IEEE Electron Device Letters*, 24(4), pp. 227–229.
5. Ramadoss, R., Lee, S., Lee, Y. C., Bright, V. M., and Gupta, K. C. (2006). RF-MEMS capacitive switches fabricated using printed circuit processing techniques, *Journal of Microelectromechanical Systems*, 15(6), pp. 1595–1604.
6. Enikov, E.T. and Lazarov, K. (2003). PCB-integrated metallic thermal micro-actuators, *Sensors and Actuators A: Physical*, 105(1), pp. 76–82.
7. Chang, S. P. and Allen, M. G. (2004). Demonstration for integrating capacitive pressure sensors with read-out circuitry on stainless steel substrate, *Sensors and Actuators A: Physical*, 116(2), pp. 195–204.
8. Palasagaram, J. N. and Ramadoss, R. (2006). MEMS-capacitive pressure sensor fabricated using printed-circuit-processing techniques, *IEEE Sensors Journal*, 6(6), pp. 1374–1375.
9. Fonseca, M. A., Allen, M. G., Kroh, J., and White, J. (2006). Flexible wireless passive pressure sensors for biomedical applications, In *Tech. Dig. Solid-State Sensor, Actuator, and Microsystems Workshop* (*Hilton Head 2006*), pp. 37–42.
10. Luque, A., Perdigones, F., Medina, D., García, J., and Quero, J. M. (2011). Capacitive pressure sensor fabricated using printed circuit board techniques,

In *Proceedings of the 8th Spanish Conference on Electron Devices*, CDE'2011, pp. 1–4.
11. Souilah, M., Chaabi, A., Perdigones, F., Quero, J. M., Flores, G., and Lain, M. R. (2016). Fabrication process for PCBMEMS capacitive pressure sensors using the Cu layer to define the gap, *IEEE Sensors Journal*, 16(5), pp. 1151–1157.
12. Rogers, J. E., Ramadoss, R., Ozmun, P. M., and Dean, R. N. (2007). A microelectromechanical accelerometer fabricated using printed circuit processing techniques, *Journal of Micromechanics and Microengineering*, 18(1), p. 015013.
13. Qiao, D., Pang, G. K., Mui, M. K., and Lam, D. C. (2009). A single-axis low-cost accelerometer fabricated using printed-circuit-board techniques, *IEEE Electron Device Letters*, 30(12), pp. 1293–1295.
14. Luque, A., Flores, G., Perdigones, F., Medina, D., García, J., and Quero, J. M. (2013). Single axis accelerometer fabricated using printed circuit board techniques and laser ablation, *Sensors and Actuators A*: *Physical*, 192, pp. 119–123.
15. Wang, X., Engel, J., and Liu, C. (2003). Liquid crystal polymer (LCP) for MEMS: processes and applications, *Journal of Micromechanics and Microengineering*, 13(5), pp. 628–633.
16. Bachman, M. and Li, G. P. (2012). Integrated MEMS in package, *Circuit World*, 38(4), pp. 184–192.
17. Boteler, D., Marsiglia, N., Lin, C.-Y., Liew, L.-A., and Lee, Y. C. (2015). Design and Fabrication of a PCB MEMS Module ith Integrated Switches and Sensor Suite, *IMAPS 11th International Conference on Device Packaging*, Fountain Hills, AZ USA.
18. Broadbent, H. A., Ketterl, T. P., and Reid, C. S. (2010). A miniature rigid/flex salinity measurement device fabricated using printed circuit processing techniques, *Journal of Micromechanics and Microengineering*, 20(8), 085008.
19. Merkel T., Graeber, M., and Pagel, L. (1999). A new technology for fluidic microsystems based on PCB technology, *Sensors and Actuators A: Physical*, 77(2), pp. 98–105.
20. Gassmann, S., Ibendorf, I., and Pagel, L. (2007). Realization of a flow injection analysis in PCB technology, *Sensors and Actuators A,* 133, pp. 231–235.
21. Lian, K., O'Rourke, S., Sadler, D., Gamboa, C., Terbrueggen, R., and Chason, M. (2009). Microfluidic systems on a printed wiring board platform, *Circuit World*, 35(3), pp. 12–20.
22. Jeong, O. C. *et al.* (2005). Fabrication of a peristaltic PDMS micropump, *Sensors and Actuators A: Physical*, 123, pp. 453–458.
23. Lee, H., Liu, Y., Ham D., and Westervelt, R. M. (2007). Integrated cell manipulation system — CMOS/microfluidic hybrid, *Lab Chip*, 7, pp. 331–337.
24. Zhu, L., Lee, C. S., and DeVoe, D. L. (2006). Integrated microfluidic UV absorbance detector with attomol-level sensitivity for BSA, *Lab Chip*, 6, pp. 115–120.

25. Wu, A., Wang, L., Jensen, E., Mathies, R., and Boser, B. (2010). Modular integration of electronics and microfluidic systems using flexible printed circuit boards, *Lab Chip*, 10, pp. 519–521.
26. Aracil, C., Perdigones, F., Moreno, J. M., Luque, A., and Quero, J. M. (2015). Portable Lab-on-PCB platform for autonomous micromixing, *Microelectronic Engineering*, 131, pp. 13–18.
27. Pittet, P., Lu, G.-N., Galvan, J.-M., Ferrigno, R., Blum, L. J., and Leca-Bouvier, B. D. (2008). PCB technology-based electrochemiluminescence microfluidic device for low-cost portable analytical systems, *IEEE Sensors Journal*, 8(5), pp. 565–571.
28. Flores, G., Aracil, C., Perdigones, F., and Quero, J. M. (2014). Low consumption single-use microvalve for microfluidic PCB-based platforms. *Journal of Micromechanics and Microengineering*, 065013.
29. Gassmann,S., Pagel, L., Luque, A., Perdigones, F., and Aracil, C. (2013). Fabrication of electroosmotic micropump using PCB and SU-8, *2013 Spanish Conference on Electron Devices (CDE)*, pp. 119–122.
30. Schumacher, S., Nestler, J., Otto, T., Wegener, M., *et al.* (2012). Highly-integrated lab-on-chip system for point-of-care multiparameter analysis, *Lab Chip*, 12(3), pp. 464–473.
31. Medoro, G., Manaresi, N., Leonardi, A., Altomare, L., Tartagni, M., and Guerrieri, R. (2003). A Lab-on-a-Chip for Cell Detection and Manipulation, *IEEE Sensors Journal*, 3(3), pp. 317–325.
32. Neuzil, P., Pipper, J., and Hsieh, T. M. (2006). Disposable real-time microPCR device: lab-on-a-chip at a low cost, *Molecular. BioSystems*, 2, pp. 292–298.
33. Perdigones, F., Aracil, C., Quero, J. M., Gutiérrez, M., Jiménez, C., and Giménez, P. (2015). Integration method of silicon sensors on SU-8-based microfluidic platforms, *Microsystem Technologies*, 21(1), pp. 155–161.
34. Berthier, J. (2012). *Micro-Drops and Digital Microfluidics.* William Andrew, MA, USA.
35. Gong, J., and Kim, C. J. (2008). Direct-referencing two-dimensional-array digital microfluidics using multilayer printed circuit board, *Journal of Microelectromechanical Systems*, 17(2), pp. 257–264.
36. Paik, P. Y., Pamula, V. K., and Chakrabarty, K. (2007). *Adaptive Cooling of Integrated Circuits using Digital Microfluidics.* Artech House, MA.
37. Millington, D. S., Sista, R., Bali, D., Eckhardt, A. E., and Pamula, V. (2014). Development of biomarker assays for clinical diagnostics using a digital microfluidics platform. *Dried Blood Spots: Applications and Techniques*, pp. 325–331.
38. http://www.integradevices.com/

Ramesh Ramadoss received his Bachelor of Engineering degree from Thiagarajar College of Engineering, Madurai Kamaraj University, India in May 1998 and PhD degree in Electrical Engineering from the University of Colorado at Boulder in May 2003. From Jun 2003 to Dec 2007, he was employed as an Assistant Professor in the Department of Electrical and Computer Engineering at Auburn University, Auburn, AL. From Jan 2008 to Mar 2012, he was employed as a Manager at FormFactor Inc., Livermore, CA. From Apr 2012 to May 2015, he was employed as a Senior Manager at Microprobe Inc. (Acquired by FormFactor Inc.), San Jose, CA. From Jun 2015 to Feb 2016, he was employed as a Technical Director at Applied Nanostructures Inc., Mountain View, CA. Currently, he is the President and CEO of IoT Technologies Inc., San Ramon, CA. He is the author or coauthor of 4 book chapters and 55 papers. He has conducted R&D projects for DARPA, NASA, US Army, US Air Force, Sandia National Labs, and Motorola Labs. In 2013, he was elevated to the grade of IEEE Senior Member. In 2014, he founded the IEEE San Francisco Bay Area MEMS & Sensors Chapter. In 2017, he was elected as the Chair of the IEEE Santa Clara Valley Section, the world's largest IEEE Section located in the heart of Silicon Valley.

Antonio Luque received the PhD degree in Electrical Engineering in 2005 from the University of Seville, Seville, Spain, where he currently holds the position of Associate Professor in the Department of Electronics Engineering. He has authored 20 journal papers, 40 conference papers, 3 book chapters, and a text book. As invited researcher and teacher, he has visited the Swiss Federal Institute of Technology Lausanne (Switzerland), Auburn University (AL, USA), Delft University of Technology (Netherlands) and Jade University (Germany). He was the Chair of the IEEE Industrial Electronics Society Technical Committee on MEMS and Nanotechnology in 2008 and 2009.

Carmen Aracil received the PhD degree in Electronics Engineering from the University of Seville, Spain, in 2011, where she works in the Department of Electronic Engineering as Researcher and Teaching Assistant. Her research interests include MEMS, BioMEMS, PCB-MEMS, and microfluidics. She has participated in numerous national and international projects related to these areas. She made predoctoral and postdoctoral stays at the Auburn University and TU Delft, respectively. She is author of 10 journal research papers, 14 conferences papers, and 2 book chapters. Currently she is Chair of Technical Committee on MEMS and Nanotechnology of IES.

Chapter 7

Single Wafer Encapsulation of MEMS Resonators

Janna Rodriguez and Thomas W. Kenny

7.1 Introduction

Nano and microelectromechanical systems (N/MEMS) resonators are currently used in a broad set of commercial and research applications. In commercial applications, resonators are used as gyroscopes (Fig. 7.1) in automotive safety systems and as timing references and frequency filters in cell phones, and computer and video game systems [8, 10, 27]. In research, resonators have been used as highly sensitive chemical and biological sensors [33] and have been investigated as future low-power replacements for CMOS switches [21]. However, despite significant progress in many MEMS resonator technologies, both commercial and research applications are limited.

In essence, MEMS resonators are extremely fragile movable structures, and effective packaging protects them while allowing them to interact with their surroundings. The specific major requirements for effective MEMS packaging are mechanical support, optimization for given space constraint limitations, compatibility with current technology and electrical connections. To illustrate, Fig. 7.1 shows the baseline of a Disk Resonator Gyroscope (DRG) that consists of 35 concentric rings that are internally connected through radial bars or spokes. The disk-shaped structure is anchored at the center and has multiple modes that vibrate with node points. These modes, also known as vibrational modes, are identified by using an eigenvalue system that determines the natural frequency of the system. After fabrication, the device operates as a planar resonating structure. For the

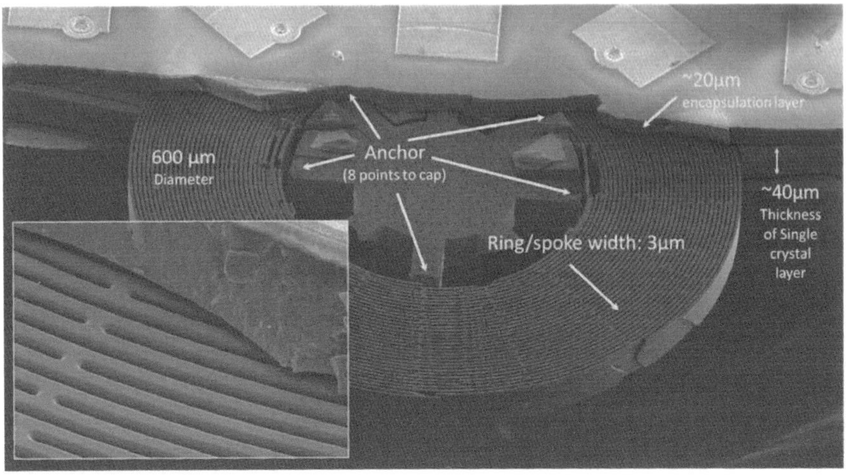

Fig. 7.1. SEM image of Epi-sealed DRG with the encapsulation partially removed [3].

case of the DRGs, these disks resonate as a degenerate pair of in-plane wineglass vibration modes.

Packaging provides the mechanical support needed by providing a shell robust enough to protect the delicate MEMS structures against the outside environment. Because these devices are very sensitive to environmental effects, even small amounts of surface contamination can significantly change device performance and permanently damage them. The support provided by packaging ensures that the MEMS devices operate in a secure and stable environment and protects these devices from mechanical shocks and vibrations, drastic temperature variations, and corrosion when they are in contact with gases. Creating a clean and stable environment is thus critical to ensuring the performance and long-term stability of MEMS resonators [30].

Packaging of a MEMS resonator must also be CMOS compatible to reduce costs and improve efficiency. Moreover, because the package is part of the complete MEMS system, and all aspects of the system must function together, the packaging must be compatible with each part of the entire system.

Frequently, packaging plays a more important role in overall performance and cost than does the inside of the MEMS device itself. Table 7.1 shows a cost breakdown for a typical silicon MEMS pressure sensor [30]. In Section 7.3, we describe packaging techniques that can occupy as much as 60% of the entire wafer, making these devices very costly. Consequently,

Table 7.1. Cost breakdown for a typical silicon pressure sensor [17].

Component	Contribution to Final Cost
Silicon Device and CMOS Circuit	35%
Packaging	45%
Calibration and Testing	20%

creating an optimal space for packaging of the MEMS devices is desirable to reduce costs.

MEMS resonators must also have the capability to be combined with integrated circuits for operations in larger electronic systems. Therefore, electrical connectivity between a device and its surrounding environment is necessary. Most of the early electrical vias were horizontal on the plane of the wafer, but as the process has improved, vertical vias have come into use.

MEMS packaging requirements depend significantly on their application, and a universal standard for MEMS packaging has not yet been established. In general, there are two different approaches for packaging: multi-chip modules and monolithic integration. In the multi-chip packaging approach, several MEMS sensors and actuators, or a combination of both, are present in a single package that forms a complex system which can perform several functions. The downside of this type of packaging is signal loss and a higher cost of packaging. The second approach, monolithic integration, is based on CMOS technology and is used when single devices for specific applications are fabricated.

Many new packaging technologies have been developed in recent years; however, there has been a growing shift away from packaging individual devices or using wafer bonding techniques toward thin-film encapsulation. This thin-film encapsulation process has the potential to reduce the amount of materials needed for most applications while increasing device throughput and yield. The thin-film encapsulation process allows devices to be encapsulated in batches, which significantly reduces the fabrication cost during large-scale production. Using this thin-film encapsulation, researchers have been able to package devices such as pressure sensors, gyroscopes, accelerometers, resonant devices, and various other MEMS devices. A full explanation for this process will be discussed in more detail in Section 7.3.2.2, where the Stanford Epi-seal packaging technique is detailed.

Having explained the importance of packaging, we next cover one of the most critical and most difficult aspects of the fabrication process, namely, pressure. We then discuss two major packaging techniques currently on the market: die-level packaging and wafer-level packaging (WLP). Toward the end of the WLP section, the Stanford Epi-seal process is reviewed. The chapter concludes with a discussion of current limitations in packaging processing as well as future trends in packaging.

7.2 Pressure and Its Importance for Frequency Stability

One of the most challenging steps in MEMS packaging is the ability to encapsulate resonators at a lower pressure, which is crucial since high pressure contributes to the damping process, which limits the quality factor of the resonator. Moreover, different device applications require different pressures inside the cavity. In general, MEMS gyros and resonators benefit from very low pressures, while accelerometers often require high pressure to achieve the required damping characteristics. In seismic accelerometers, the gas damping has to be reduced to a minimum requirement of 1 Pa or lower pressure inside the chamber.

In addition to the different pressures required by specific applications, MEMS resonators require a constant pressure over a long period [12]. Achieving both a specific pressure inside the cavity and long-term stability requires a good design and excellent packaging fabrication. During the design phase, the designer can optimize the resonator design to limit the influence of the least attractive dissipation mechanisms, such as gas damping. During the encapsulation phase, a low cavity pressure is essential in reducing gas damping. Modeling gas damping is challenging because the continuum model becomes less applicable as the mean free path approaches the characteristics of the resonators. Different encapsulation processes result in different cavity pressures for the resonators. In the Stanford Epi-seal process, for example, the cavity has been measured to be less than 1 Pa at room temperature. Several models have been used to predict pressure inside resonators; two of the most commonly used are the Christian Model and a modified version of it, the Bao Model [5, 9]. Bao's model is simplified in equation (7.1), which can be further simplified to equation (7.2). This equation (7.2) shows a relationship between gas damping and pressure commonly used to understand the performance in MEMS resonators, with a range of pressures from 0.1 Pa to 100 kPa.

$$Q_{\text{gas}} = C\frac{\sqrt{K_b T}}{P} \tag{7.1}$$

$$Q_{\text{gas}} \propto \frac{1}{P} \tag{7.2}$$

C: Constant involving the resonator geometry
K_b: Boltzmann Constant
T: Temperature
P: Pressure

Various techniques to determine the pressure inside the cavity are available. One effective technique is to design and test MEMS resonators with a quality factor that is highly sensitive to pressure. For example, when a device is packaged with the thin film encapsulation method, one way to determine the pressure inside the encapsulation cavity is by measuring the quality factor at various pressures after the encapsulation has been broken and then subsequently measuring the quality factor, after the encapsulation has been broken for one of the devices. At lower pressures (below 1 Pa), the measured Q becomes pressure independent, which indicates that other damping mechanisms become dominant over gas damping. The measured Q, per equation (2) is inversely proportional to pressure, because gas damping is dominant in the system. Measuring the quality factor allows us to predict the gas damping of the encapsulated devices (in Epi-seal) from the extrapolation of the Q-pressure relationship (Fig. 7.2). To illustrate

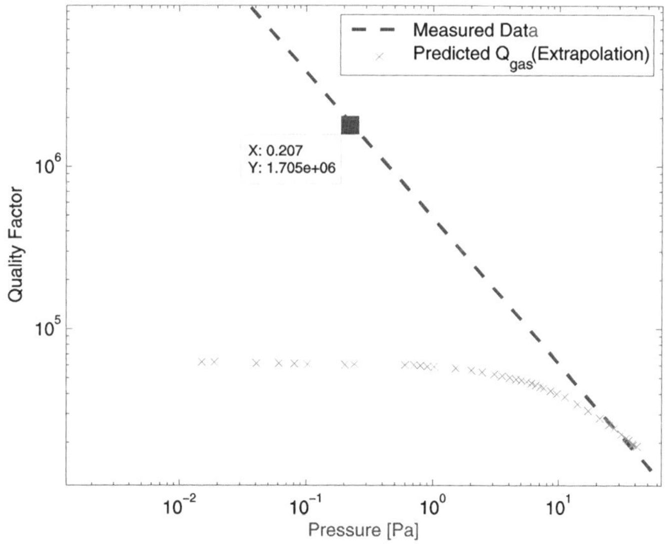

Fig. 7.2. Quality factor of a DRG tested at different pressures showing an approximation of the pressure inside the cavity fabricated by the Stanford Epi-seal process and the extrapolation of Q_{gas} [29].

this, we can look at the results of testing a device, shown in Fig. 7.2. This device was a DRG encapsulated using the Epi-seal process, with a cavity pressure of 0.25 Pa at room temperature. The gas damping contribution was extrapolated to be ~2e6. The thin-film encapsulation process used to package this device was created in a wafer that contains multiple resonators, and the pressure of all other resonators on the wafer could thus be determined. These results are consistent with other research that has demonstrated that significantly reducing the effect of gas damping requires that the pressure in the encapsulation cavity be less than 0.1 Pa. [3]

In the introduction to this chapter, we spoke about clean and stable packaging environments being essential for the stability of the resonant frequency of the MEMS resonators because frequency stability depends directly on the quality of the packaging environment. Small changes in pressure over a long period result in a noticeable shift in the resonant frequency over the usable temperature range. Moreover, during the encapsulation process, gas molecules can absorb onto or desorb from the MEMS devices. These gases may introduce moisture that could defeat the purpose of vacuum packaging, and may also cause corrosion or stiction of the packaged components. However, if MEMS devices are packaged at very high temperatures, in extremely clean environments and with encapsulation pressures lower than 0.01 Pa, unwanted molecule absorption could be significantly reduced. Since accomplishing this encapsulation pressure with all these elements present simultaneously is difficult, different techniques have been developed to mitigate this problem. One of the most common techniques is the use of noble gases such as He, Ne, Ar, or Kr as the residual gas in the package, because these gases do not adhere to the surface of the MEMS resonator at room temperature. Another approach is to capture these unwanted molecules by using a chemically-active film, or a "getter," which is added as part of the fabrication process.

7.3 Fabrication Techniques and the Packaging Process of MEMS Resonators

In this chapter, the packaging techniques for MEMS resonators are separated into two categories: die-level and WLP (Fig. 7.3). The designer needs to consider which packaging process is most suitable for a specific application. There are three major differences between the two processes: the dicing step, the hermeticity, and the cost.

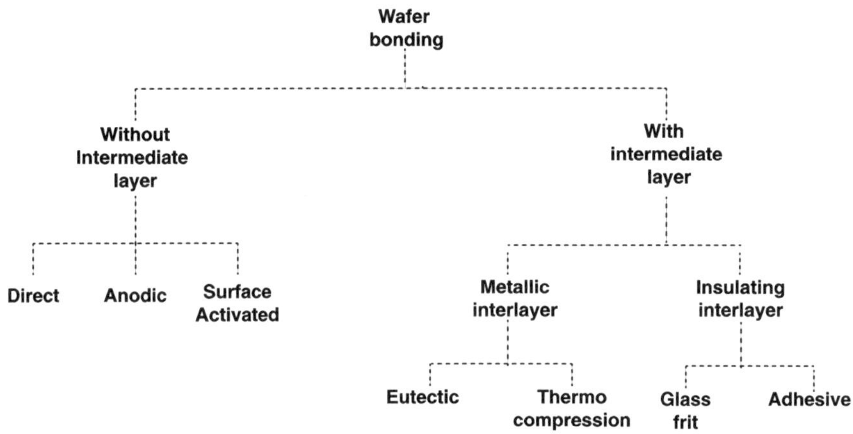

Fig. 7.3. Overall classification of MEMS resonator packaging.

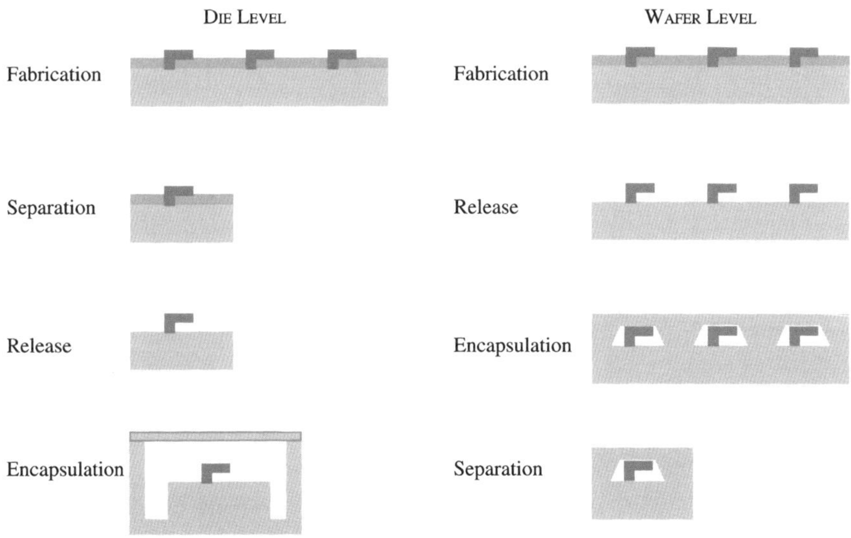

Fig. 7.4. Illustration of the encapsulation and separation process for the die-level process vs. the wafer-level process.

The first approach, die-level packaging, is the oldest method of packaging. It is performed individually for each die on the wafer (Fig. 7.4). For example, for vacuum-can packaging, the die is first bonded to a base, and then the vacuum-can is adhered to the base, around the die. This process is repeated for all dies that the wafer has.

The second packaging technique, WLP, is currently one of the most common packaging techniques. Because it is done for the entire wafer at once, it offers several advantages: it makes much smaller packages possible, it is cost-effective, it improves performance, and it can be done by either bonding a wafer cap or by a thin-film deposition (Fig. 7.4). One successful encapsulation process currently implemented by SiTime Inc is the Stanford Epi-seal wafer-level thin-film encapsulation, the subject of a detailed discussion in Section 7.3.2.2. With either method, the final package must include the basic MEMS packaging characteristics, such as environmental interface, hermeticity, protection, and electrical interface. These two techniques are defined in the following sections.

7.3.1 *Die-Level Packaging*

Die-level packaging is a packaging method that requires separation of the die from the wafer followed by individual packaging of each die. There have been varied implementations of this packaging technique, most commonly using metals, ceramics, and plastics, each of which has benefits and limitations [25]. This packaging method also involves many hurdles that must be overcome to produce high-quality and reliable resonators, specifically, high cost, the need for extensive individual handling, reduced yield, and a process that results in variations in the resonator frequency. With this kind of packaging, each resonator is individually encased in its own environment, so reliability and thus repeatability can be a problem. Reliability for MEMS devices is commonly defined as the probability that a device will perform its required task for a set amount of time. Repeatability is defined as a set of the measurements taken without variation on the same resonator, under the same conditions. These two essential characteristics become more difficult to control and measure when resonators are being mass-produced.

One critical part of die-level packaging is the die separation step. The released resonators are very fragile and can easily become damaged from contamination or handling. Four of the most common methods that are used to separate the die are wafer saw, laser, plasma, and scribe.

Wafer Saw

Traditionally, an abrasive wafer saw is used to separate the die from the wafer as illustrated in Fig. 7.5. A wafer saw is typically diamond coated and uses this abrasiveness to remove wafer material and separate the dies

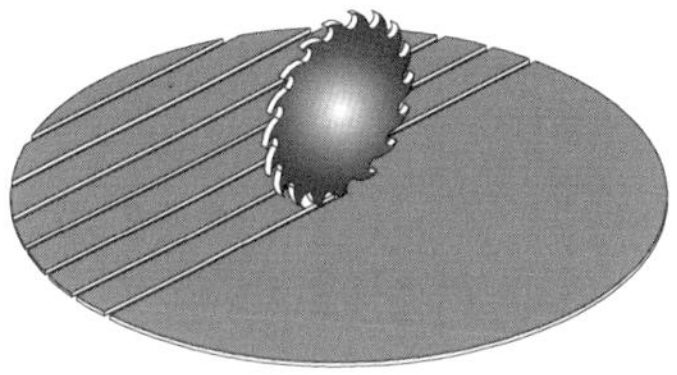

Fig. 7.5. Illustration of wafer saw cutting.

from one another. Wafer material is removed along dicing lines (or streets) and the street width is usually always larger than the wafer saw width. For example, creating a street width of 50 to 76 μm requires a recommended wafer blade thickness of 20 to 30 μm [28]. This method has four main drawbacks. First, because of the mechanical force required to separate the die in this method, damage to the die can occur such as chipping, die breaking, or delamination of the bonded layers [28]. This is most likely to occur with thinner wafers. The second drawback is that wafer saw creates heat as it cuts, requiring a coolant (usually water) to be sprayed on the wafer [32]. The third drawback is possible contamination. Particles from the wafer saw process, as well as the coolant, can contaminate the resonators and affect their performance [28]. To combat this, some have proposed that the wafer be diced while the wafer is attached, face-down, to sufficiently thick dicing tape. For this approach, cut-outs are placed in the dicing tape for each resonator and then aligned with the wafer so each resonator is protected. The fourth main drawback is that indications or alignment marks are necessary on the back side of the wafer to align the wafer saw because cutting occurs on the back side of the wafer [28, 32].

Laser Dicing

To avoid the drawbacks of wafer sawing, a laser can be used to facilitate separation of the die from the wafer. With laser dicing, a high-power laser is used with a wavelength that is absorbed by the silicon substrate. Die separation can be achieved in one of the following three ways: cutting trenches, cutting a series of holes, or stealth dicing. In the first approach, a laser is used to cut trenches. A continuous groove cuts the part through the thickness of the wafer partially (Fig. 7.6). This cutting weakens the wafer along this line and will allow the die to break off and separate if a bending stress is applied at this location. In the second approach, a laser is used to create a series of holes down the length of the wafer. As with the trench method, in this second approach, a laser weakens the wafer and allows the die to break off and separate at this location. However, both of these methods have a drawback in that when the laser is cutting, molten silicon is ejected from the cut location and can be deposited elsewhere. This silicon has the potential to damage the resonator and/or change its resonant frequency. Moreover, this second approach can lead to part-to-part variations in the resonators. The third approach, stealth dicing, avoids these complications by using a dual-focused laser beam focused on a point internal to the wafer. In this approach, an area of the wafer is not exposed to the outside environment, so molten silicon is not ejected and will not create debris. However, applying a bending force at the ablated location will cause the die to break off and separate from the wafer. All three of these laser dicing processes are dry processes, so no coolant is necessary.

Plasma Dicing

Plasma etching is a dicing technique that uses a chemical reaction to separate the dies from each other. The wafer is placed in a vacuum chamber with an erosive gas, and then the wafer is exposed to plasma discharge where the die separation is to take place (Fig. 7.7). This plasma discharge accelerates the chemical reaction that etches the wafer until die separation is complete [28]. Plasma etching is a dry process, so no coolant is required. Moreover, the plasma etching process occurs on the whole wafer at once, so

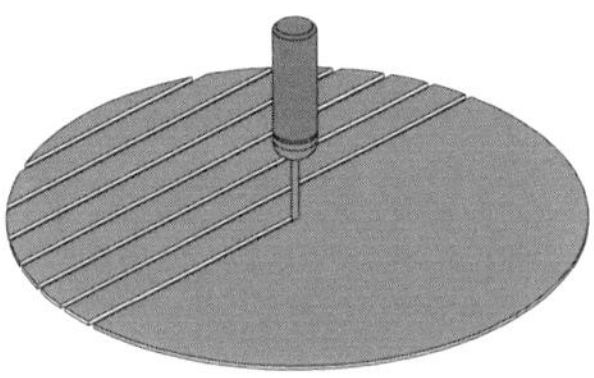

Fig. 7.6. Illustration of laser cutting.

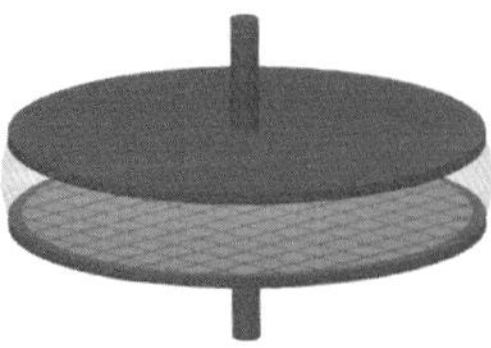

Fig. 7.7. Illustration of plasma cutting.

processing densely packed wafers takes the same amount of time as with less densely packed wafers. Even though the plasma dicing machines are relatively expensive, the cost to plasma dice can be less than traditional wafer saw dicing if the individual dies on the wafer are less than several millimeters because dicing with a wafer saw on such small dies would be much slower.

Scribe and Break

The scribe and break method relies on a diamond-tipped scoring device as illustrated in Fig. 7.8. The scoring tool is scraped along the wafer in the areas that need to be separated, and a bending force is applied that causes the die to break off from the rest of the wafer. Experience has shown that the sharpness of the cut is much more important than the depth of cut, and, in fact, typical depths of cut range only from 3 to 5 μm deep [4], resulting in the narrowest kerf width out of all the separation methods.

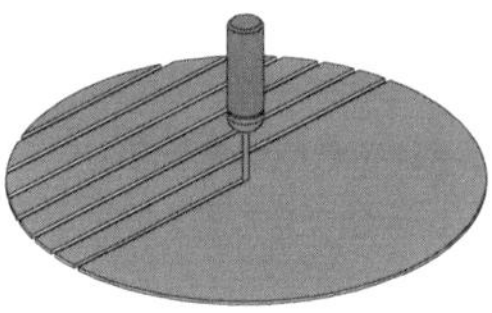

Procedure: Sharp tool scraped against surface
Speed: Medium
Fracture Toughness: Mechanically damaged
Dryness: Dry process

Fig. 7.8. Illustration of scribe cutting.

Some of the limitations of the die-level process are that it decreases reliability and consistency. Because of the lack of standard processes, companies have developed their own processes, which impose additional development time, the acquisition of non-standard tools, and an increase in cost.

7.3.2 *Wafer-Level Packaging*

In contrast to die-level packaging, wafer-level packaging (WLP) encloses all MEMS dies on a wafer at the same time before the individual dies are separated. In this way, the delicate MEMS devices do not risk coming into contact with contaminants during the die separation step because they are already protected. Two methods can be used to package a MEMS device on the wafer level: wafer-bonding and thin-film encapsulation. With wafer-bonding, a secondary wafer is created with cavities for the MEMS devices and then bonded onto the first wafer, which protects all MEMS devices at once; note that a more detailed description of wafer-bonding techniques is described in other chapters. With thin-film encapsulation

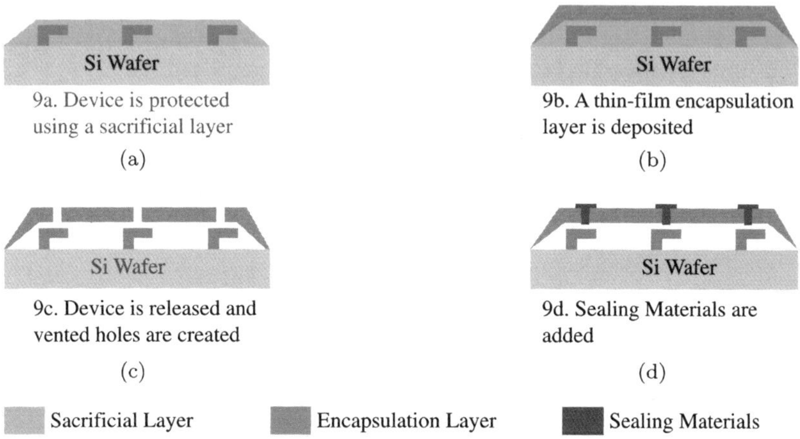

Fig. 7.9. Illustration of the thin-film encapsulation process: (a) Device is protected using a sacrificial layer, (b) A thin-film encapsulation layer is deposited, (c) Vent holes are created and device is released, and (d) Sealing materials are added.

(Fig. 7.9), thin-film layers are deposited and selectively removed to build the packaging around the MEMS devices without the addition of a secondary wafer, a process detailed in Section 7.3.2.1 below. One successful example of a thin-film encapsulation technique is the Stanford Episeal packaging process (Fig. 7.10). This packaging technique, also detailed below, has great potential because it can protect the devices before separation and allows the use of standard and economical integrated circuit packages.

Thin Film

The thin-film process for MEMS packaging is an attractive solution for MEMS devices because it offers several advantages: a reduced area required for packaging, a contact pad to access the buried devices, the fabrication is done with standard equipment, and the need for only a single wafer, which eliminates the need for bonding tools. The thin-film encapsulation is also called "zero-level" packaging since the device is encapsulated as part of the fabrication process. It is CMOS compatible and readily available in most foundries. In general, thin-film encapsulation tends to be less flexible than the use of a bonded cap, and the MEMS device's shape and size being packaged has a large impact on its ability to be packaged, as we explain below.

The standard thin-film process starts by patterning and etching the structures into the device layer. This is followed by adding a sacrificial layer that will later be removed until a planar surface is created (Fig. 7.9a). This sacrificial layer can be deposited by chemical vapor deposition (CVD), evaporation, sputtering or atomic layer deposition (ALD). Although this layer is usually SiO_2, materials such as silicon nitride, silicon carbide, phosphorous silicate glass (PSG), photoresist and polycrystalline silicon have also been used [13–15, 22, 23]. The next step is the deposition of a capping layer done by either a dry or a wet process (Fig. 7.9(b)). Various materials such as polysilicon, epitaxial polysilicon, silicon nitride, silicon dioxide, aluminum, nickel, and gold have been explored as encapsulation layers. The capping layer is usually pattered with release vents to facilitate the removal of the sacrificial layer (Fig. 7.9(c)). Once the device is capped, the next part of the process involves the release of the MEMS resonator structure. This release process is typically a dry process that is limited by the number, size, and location of the release ports or vents. Moreover, the release vents have to be made small enough, or placed far away from the structure, to prevent the final sealing material from depositing onto the functional devices and affecting their operation. At this point, the resonator is free to move. Next, silicon is grown to seal the vents. Last, electrical connections are made by etching the cap layer and depositing aluminum to create pads to allow electrical access to the resonator (Fig. 7.9(d)).

This encapsulation technique is comparable to micro-machining and CMOS integration. Therefore, all procedures and equipment currently in place could be used for the encapsulation process, which reduces cost and increases yield. The drawback to this technique is the limited span that can be capped with the relatively thin film without introducing supporting posts within the MEMS structures. To address this shortcoming, the Stanford Epi-seal thin-film process allows the device to be buried under a thick epitaxial polysilicon (epi-poly) encapsulation with electrical routing and contacts on the planar wafer surface. The wafer can be diced as is and mounted in a package, while maintaining the ultra-clean cavity within the encapsulation. The Stanford Epi-seal technique is described in detail below.

Stanford Epi-seal process

The Stanford Epi-seal packaging process, a type of thin-film encapsulation packaging, is a wafer-scale process that has high stability, low cavity pressure, and is contamination-free. The Epi-seal process has proven to be a good solution for the fabrication of MEMS resonators, and the long-term frequency stability it offers is far superior to other MEMS resonators [19]. As mentioned above, the Stanford Epi-seal process is also compatible with a traditional CMOS process. This compatibility is highly valued since it allows MEMS devices to be brought to market in high volumes. The Epi-seal process was developed through a collaboration between the Micro Structures and Sensors Lab at Stanford University and The Bosch Research and Technology Center (RTC) in Palo Alto, CA. The initial team of collaborators included Markus Lutz, Gary Yama, Aaron Partridge, Rob Candler, Bongsang Kim, Matt Hopcroft, WooTae Park and Thomas Kenny [6, 7, 16, 18].

As noted above, the key components of the Stanford Epi-Seal Process are high stability, low cavity pressure, and is contamination-free. This process has been proven to be robust and capable of withstanding wafer dicing, wire bonding, and plastic injection molding, among others. The use of an epitaxial grown silicon cap layer makes it possible to maintain regions of single-crystal silicon (SCS) necessary for CMOS integration. The vertical electrical interconnections on the top of the devices results in footprints only slightly larger than the device itself. Moreover, the hermeticity of the package allows for excellent long-term stability. In the following 12 steps (Fig. 7.10), we provide an overview of the Stanford Epi-seal process.

Step 1: The process begins with a highly doped silicon-on-insulator (SOI) substrate as a base layer (Fig. 7.10(a)), a $2\,\mu$m thick buried oxide layer and a $40\,\mu$m thick SCS. This substrate has been previously aligned in one of the several relative directions known as crystal orientation, defined by the Miller index. This wafer alignment and doping are what later on will determine the anisotropic or isotropic properties, the design symmetry, and the mode matching of the devices built into this wafer [2, 29]. In the Epi-seal process, devices have been built using (111) and (100) orientations with either SCS or polysilicon (Fig. 7.11).

Step 2: Using deep reactive ion etching (DRIE) (Fig. 7.12), instrumentation trench features are defined according to their geometry by etching them into the device layer of the SOI wafer (Fig. 7.10(b)). These trenches define the resonator structure and transduction gaps. These transduction gaps are usually between $0.7\,\mu$m and $1.5\,\mu$m; and this allowable width is limited by the next step, which is the oxide deposition (Fig. 7.10(c)).

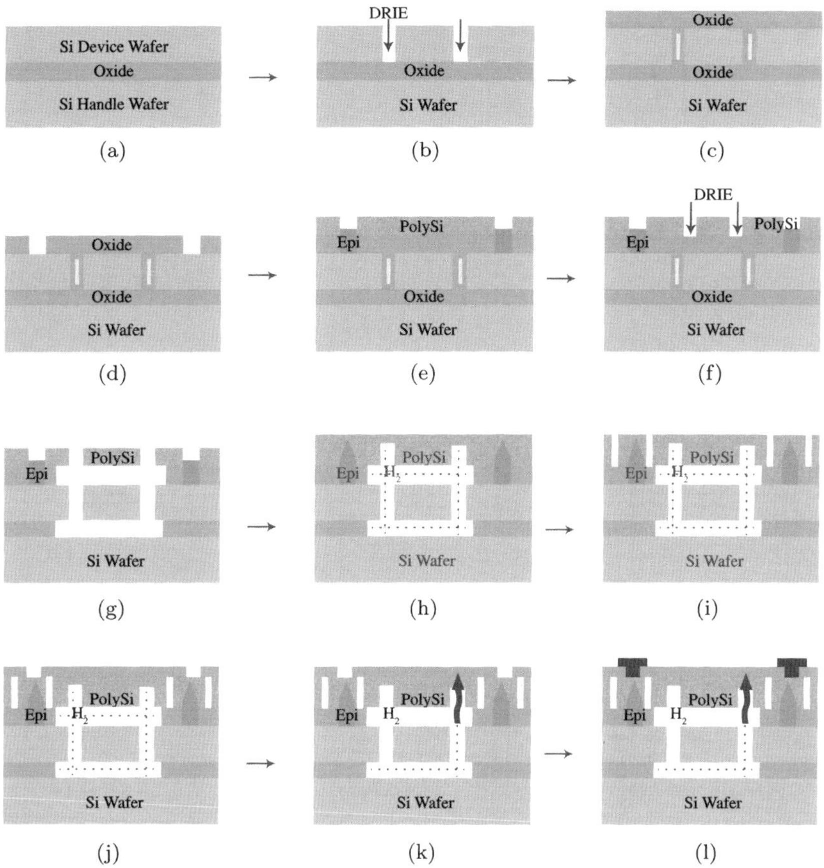

Fig. 7.10. The Stanford Epi-seal process: (a) Start with SOI wafer, (b) Defining device, (c) Deposit, (d) Etch oxide, (e) Epi-poly cap, (f) Etch vents, (g) HF vapor etch oxide for releasing the device, (h) Epi-seal, (i) Etch isolation in cap, (j) Deposit and pattern oxide, (k) Diffuse out H_2, and (l) Sputter and pattern Al.

Step 3: Oxide is deposited to fill the trenches and to be used as a top spacer layer (Fig. 7.10(c)). The deposition takes place using a low-pressure chemical vapor deposition (LPCVD), and the top spacer layer is usually about $2\,\mu$m thick, which is more than enough to cover or bridge these trenches and achieve a planar surface (Fig. 7.13).

Step 4: Contact holes are subsequently etched into this oxide. These contact holes provide electrical access to the device and electrodes, creating a mechanical anchor (Fig. 7.10(d)).

Fig. 7.11. Due to silicon migration at high temperatures, polysilicon and SCS sidewall profiles have a very different final surface finish. As shown, encapsulated polysilicon devices (top-left), with a zoomed-in view (bottom-left), experience significant surface roughness, whereas the encapsulated SCS device (top-right), with a zoomed-in view (bottom-right), ends up with smooth surface finish [31].

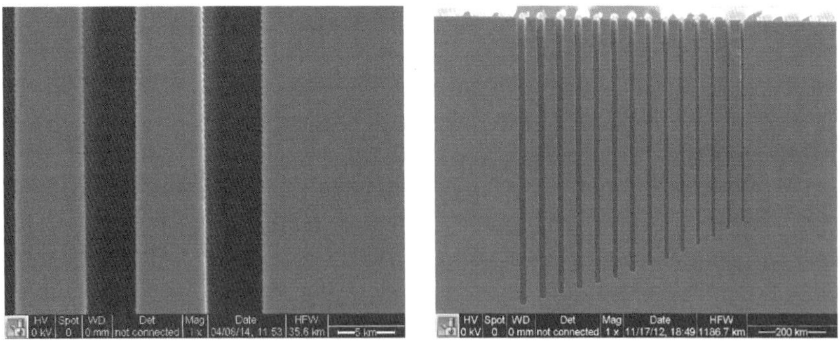

Fig. 7.12. To the left: SEM of different variations of width from 0.4 μm to 2 μm. To the right: Zoomed-in view of the trenches where scallops are visible. [25]

Step 5: To obtain a uniform deposition across the wafer, a thin seed layer (~100 nm) of amorphous polysilicon is deposited (Fig. 7.10(e)), followed by growing ~6 μm of polysilicon on top of the sacrificial oxide. This last process is called epi-poly because of the epitaxial recipe used for the

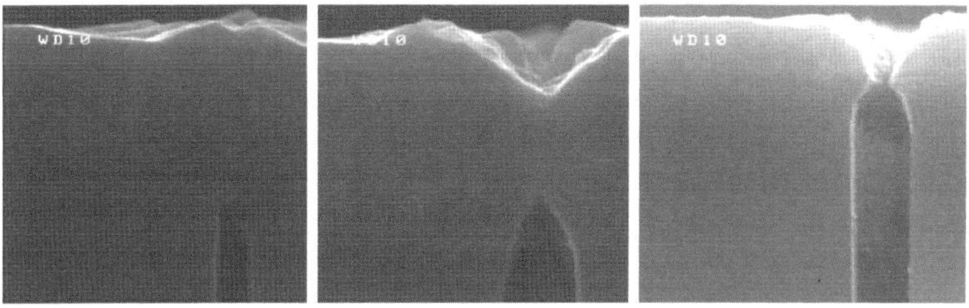

(a) Trench size 2 μm (b) Trench size 4 μm (c) Trench size 6 μm

Fig. 7.13. SEM images collected during the Epi-seal fabrication process, which shows a roughly 3 μm fill using a LPCVD Tetraethyl Orthosilicate TEOS to deposit oxide which results in planar and non-planar surfaces, depending on their gap size.

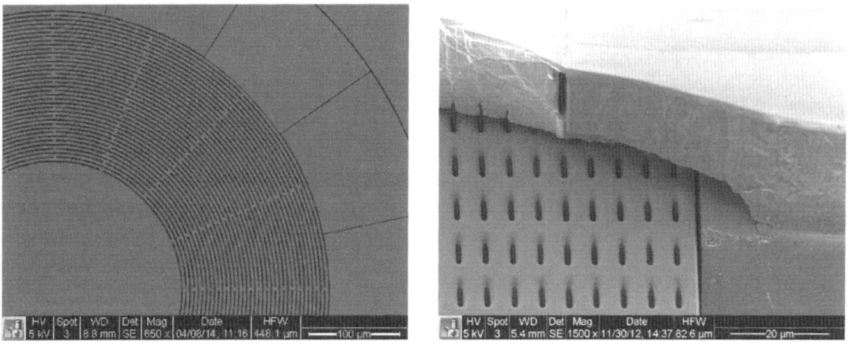

Fig. 7.14. SEM image of a DRG wine glass mode resonator (left) and a length extensional bulk-mode resonator (right) with release etch-holes in the device layer.

deposition of polysilicon. Epi-poly silicon is a low-stress material that can be grown in thick layers and has been demonstrated to be excellent for MEMS devices [13].

Step 6: By means of a DRIE process, vent holes are created (Fig. 7.10(f)). Since these vents must be sealed after the device is released, the width of these vents is typically $<0.7\,\mu$m (Fig. 7.14).

Step 7: Vapor-phase hydrofluoric acid (HF) is used to etch the sacrificial oxide and release the device structure. HF etching is highly sensitive to moisture; consequently, the release process is performed in cycles. The release time depends on the size of the structures to be released, and compensation for larger structures should be taken into consideration during the design process (Fig. 7.10(g)). For instance, larger structures should be designed with larger anchor sizes.

Step 8: After the HF vapor step is completed, a thick ~15 μm additional encapsulation layer is deposited epitaxially to seal the vent holes and create a hermetic cavity (Fig. 7.10(h)). This is done in the epitaxial silicon reactor, which will remove all the SiO_2. Note that depending on the material used for the wafer, this process can result in different surface finishes on the sidewalls. After going through the epitaxial silicon reactor, the surface finish for polysilicon is rough whereas the surface finish for SCS is smooth [18]. Then the wafer surface is planarized using chemical mechanical polishing (CMP) in preparation for the subsequent step.

Step 9: Once the device has been released, electrical isolation trenches are patterned in the cap layer (Fig. 7.10(i)).

Step 10: These trenches are bridged using a low temperature deposition of passivation oxide ~2 μm thick, and contact holes are etched into the LTO for electrical access. (Fig. 7.10(j)).

Step 11: The next step is a hydrogen anneal that is performed to prepare the device for sealing (Fig. 10(k)). This annealing will remove any native oxide on the device surface. At least 1 μm of silicon is deposited to seal the vents, and, subsequently, an extra 10–20 μm can be deposited to strengthen the cap layer. During deposition, at least 90% of the gas flowing is hydrogen. The other gasses in the chamber are dichlorosilane ($SiCl_2H_2$), phosphine (PH_3), and hydrochloric acid (HCl). The final pressure inside the cavity is less than 1 Pa.

Step 12: Once the device is sealed, electrical isolation trenches are patterned in the cap layer to make isolated electrical contacts to the buried electrodes (Fig. 7.10(l)). These trenches cannot be defined over the cavity, since this would vent the cavity. These trenches are bridged over by a deposited low temperature oxide. Contact vias are etched into this layer, and aluminum is deposited and patterned.

Last, the dies can be separated easily using a standard wafer separation technique with little or no reduction in yield (Fig. 7.15). MEMS devices can be wire bonded and molded into their final standard electrical package using standard tools, such as those used for CMOS devices. The Kenny Group at Stanford University has used this process as a "baseline process" to support many generations of improved resonators [1, 11, 19, 24, 26].

Some limitations, described below, of the Epi-seal process are gap sizes, limited released regions, out-of-plane actuation and sensing, and limits to the types of materials that can be used. Some of these limitations have

Fig. 7.15. Completed 2 × 2 mm die which could contain as many as 12 encapsulated devices at once [25].

been addressed by modifications made to some of the steps, as shown in Fig. 7.10.

Wide Gaps vs. Narrow Gaps

Gap size in MEMS resonators is critical for transduction, with the motional impedance variant to the fourth power of the gap size. In the Epi-seal process, gap sizes range from 0.7 μm up to 2 μm. The gap size limits many potential applications, which require more movable structures such as comb drives. Since the Epi-seal process relies on a sacrificial layer followed by a structural cap layer, one of two conditions that involve the device layer gaps must be met. Either the open features must be bridged over with the sacrificial material, or they must be filled with the sacrificial material. In the case of the Epi-seal process, the solution is a combination of both of these. When the gap size is limited to 2 μm, the gaps are both partially filled and bridged over after only 2 μm of oxide deposition. Even though a keyhole is left in the trench itself, the topmost surface of the oxide film is nearly planar and ideally suited for the epitaxial silicon cap (Fig 7.13(a)). However, when gap sides are wider, the sacrificial layer does not fully bridge the gap, and the subsequent deposit of the structural cap layer will also have the same topology as the larger device feature, leading to epitaxial silicon depositions that reach down into these gaps and yield an upper surface with this same topology. These larger features do not yield cleared, open features upon completion of the encapsulation process, and hence lead to poor yield (Figs. 7.13(b) and 7.13(c)). Moreover, when the oxide layer does not completely bridge over all trenches, there is the possibility of electrical shorting during the silicon deposition step. Therefore, trenches with a width greater than 2 μm, cross-shaped trenches, and acute-angled structures cannot be handled by the standard process. Efforts have been made to increase the gap size within the Epi-seal process using a modified version of the standard Epi-seal process [12].

Limited Released Regions

Because the oxide is released from vents at the top (Fig. 7.10(g)), the buried oxide layer can only be released via trenches in the device layer. This limits the released regions in the device layer. Thin beam devices can be released using vapor HF to remove the oxide underneath within a reasonable amount of time; however, this process cannot be used for devices with larger area since, because of their size, these devices would not be released. This problem is mitigated by the use of etch holes (Fig. 7.14). In certain bulk mode resonators, however, the use of etch holes have been shown to reduce the quality factor and proof mass.

Out-Of-Plane Actuation and Sensing

One of the limitations of the standard process is the lack of out-of-plane electrodes, which limits the type of devices that can be fabricated. For instance, this feature is highly desirable for 3-axis sensors, such as accelerometers, gyroscopes, and magnetometers.

Limited Materials

The Epi-seal fabrication is done at very high temperatures (~1100°C) with CMOS compatible tooling. High temperatures ensure cleanliness of the cavity, but at the same time these temperatures limit the types of materials available for fabrication. Moreover, because the fabrication is done using CMOS tooling, only a specific set of materials are allowed inside these fabrication facilities.

7.4 Conclusion

This chapter has reviewed the packaging process for MEMS resonators with a particular emphasis on the Epi-seal process. This process has been successfully demonstrated on the wafer-scale and with a low cavity pressure (<1 Pa), which has allowed its use to build resonators, accelerometers, gyroscopes, magnetometers, switches, and pressure sensors, just to mention a few. The silicon surface in the cavity is free from native oxide and is smoothed by silicon migration. Because the Epi-seal process has high stability, low cavity pressure, and is contamination free, it is ideal for resonant structures.

References

1. Ahn, C. H., Hong, V. A., Park, W., Yang, Y., Chen, Y., Ng, E. J., *et al.* (n.d.). On-chip ovenization of encapsulated disk resonator gyroscope (DRG), Presented at the 2015 Transducers — 2015 *18th International Conference on Solid-State Sensors, Actuators and Microsystems*, pp. 39–42.
2. Ahn, C. H., Ng, E. J., Hong, V. A., Yang, Y., Lee, B. J., Ward, M. W., and Kenny, T. W. (n.d.). Geometric compensation of (100) single crystal silicon disk resonating gyroscope for mode-matching, Presented at the 2015 Transducers — 2015 *18th International Conference on Solid-State Sensors, Actuators and Microsystems*, pp. 1723–1726.
3. Ahn, C. H. (2016). Disk Resonator Gyroscope within a Wafer-Scale Encapsulation Process, PhD Thesis, July.
4. Acker. (2001). The backend process: step11-scribe and break, *Advanced Packaging.*
5. Bao, M. and Yang, H. (2007). Squeeze film air damping in MEMS. *25th Anniversary of Sensors and Actuators a: Physical*, 136(1), pp. 3–27.
6. Candler, R. N., Park, W.-T., Hopcroft, M., Kim, B., and Kenny, T. W. (n.d.). Hydrogen diffusion and pressure control of encapsulated MEMS resonators, Presented at the 2015 Transducers — 2015 *18th International Conference on Solid-State Sensors, Actuators and Microsystems*, Vol. 1, pp. 920–923.
7. Candler, R. N., Park, W.-T., Li, H., Yama, G., Partridge, A., Lutz, M., and Kenny, T. W. (2007). Single wafer encapsulation of MEMS devices. *IEEE Transactions on Advanced Packaging*, 26(3), pp. 227–232.
8. Chandrahalim, H., Weinstein, D., Cheow, L. F., and Bhave, S. A. (2006). Channel-select micromechanical filters using high-k dielectrically transduced MEMS resonators, Presented at the *19th IEEE International Conference on Micro Electro Mechanical Systems*, pp. 894–897.
9. Christian, R. G. (1966). The theory of oscillating-vane vacuum gauges. *Vacuum*, 16(4), pp. 175–178.
10. Eloy, J. C., Mounier, E., and Roussel, P. (2005). Status of the inertial MEMS-based sensors in the automotive, In J. Valldorf & W. Gessner (Eds.), *Advanced Microsystems for Automotive Applications 2005* (pp. 43–48). Berlin, Heidelberg: Springer Berlin Heidelberg. http://doi.org/10.1007/3-540-27463-4_3
11. Graham, A. B., Messana, M. W., Hartwell, P. G., Provine, J., Yoncoka, S., Melamud, R., *et al.* (n.d.). A Method for wafer-scale encapsulation of large lateral deflection MEMS devices, *Journal of Microelectromechanical Systems*, 19(1), pp. 28–37.

12. Graham, A. B. (2010). Methods for the wafer-scale encapsulation of MEMS, PhD Thesis, August.
13. Guckel, H., Rypstat, C., Nesnidal, M., Zook, J. D., Burns, D. W., and Arch, D. K. (n.d.). Polysilicon resonant microbeam technology for high performance sensor applications, Presented at the T*echnical Digest IEEE Solid-State Sensor and Actuator Workshop*, pp. 153–156.
14. He, R., and Kim, C.-J. (2006). Alow temperature vacuum package utilizing porous alumina thin film encapsulation, Presented at the *19th IEEE International Conference on Micro Electro Mechanical Systems*, pp. 126–129.
15. Jones, D. G., Azevedo, R. G., Chan, M. W., Pisano, A. P., and Wijesundara, M. B. J. (2007). Low temperature ion beam sputter deposition of amorphous silicon carbide for wafer-level vacuum sealing, Presented at the *19th IEEE International Conference on Micro Electro Mechanical Systems*, pp. 275–278.
16. Joyce, B. D. and Baldrey, J. A. (1962). Selective epitaxial deposition of silicon, *Nature*, 195(4840), pp. 485–486. http://doi.org/10.1038/195485a0
17. Jung, E. (2003). Packaging Options for MEMS devices, *MRS Bulletin*, 28(1), pp. 51–54. http://doi.org/10.1557/mrs2003.17
18. Kim, B., Candler, R. N., Hopcroft, M., Agarwal, M., Park, W.-T., Li, J. T., and Kenny, T. (2004). Investigation of MEMS resonator characteristics for long-term and wide temperature variation operation, Presented at the *ASME 2004 International Mechanical Engineering Congress and Exposition*, ASME, pp. 413–416. http://doi.org/10.1115/IMECE2004-61727
19. Kim, B., Candler, R. N., Melamud, R., Hopcroft, M. A., Yoneoka, S., Lee, H. K., *et al.* (2009). Hermeticity and diffusion investigation in polysilicon film encapsulation for microelectromechanical systems. *Journal of Applied Physics*, 105(1), p. 013514. http://doi.org/10.1063/1.3054366
20. Kim, B., Melamud, R., Hopcroft, M. A., Chandorkar, S. A., Bahl, G., Messana, M., *et al.* (2008). Si-SiO_2 Composite MEMS resonators in CMOS compatible wafer-scale thin-film encapsulation, Presented at the 2008 *IEEE International Frequency Control Symposium*, pp. 1214–1219.
21. Lin, Y., Li, W.-C., Ren, Z., and Nguyen, C. T. C. (2008). A resonance dynamical approach to faster, more reliable micromechanical switches, Presented at the 2008 *IEEE International Frequency Control Symposium*, pp. 640–645.
22. Metzger, L., Fischer, F., and Mokwa, W. (2007). Polysilicon sacrificial layer etching using ClF_3 for thin film encapsulation of silicon acceleration sensors with high aspect ratio, *25th Anniversary of Sensors and Actuators A: Physical*, 133(1), pp. 259–265.
23. Najafi, K. (2003). Micropackaging technologies for integrated microsystems: Applications to MEMS and MOEMS. In J. A. Yasaitis, M. A. Perez-Maher, & J. M. Karam (Eds.), Presented at the *Micromachining and Microfabrication*, SPIE, Vol. 4979, pp. 1–19. http://doi.org/10.1117/12.484953
24. Ng, E. J., Chiang, C. F., Yang, Y., Hong, V. A., Ahn, C. H., and Kenny, T. W. (n.d.). Ultra-high aspect ratio trenches in single crystal silicon with epitaxial gap tuning,) Presented at the 2015 Transducers — 2015 18th *International Conference on Solid-State Sensors, Actuators and Microsystems*, pp. 182–185.

25. Ng, E. J. (2015). Temperature-Compensated Silicon MEMS Resonators with High Quality Factors and Low Motional Impedance, PhD Thesis, July.
26. Nitzan, S., Ahn, C. H., Su, T. H., Li, M., Ng, E. J., Wang, S., *et al.* (2013). Epitaxially-encapsulated polysilicon disk resonator gyroscope, Presented at the 19th *IEEE International Conference on Micro Electro Mechanical Systems*, pp. 625–628.
27. Partridge, A., Lutz, M., Kim, B., Hopcroft, M., Candler, R. N., Kenny, T. W., Petersen, K., and Esashi, M. (2005). MEMS Resonators: Getting the Packaging Right, in *SEMICON-Japan*.
28. Roberts, C. M., Long, L. H. and Ruggerio, P. A. (1994). *Method for Separating Circuit Dies from a Wafer*. Google Patents.
29. Rodriguez, J., Yang, Y., Ahn, C. H., Chen, Y., Ng, E. J., Hong, V. A., *et al.* (n.d.). Damping mechanisms in light and heavy-doped dual-ring and double-ended tuning fork resonators (DETF), Presented at the 2015 Transducers — 2015 18th *International Conference on Solid-State Sensors, Actuators and Microsystems*, pp. 2005–2008.
30. Senturia, S. D. (2001). *Microsystem Design*. Norwell, MA, USA: Kluwer Academic Publishers.
31. Stehle, J., Hong, V. A., Feyh, A., O'Brien, G. J., Yama, G., Ambacher, O., Kim, B., and Kenny, T. W. (2014). Silicon migration of through-holes in single- andpoly-crystalline silicon membranes, *Solid-State Sensors, Actuators, and Microsystems Workshop Hilton Head Island*, pp. 1–4.
32. Sparks, D. and Hubbard, T. (2004). Micromachined needles and lancets with design adjustable bevel angles, *Journal of Micromechanics and Microengineering*, 14(8), p. 1230.
33. Waggoner, P. S., and Craighead, H. G. (2007). Micro- and nanomechanical sensors for environmental, chemical, and biological detection. *Lab on a Chip*, 7, pp. 1238–1255. http://doi.org/10.1039/B707401H

Janna Rodriguez received the BS degree in Mechanical Engineering from the University of California, Merced, in 2012, and the MS degree in Mechanical Engineering from Stanford University in 2016, where she is currently pursuing her PhD degree in Mechanical Engineering. Her research focuses on understanding the different dissipation mechanisms that affect the quality factor through the use of models and wide-range temperature measurements.

Thomas W. Kenny received the BS degree from the University of Minnesota, Minneapolis, in 1983, and the MS and PhD degrees from the University of California at Berkeley in 1987 and 1989, respectively, all in physics. From 1989 to 1993, he was with the Jet Propulsion Laboratory, National Aeronautics and Space Administration, Pasadena, CA, where his research focused on the development of electron-tunneling high-resolution microsensors. In 1994, he joined the Department of Mechanical Engineering, Stanford University, Stanford, CA, where he directs microsensor-based research in a variety of areas, including resonators, wafer-scale packaging, cantilever beam force sensors, microfluidics, and novel fabrication techniques for micromechanical structures. He is the Founder and CTO of Cooligy (now a division of Emerson), a microfluidics chip cooling component manufacturer, and the Founder and a Board Member of SiTime Corporation (now a division of MegaChips), and a Developer of timing references using MEMS resonators. He is the Founder and a Board Member of Applaud Medical, developing noninvasive therapies for kidney stones. From 2006 to 2010, he was on leave to serve as a Program Manager at the Microsystems Technology Office, Defense Advanced Research Projects Agency, starting and managing programs in thermal management, nanomanufacturing, manipulation of Casimir forces, and the Young Faculty Award. He has authored or co-authored over 250 scientific papers and holds 50 issued patents, and has been an Advisor to more than 50 graduated PhD students from Stanford University. He is currently the Richard Weiland Professor of Mechanical Engineering and the Senior Associate Dean of Engineering for Student Affairs. He was the General Chairman of the 2006 Hilton Head Solid State Sensor, Actuator, and Microsystems Workshop, and the General Chair of the Transducers 2015 meeting in Anchorage.

Chapter 8

Heterogeneous Integration and Wafer-Level Packaging of MEMS

Masayoshi Esashi and Shuji Tanaka

8.1 Introduction

Microelectromechanical systems (MEMS) are used in combination with integrated circuits (IC) or large-scale integration (LSI) in many cases. This heterogeneous integration (heterointegration) requires process flexibility to fabricate high-performance MEMS in order to allow multiple elements with interconnections and reduced stray capacitance and stray inductance [1]. The heterogeneous integration described in this chapter is a wafer-level transfer of multiple MEMS on a carrier wafer to a LSI wafer. This method fulfills the requirements described earlier. The heterointegration devices should be packaged to be protected from the environment. For this reason, wafer-level packaging (WLP) [2] is needed as described later in Section 8.3. The concept of the heterointegration combined with WLP is shown in Fig. 8.1. MEMS or thin films of functional material are formed on a Si carrier wafer and transferred on a LSI wafer by adhesive bonding and succeeding removal of the carrier wafer. Another bonding method called bump bonding can be also used as discussed in Fig. 8.3(c). The LSI wafer with transferred MEMS is covered by a low-temperature co-fired ceramic (LTCC) wafer with through-via interconnection for the purpose of the WLP. Finally, the covered wafer is diced to obtain the packaged heterointegration chips.

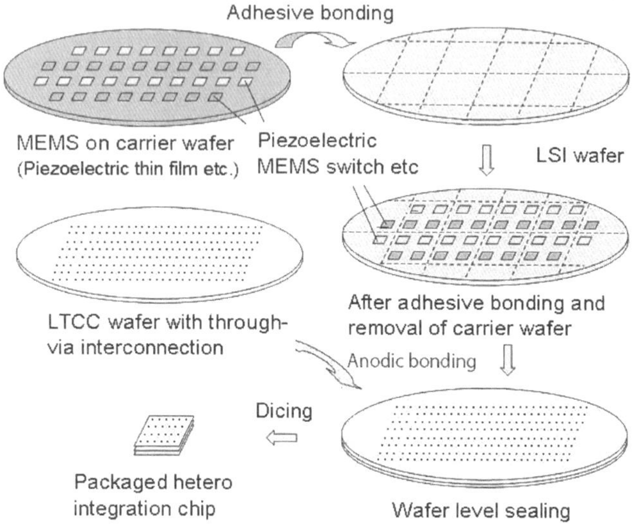

Fig. 8.1. Concept of the heterogeneous integration by the wafer-level transfer and the WLP.

8.2 Heterointegration

Three methods of the heterointegration are shown in Fig. 8.2: (a) surface micro-machining, (b) assembly of packaged MEMS and LSI, and (c) the transfer of MEMS to LSI.

Surface micro-machining, shown in Fig. 8.2(a), has been used to fabricate mirror array for display [3], infrared imager, and other devices. This method is suitable especially for array MEMS on LSI. Sacrificial layer made of, for example, photoresist is patterned and then MEMS structures are formed on it. Finally, the sacrificial layer is etched out to make MEMS structures on LSI and the wafer is diced to each chip. High-density interconnections between the MEMS and the LSI can be achieved in this process. However, the fabrication of MEMS has to be carried out on the LSI wafer and hence there are limitations such as temperature not to damage the LSI during MEMS fabrication.

The assembly of packaged MEMS and LSI in Fig. 8.2(b) has been used, for example, in capacitive sensors for accelerometers and gyroscopes. Packaged MEMS chips are fabricated by protecting the fabricated MEMS wafer with a cover wafer and by dicing the wafers. LSI chips are located on or besides the packaged MEMS chips and the interconnection between them are made by wire bonding. Since MEMS are packaged, conventional plastic

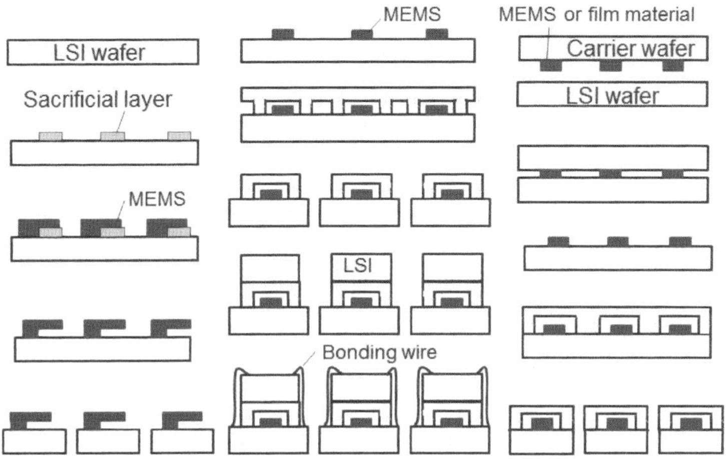

Fig. 8.2. Heterogeneous integration processes. MEMS and LSI (WLP): (a) Surface micro-machining; (b) assembly of packaged MEMS and LSI; and (c) transfer of MEMS to LSI.

molding can be applied to protect the assembled device. The MEMS fabrication process has much freedom and the LSI fabrication can be a standard one. The number of interconnections is limited because of wire bonding. Stray capacitance and the parasitic inductance of the bonding wire limit its application. The capacitance of the capacitive sensor should not be very small. For this reason, special deposition method called epitaxial poly-Si is used to deposit a thick film ($\sim 15\,\mu$m) to increase the capacitance of the sensor [4].

The transfer of MEMS to LSI, shown in Fig. 8.2(c), enables new heterogeneous integration of MEMS on LSI. MEMS or films of functional materials such as piezoelectric material are fabricated on a carrier wafer. They are transferred on a LSI wafer by bonding and dissolving out the carrier wafer as explained in Fig. 8.1. The LSI wafer, which has MEMS on it, is diced to each chip or diced after being covered with a lid wafer in case of the WLP. The carrier wafer can be also used as the cover to enable the WLP (Nasiri fabrication) [5].

The wafer-level transfer methods of MEMS (or film of functional material) can be categorized as (a) film transfer, (b) device transfer (via-last), and (c) device transfer (via-first) as shown in Fig. 8.3 [6]. Each process will be explained in the following and examples of each methods will be shown in Section 8.2.1 through 8.2.3, respectively.

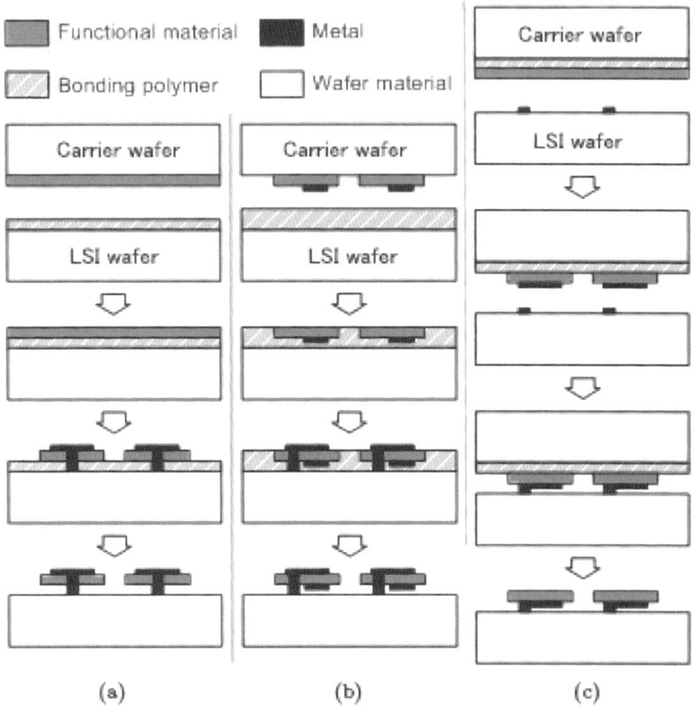

Fig. 8.3. Wafer-level transfer of heterogeneous MEMS to LSI: (a) Film transfer, (b) Device transfer (via-last), and (c) Device transfer (via-first).

Figure 8.3(a) shows the process sequence of the film transfer method. A film of the functional materials such as diamond film or lead zirconate titanate (PZT) film is formed on a carrier wafer by chemical vapor deposition (CVD), sputtering, or other deposition methods. The film is transferred to the LSI wafer by adhesive bonding or other method and by succeeding removal of the carrier wafer. This removal can be done by etching out the Si of the carrier wafer or by etching out a sacrificial layer between the film and the carrier wafer. MEMS are fabricated using the film, and electrical and mechanical connections to the LSI wafer are made by using an electroplating or other deposition methods of metal. The adhesive layer is etched out and we can get the MEMS on LSI. Since the MEMS fabrication is carried on the LSI wafer, the process is limited not to damage the LSI.

Figure 8.3(b) shows the process sequence of the device transfer (via-last) method. The MEMS are fabricated on a carrier wafer and the wafer is bonded to the LSI wafer using the adhesive bonding. After removing the carrier wafer, electrical and mechanical connections (vias) to the LSI wafer are made using the electroplating and finally the adhesive layer is etched out. This process is called via-last because vias are made after bonding.

Figure 8.3(c) is the process sequence of the device transfer (via-first). The MEMS is fabricated on a carrier wafer and bumps for electrical interconnection (via) are formed on the LSI wafer. These wafers are bonded using the bumps and finally the carrier wafer is taken off by removing the bonding polymer. Since we can use another method, the bonding polymer is not always necessary. This process is called via-first because vias are made before bonding.

The transfer of the MEMS devices to the carrier wafer is carried out in batch process as shown in Fig. 8.3. However, selective transfer process is required for different-sized MEMS chips from LSI chips [7]. The selective transfer process is shown in Fig. 8.4 and explained in the following. Au pads are formed on the MEMS wafer for Au–Au bump bonding (Fig. 8.4(a)). The MEMS wafer is bonded to a glass carrier wafer using polymer (acrylic resin) and grooves are made on the MEMS wafer by dicing (Fig. 8.4(b)). Au bumps are formed on the LSI wafer (Fig. 8.4(b)). The MEMS wafer

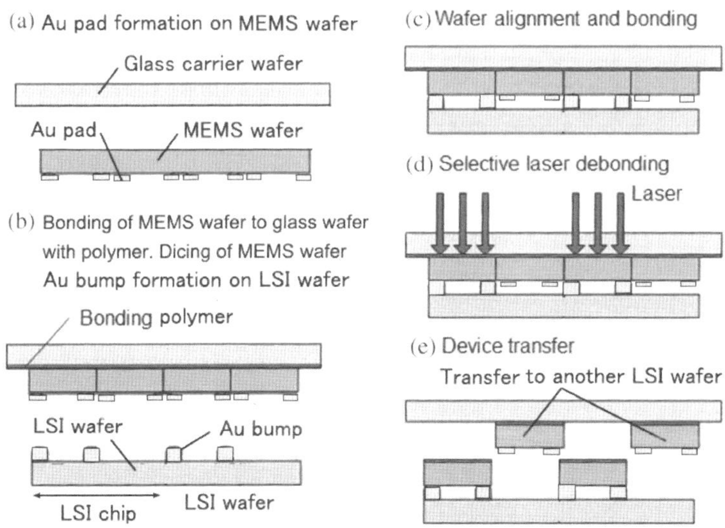

Fig. 8.4. Selective transfer process: (a) Au pad formation on MEMS wader, (b) bonding of MEMS wafer to glass wafer with polymer. Dicing of MEMS wafer and Au bump formation on LSI wafer, (c) wafer alignment and bonding, (d) selective laser bonding, and (e) device transfer.

on the carrier wafer is aligned with the LSI wafer and these wafers are bonded by pressing them uniformly at 180°C after Ar plasma activation (Fig. 8.4(c)). Selective debonding is achieved by irradiating the interfacial acrylic resin using Nd:YVO_4 third harmonic laser ($\lambda = 355\,\text{nm}$) through the glass carrier wafer (Fig. 8.4(d)) [8]. The acrylic resin is carbonized to lose adhesion and the MEMS device is transferred to the surface of the LSI wafer (Fig. 8.4(e)). The MEMS dies remaining on the glass carrier wafer can be transferred to another LSI wafer.

8.2.1 *Film Transfer*

A 20×20 array of boron-doped diamond (BDD) electrodes is formed on a CMOS LSI for simultaneous multipoint amperometric detection of biochemical substances and for real-time bioimaging [9]. The fabrication process of the BDD electrode array on LSI is shown in Fig. 8.5. Metallization of Cr/Pt/Au/Pt/Cr is carried out on the LSI wafer (Fig. 8.5(a)). Diamond is doped with boron for electrical conduction. The BDD is prepared by microwave chemical vapor deposition (MW-CVD) at 800°C on a carrier Si wafer. The wafer-level transfer of the BDD film to the LSI is carried out by film transfer explained in Fig. 8.3(a). Benzocyclobutene (BCB) is used as the bonding polymer (Fig. 8.5(b) and (c)). The BDD is exposed by dry-etching of the Si carrier wafer in SF_6 plasma (Fig. 8.5(d)). The BCB is selectively etched by plasma ($SF_6 + O_2$) using patterned metal as a mask (Fig. 8.5(e)). Cr and Au are deposited and patterned for an electrical interconnection between the LSI pad and the BDD (Fig. 8.5(f)). Finally, it is covered with thick photoresist (SU-8) and windows are opened on the BDD (Fig. 8.5(g)).

The structure and photographs of the BDD on CMOS LSI are shown in Fig. 8.6(a). The 20×20 (400) array of operational amplifier I–V converters are fabricated on the chip. Current is detected by integrating it in a feedback capacitor for some period depending on the current level.

The cyclic voltammogram (CV) of the BDD compared with that of Au is shown in Fig. 8.6(b). The voltage to generate oxygen gas by oxidation of hydroxide ion at positive voltage and also that to generate hydrogen gas by reduction of hydrogen ion at negative voltage are large in the case of BDD compared with Au, because diamond does not have a catalytic property. The CV of the BDD in histamine solution is shown in Fig. 8.6(b). Histamine could be detected amperometrically using the BDD with a high oxidation potential around 1.1 V, which is too high for a conventional Au

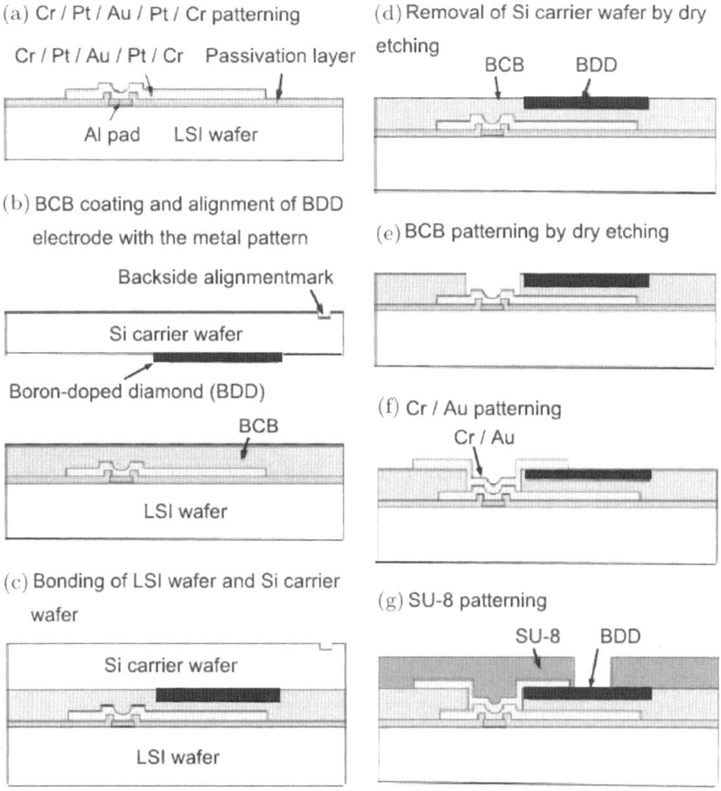

Fig. 8.5. Fabrication process of BDD electrode array on LSI for electrochemical biosensing: (a) Cr/Pt/Au/Pt/Cr patterning, (b) BCB coating and alignment of BDD electrode with the metal pattern, (c) Bonding of LSI wafer and Si carrier wafer, (d) Removal of Si carrier wafer by dry etching, (e) BCB patterning by dry etching, (f) Cr/Au patterning, and (g) SU-8 patterning.

electrode to detect. The diffusion behavior of histamine in a solution was imaged in real time by parallel measurement of the oxidation current at 400 points (Fig. 8.6(c)). This CMOS-based amperometric sensor array has been successfully used for drug screening being applied to cancer cells on the array.

Surface acoustic wave (SAW) filter plays an important role in our communication systems. A one-chip band width-tunable SAW filter was developed by the wafer-level transfer of $BaSrTiO_3$ (BST) film to a $LiTaO_3$ wafer [10]. The BST film is needed for variable capacitors. The fabrication process is shown in Fig. 8.7. Pt, BST, Pt, and Au are successively

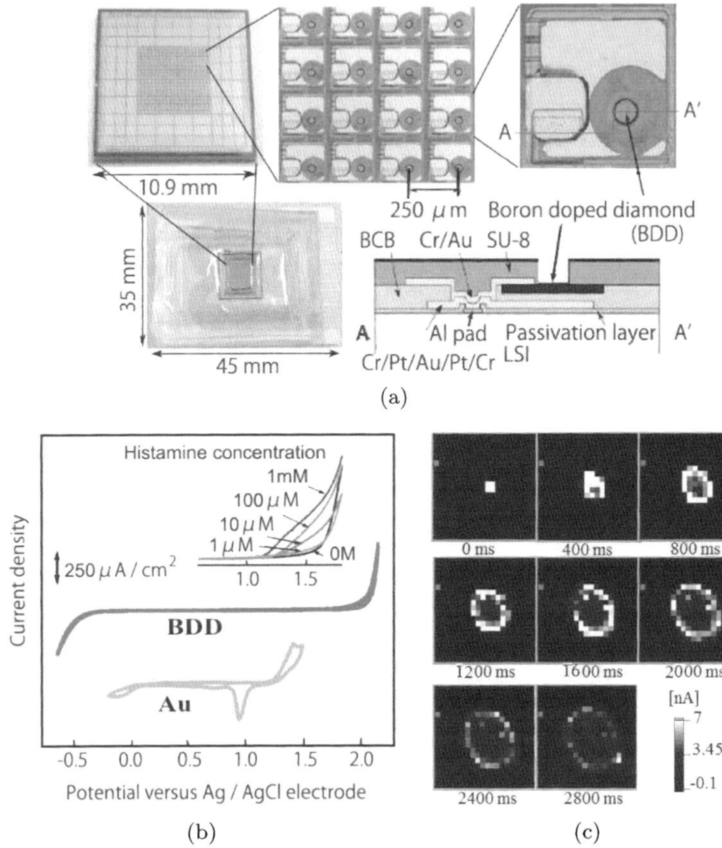

Fig. 8.6. 20 × 20 array of BDD electrodes on LSI for the electrochemical detection of biochemical substances: (a) Structure, (b) Cyclic voltammogram (CV) on Au, and (c) Diffusion of dropped histamine. BDD in 0.5M H_2SO_4 with CV on BDD in 0.1 mM histamine in Dulbecco's phosphate buffer solution.

deposited on a sapphire wafer (Fig. 8.7(a)). Au, Pt, and BST are etched using photoresist as a mask (Fig. 8.7(b)). The Pt layer is irradiated with Nd:YVO_4 laser (wavelength 355 nm) through the sapphire wafer. The purpose of this process step is to weaken the adherence between Pt and BST and this is called laser pre-irradiation (Fig. 8.7(c)). Using Au–Au bonding, the sapphire wafer is bonded to the $LiTaO_3$ wafer, which has inter digital transducer (IDT) electrodes for SAW device (Fig. 8.7(d)). The patterned BST, Pt, and Au layers are transferred to the $LiTaO_3$ wafer by separating the sapphire wafer from the $LiTaO_3$ wafer (Fig. 8.7(e)). Finally,

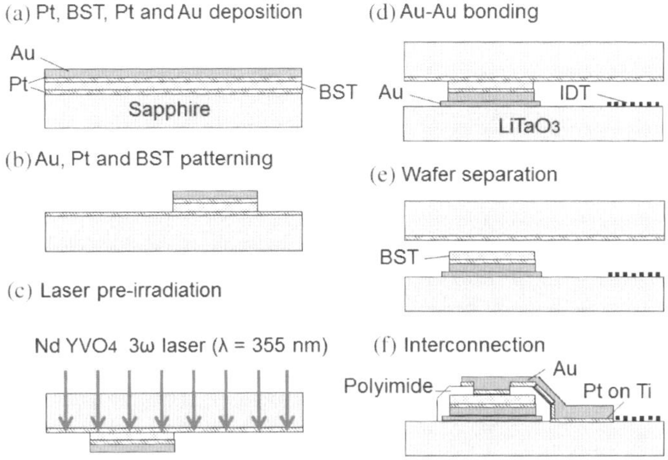

Fig. 8.7. Fabrication process of tunable SAW filter using the selective transfer process: (a) Pt, BST, Pt and Au deposition, (b) Au, Pt and BST patterning, (c) Laser pre-irradiation, (d) Au-Au bonding, (e) Wafer separation, and (f) Interconnection.

for interconnection, the side of the BST is insulated with polyimide and metalized with Ti, Pt, and Au (Fig. 8.7(f)).

The circuit diagram (Fig. 8.8(a)), the photograph (Fig. 8.8(b)), and the characteristics (Fig. 8.8(c)) of the bandwidth-tunable SAW filter are shown. The Y_p and the Y_s are the SAW filters and the bandwidth is tunable by connecting variable capacitors C_p and C_s made of the BST in the circuit diagram. Owing to the nonlinear dielectric constant of ferroelectric BST, the capacitance decreases to approximately 50% by applying 5 V. The 3-dB bandwidth of the filter is tuned between 3.25 and 6.25 MHz by applying DC voltage to the BST variable capacitors, while the center frequency is constant at 1.004 GHz. The prototyped filter installed in a cognitive radio demo system based on IEEE802.11af draft successfully demonstrated wireless LAN communication on the digital TV band.

8.2.2 *Device Transfer (Via-Last)*

PZT-actuated MEMS switches on LSI were fabricated as shown in Fig. 8.9 [11]. The device transfer (via-last) explained in Fig. 8.3(b) is used for the fabrication. The piezoelectric MEMS switch works at lower driving voltage and occupies a smaller area than electrostatic MEMS switch. PZT was deposited on a Si carrier wafer by a sol–gel method. A symmetrical structure made of two PZT layers is formed in order to prevent bending

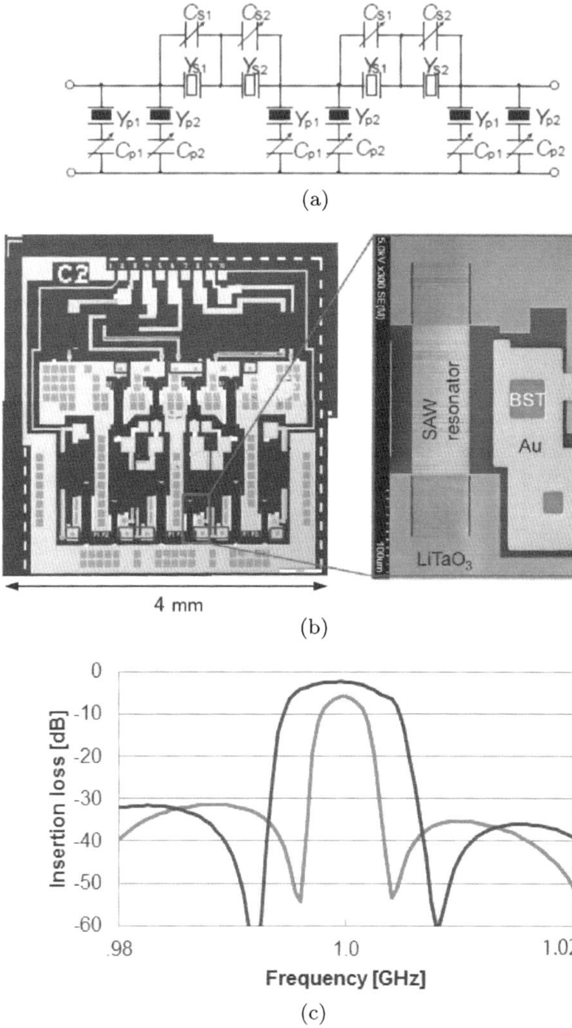

Fig. 8.8. Circuit diagram, photograph, and characteristics of the tunable SAW filter: (a) circuit diagram (resistors to apply voltage to capacitors are not shown), (b) photograph, and (c) characteristics.

due to intrinsic and thermal stress (Fig. 8.9(a)). The flipped carrier wafer is bonded to the LSI wafer using a polymer adhesive (Fig. 8.9(b) and (c)). The two PZT layers are transferred to the LSI wafer by etching out the Si carrier wafer (Fig. 8.9(d)). After connecting the MEMS and the LSI using an electroplated metal (Figs. 8.9(e) and (f)), the polymer is removed by

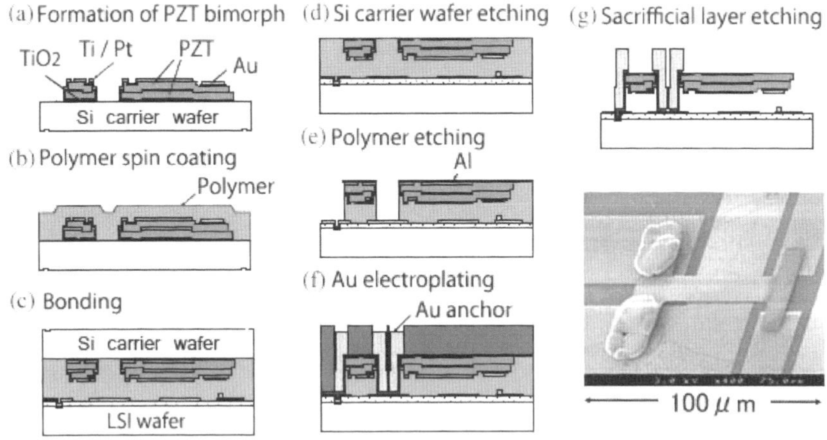

Fig. 8.9. PZT MEMS switch on LSI: (a) Formation of PZT bimorph, (b) Polymer spin coating, (c) Bonding, (d) Si carrier wafer etching, (e) Polymer etching, (f) Au electroplating, and (g) Sacrificial layer etching.

O_2 plasma to release the MEMS switches (Fig. 8.9(g)). The deflection of the PZT bimorph beam obtained was $5\,\mu$m at 8 V. The circuit of the LSI was not damaged during the fabrication process and the operation of the switch control circuit was confirmed.

Distributed tactile sensors (tactile sensor network) are needed on the skin of nursing care robots, rehabilitation robots, etc. to ensure their collision safety and enable body communication. Our tactile sensor network acquires sensing data from tactile sensors on the skin by autonomous data transmission (event driven). The tactile sensor chips are connected to a flexible cable with common 4-wire bus for power supply, ground, and two signals as shown in Fig. 8.10 [12]. The capacitive tactile force sensor is formed on a communication LSI by adhesive bonding of a MEMS wafer using benzo cyclo butenel (BCB). The interconnection pads to the flexible cable are located at the backside of the sensor chip via through silicon groove (TSG).

The fabrication process of the tactile sensor is shown in Fig. 8.11. V groves are formed on the LSI wafer using a dicing saw (Fig. 8.11(a)). The groves are insulated by depositing SiO_2 and then metal interconnection is made between boding pads and the grooves by Ti and Au (Fig. 8.11(b)). Polymer (BCB) is coated (Fig. 8.11(c)) and after making via in the polymer Al pattern is made for the sensing capacitor (Fig. 8.11(d)). Si MEMS wafer having diaphragms for sensing capacitors are bonded to

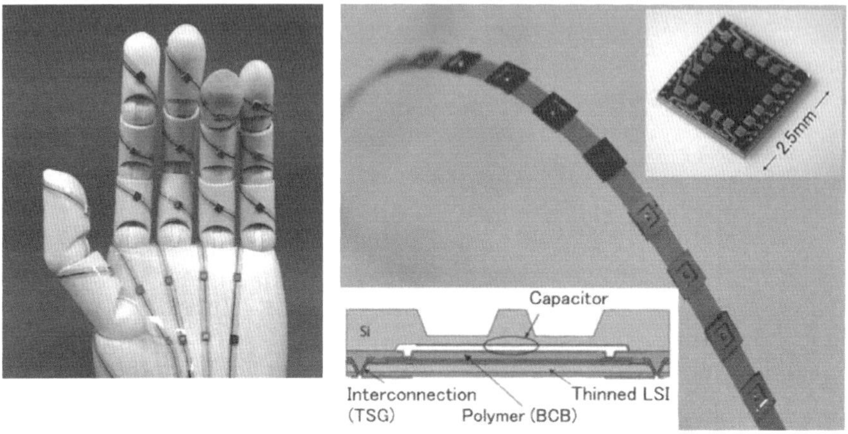

Fig. 8.10. Tactile sensor network.

the polymer (Fig. 8.11(e)). The LSI wafer is thinned from the backside by grinding and polishing. This makes the bottom of the V grove exposed (Fig. 8.11(f)). BCB is coated on the back side (Fig. 8.11(g)) and the BCB and the SiO_2 are etched to expose the metal connected from the front side (Fig. 8.11(h)). Finally pads on the backside are formed by Ti and Au (Fig. 8.11(i)).

The function of the communication LSI for event-driven data transmission at a clock rate of 45 MHz was confirmed. An example of packet communication of the tactile sensor network is shown in Fig. 8.12. The host computer recognizes the sensor position and the applied force from the sensor ID and the force data in the packet signal. The force is detected as a capacitance change, which is converted to a digital output and the linear relationship between the force and the digital output is shown in Fig. 8.12.

8.2.3 *Device Transfer (Via-First)*

Wafer-level selective transfer from one carrier wafer to multiple target wafers are needed for cost-effective integration of dies with different sizes. Multiple SAW resonators are formed on a LSI chip as shown in Fig. 8.13(a) [13]. The selective transfer technology applied is the one that has been explained in Fig. 8.4. The SAW resonators fabricated on a $LiNbO_3$ wafer is bonded to a glass carrier wafer using a polymer. After dicing the $LiNbO_3$ wafer on the glass carrier wafer, it is bonded to the CMOS wafer

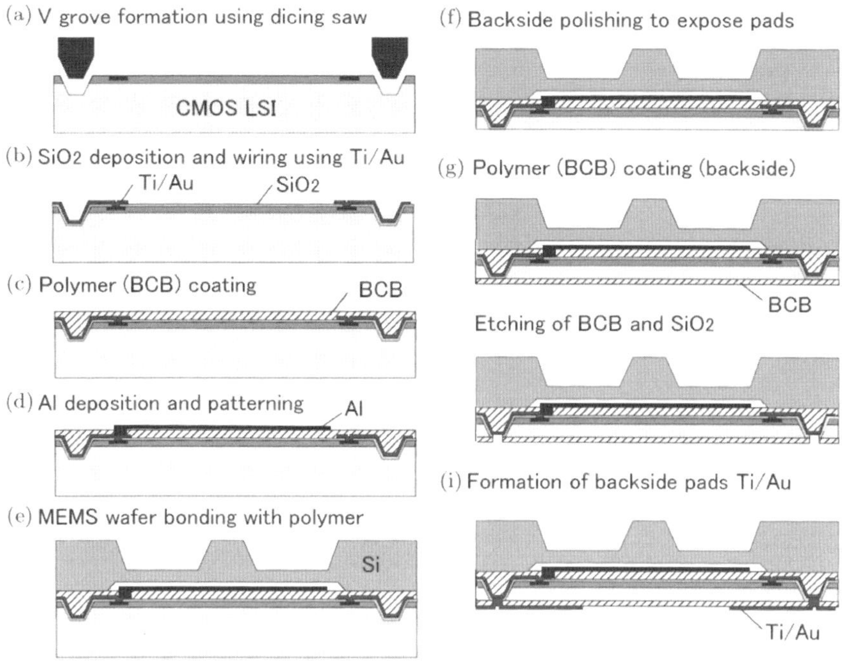

Fig. 8.11. Fabrication process of the tactile sensor: (a) V grove formation using dicing saw, (b) SiO_2 deposition and wiring using Ti/Au, (c) Polymer (BCB) coating, (d) Al deposition and patterning, (e) MEMS wafer bonding with polymer, (f) Backside polishing to expose pads, (g) Polymer (BCB) coating (backside), (h) Etching of BCB and SiO_2, (i) Formation of backside pads Ti/Au.

using Au bumps. The selective laser debonding is applied and the SAW devices are transferred to the LSI chip. Finally, polymer underfilling fixes the SAW chips on the LSI chip.

Frequency tuning characteristics of two integrated oscillators using different SAW filters on a same LSI chip are shown in Fig. 8.13(b). The phase noise measured was low, which is one of the advantages of the heterogeneous integration device owing to the small stray capacitance and inductance.

Digital fabrication of LSI-based on massless lithography is expected for cost-effective small-volume production and short-term development. Extremely high through put direct electron beam (EB) lithography is required because the latest LSI wafer has 10^{12} (1 tera) nanoscale transistors on it. For this purpose, a massive parallel EB exposure system with an active matrix nanocrystal silicon (nc-Si) emitter has been developed using the device transfer (via-first) method shown in Fig. 8.3(c). As shown

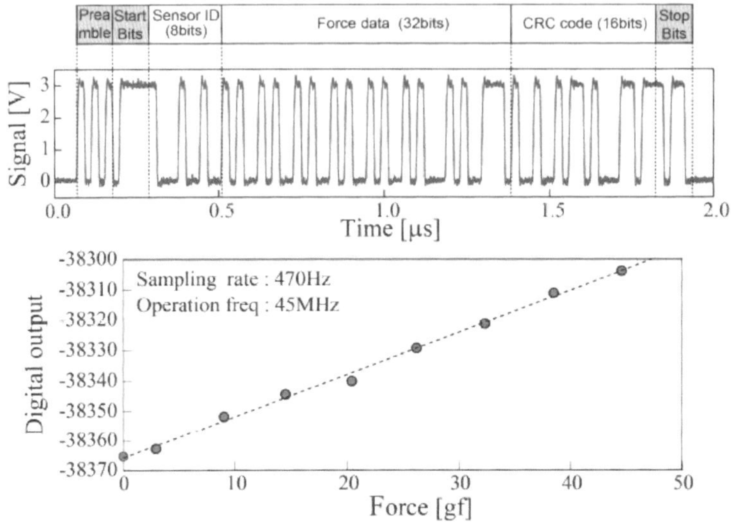

Fig. 8.12. Packet communication signal and the digital output versus force.

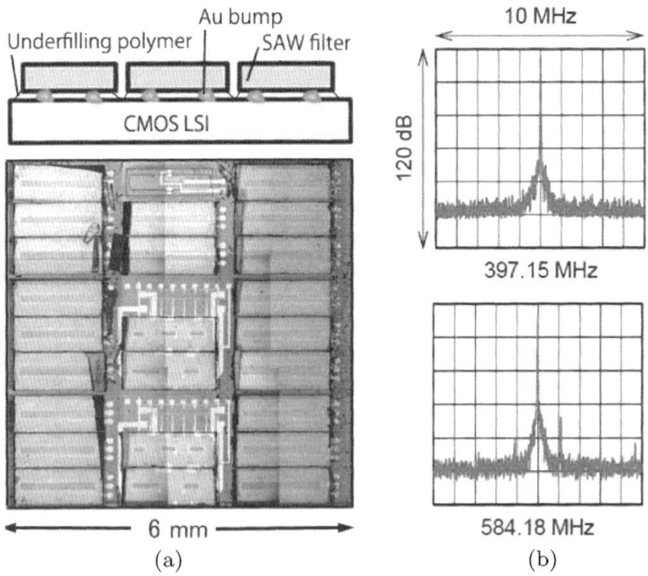

Fig. 8.13. Multi-SAW filters on a LSI chip by the selective transfer process: (a) Structure and photograph, and (b) Characteristics.

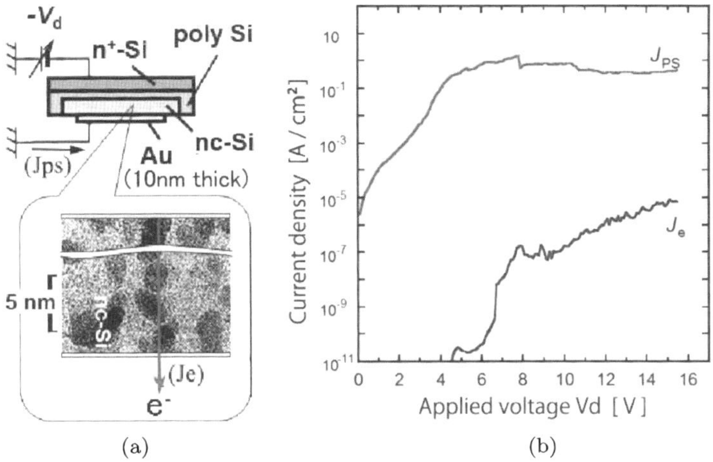

Fig. 8.14. Nanocrystal Si (nc-Si) electron emitter: (a) Principle of nc-Si emitter, and (b) The emission current (J_e) and the current to the thin Au layer (J_{ps}) versus applied voltage (V_d).

in Fig. 8.14(a), the nc-Si emitter consists of cascaded tunnel junctions and accelerated ballistic electrons are emitted through a thin (10 nm thick) Au layer. The emission current (J_e) and the current to the thin Au layer (J_{ps}) plotted versus applied voltage (V_d) are shown in Fig. 8.14(b). The emitted current by the ballistic electrons is obtained at a voltage as low as 10 V [14]. The nc-Si is formed by anodizing Si in a HF solution and by following electro chemical oxidation (ECO). The low-voltage electron emission is needed for making high-density arrayed active matrix emitter because the size of the transistor depends on the voltage required. The active matrix LSI developed for 100×100 array is $100\,\mu$m pitch on a 10 mm square chip.

A planer-type nc-Si emitter array with through silicon via (TSV) was fabricated using the device transfer (via-first) method as shown in Fig. 8.15 [15]. The Si wafer is thermally oxidized after deep reactive ion etching (deep RIE(DRIE)) to make vias (Fig. 8.15(b)). The vias are filled with poly-Si deposition and planarization (polishing) (Fig. 8.15(c)). Columnar poly-Si is deposited for the purpose of nc-Si and poly-Si is patterned (Fig. 8.15(d)). Si_3N_4 is deposited and patterned (Fig. 8.15(e)). The next process step is the formation of the nc-Si. The poly-Si is anodized in HF(55%) + C_2H_5OH (1:1) solution, followed by electro chemical oxidation

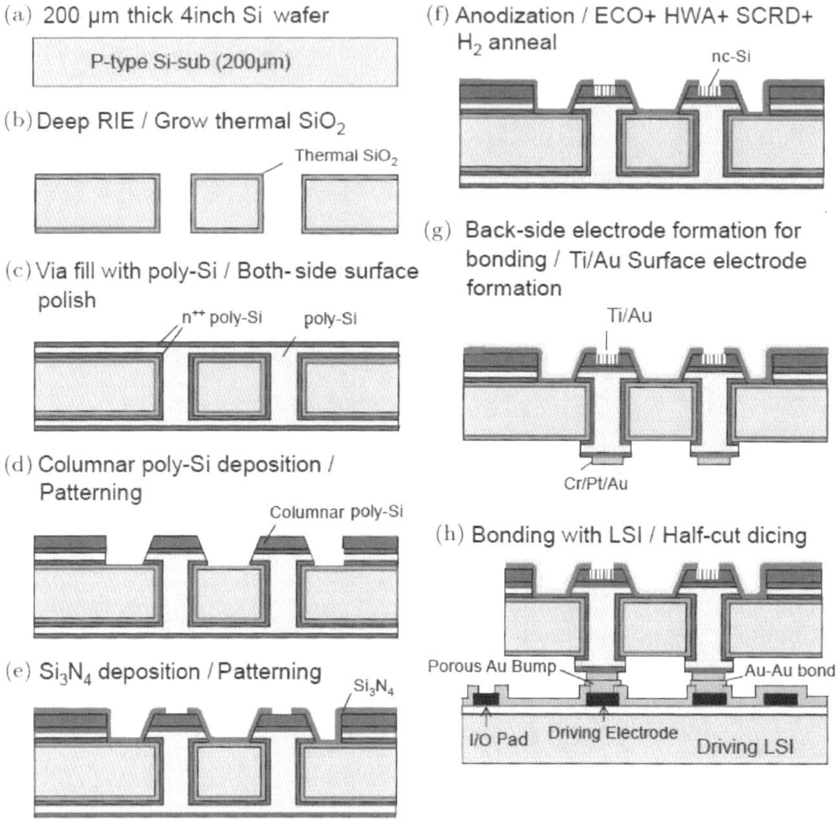

Fig. 8.15. Fabrication process of nc-Si emitter array with TSV: (a) 200 μm thick 4 inch Si Wafer, (b) Deep RIE/Grow thermal SiO_2, (c) Via fill with poly-Si/Both-side surface polish, (d) Columnar poly-Si deposition/Patterning, (e) Si_3N_4 deposition / patterning, (f) Anodization/ECO + HWA + SCRD + H_2 anneal, (g) Back-side electrode formation for bonding / Ti/Au Surface eiectrode formation, (h) Bonding with LSI/Half-cut dicing.

(ECO) in ethylene glycol and potassium nitrate, high-pressure water vapor annealing (HWA), supercritical rinsing and drying (SCRD), and annealing in H_2 (Fig. 8.15(f)). Back-side electrode for the bump bonding and the thin surface electrode (Ti (1 nm) + Au (9 nm)) are formed (Fig. 8.15(g)). Finally, the wafer is bonded to the LSI wafer with bumps and bonding pads of the LSI is exposed by half-cut dicing (Fig. 8.15(h)). Figure 8.16 shows cross sectional photographs of the fabricated nc-Si electron emitter.

Fig. 8.16. Cross sectional photographs of the nc-Si electron emitter with TSV.

A preliminary experiment was conducted as shown in Fig. 8.17. The 1:1 exposure system with a magnetic focusing shown on the right was used for this experiment. Photographs of the nc-Si emitter array (top view) and a resist pattern exposed to the electron from the nc-Si emitter are shown on the left. The 12 μm square exposed pattern corresponding to the nc-Si emitter size was formed on the resist successfully. LSI for the active matrix emitter driver was developed and the basic operation of the LSI was confirmed. The LSI receives external writing bitmap data and switches 100×100 electron beam lets on and off. A reduction electron optics was designed to reduce the electron beamlets by an objective lens with a factor of 100 at an acceleration voltage of 5 keV.

8.3 Wafer-Level Packaging

Packaged chip-size MEMS have been fabricated by WLP [2]. The MEMS on the Si wafer are encapsulated by glass wafer using the anodic bonding, and finally the wafer is diced to chips as shown in Fig. 8.18. This process makes batch assembly possible and has advantages of elimination of assembly equipment, reduced package cost, small size, high yield, and high reliability.

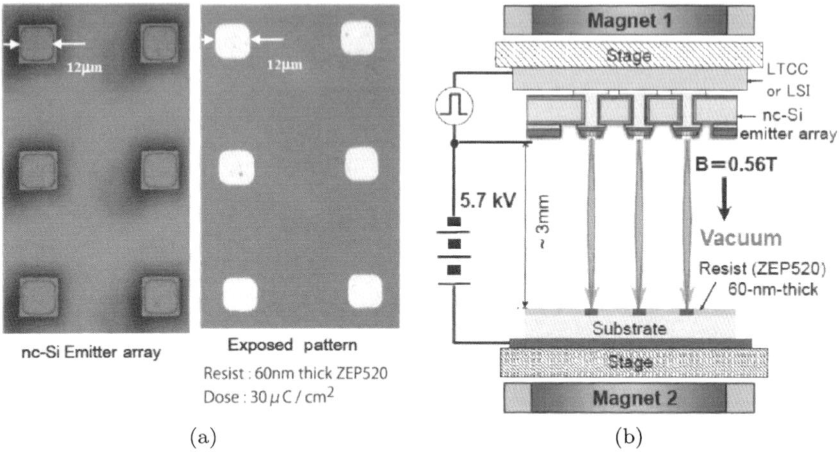

Fig. 8.17. Photographs of nc-Si emitter array and exposed pattern and 1:1 exposure system: (a) nc-Si emitter array and exposed pattern, (b) 1:1 exposure system.

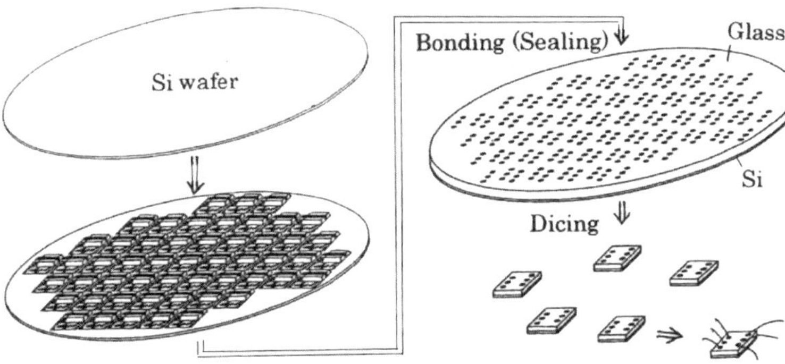

Fig. 8.18. Concept of the WLP.

Three products to which the WLP was successfully applied are shown in the following. The fabrication process and the photograph of the integrated capacitive pressure sensor are shown in Fig. 8.19 [13]. First, the Si wafer is etched and holes are made in the glass (Fig. 8.19(a)). CMOS circuits to detect capacitance are formed on the Si wafer. Metal is deposited and patterned on the glass wafer (Fig. 8.19(b)). The Si wafer and the glass wafer are anodically bonded. Negative voltage (−500 V) is applied to the glass wafer at around 400°C, which causes electrostatic force to

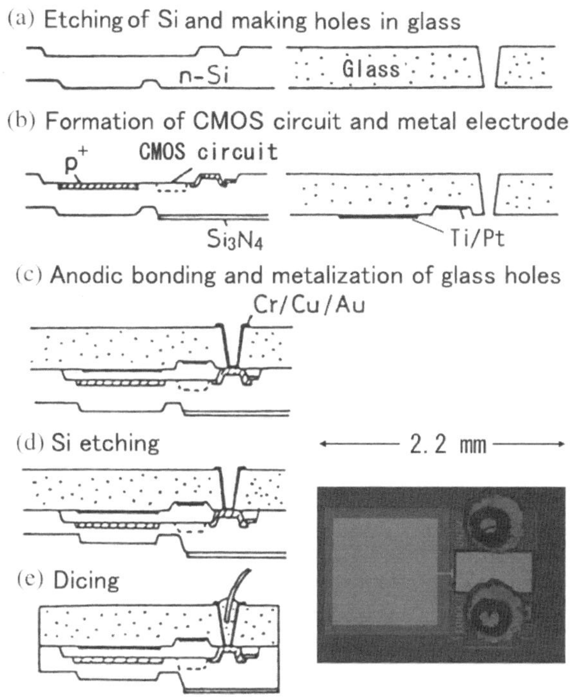

Fig. 8.19. Fabrication process of integrated capacitive pressure sensor: (a) Etching of Si and making holes in glass, (b) Formation of CMOS circuit and metal electrode, (c) Anodic bonding and metalization of glass holes, (d) Si etching, (e) Dicing.

bond at the glass–Si interface. Holes in the glass are metalized by sputter deposition (Fig. 8.19(c)). The Si wafer is selectively etched to make thin diaphragm (Fig. 8.19(f)). Finally, the wafer is diced to get packaged chips (Fig. 8.19(e)).

Vacuum cavity is needed for MEMS devices such as resonators (gyroscope etc.), infrared imagers, and capacitive vacuum sensors. Oxygen gas is generated during the anodic bonding process by the electrochemical decomposition of glass at the Si–glass interface. Figure 8.20 shows the evidence of the oxygen generation [17]. Si with a thin diaphragm is anodically bonded to a glass in vacuum. When exposed to the atmosphere, samples with a large cavity volume showed a deflated diaphragm. On the other hand, the diaphragm with a small cavity volume inflated because of the generation of oxygen.

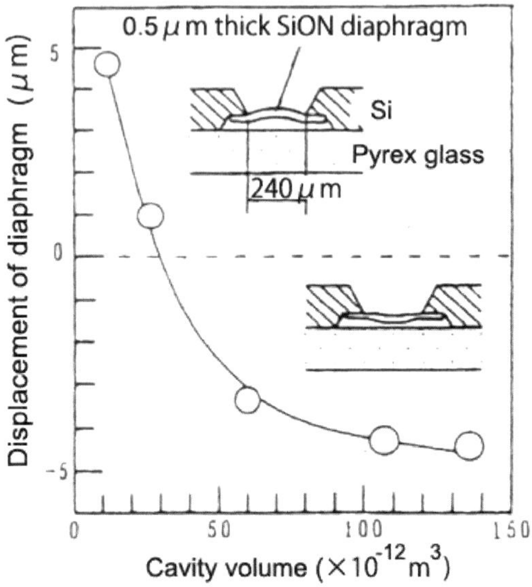

Fig. 8.20. Evidence of electrochemical oxygen generation during anodic bonding process.

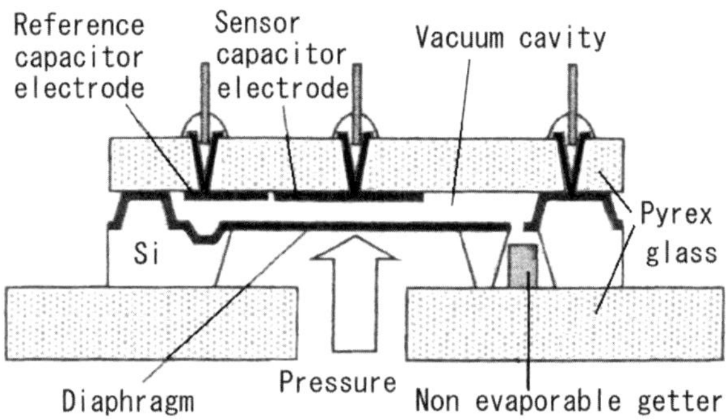

Fig. 8.21. Si diaphragm capacitive vacuum sensor with a vacuum cavity.

It was found that a vacuum cavity can be achieved by using nonevaporable getter in the cavity during the anodic bonding in vacuum [18]. The getter is activated at the temperature of the anodic bonding and adsorbs the oxygen gas generated. Figure 8.21 is a capacitive vacuum sensor, which has a thin Si diaphragm and a vacuum cavity for the purpose of the reference pressure [19].

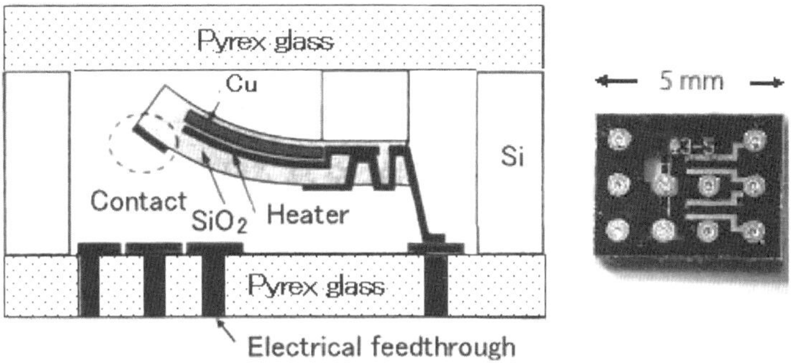

Fig. 8.22. Thermal MEMS switch using electrical feedthrough in glass.

Figure 8.22 is a thermal MEMS switch used for LSI tester [20]. It is fabricated using the WLP. The MEMS switch has a cantilever, which has a heater in it, which bends by the principle of bimetal and makes electrical contact. The MEMS switch requires glass wafer having electrical feedthrough in it to improve its high-frequency characteristics.

8.3.1 *LTCC with Feedthrough for WLP*

The glass having electrical feedthrough used in Fig. 8.22 can be replaced by low-temperature cofired ceramic (LTCC) [21]. The coefficient of thermal expansion (CTE) of the LTCC is close to that of Si as shown in Fig. 8.23(a) and hence it can be anodically bonded to Si. The photograph of the bonded wafer is shown in Fig. 8.23(b). The WLP process using the LTCC with the electrical feedthrough is shown in Fig. 8.23(c). The through holes are made by puncturing the soft green sheet (Fig. 8.23(c) (1) and (2)) and the holes are filled with a gold paste (Fig. 8.23(c) (3)).

After making metal patterns on the green sheets, they are laminated (Fig. 8.23(c) (4)) and sintered to make a ceramic (Fig. 8.23(c) (5)). The lateral dimension can be controlled by fixing on a plate to minimize shrinkage during sintering. Finally, the LTCC wafer is anodically bonded to the Si wafer at around 400°C (Fig. 8.23(c) (6)). The laminated LTCC with electrical feedthrough is less expensive and more air-tight than the glass.

Electrical interconnections between the electrical feedthrough (Au vias) in the LTCC and the pads on the Si are needed. These can be made not only at the anodic bonding interface as shown in Fig. 8.23 but also inside the cavity by metal bonding as shown in Figs. 8.24(a)–(c). Figure 8.24(a)

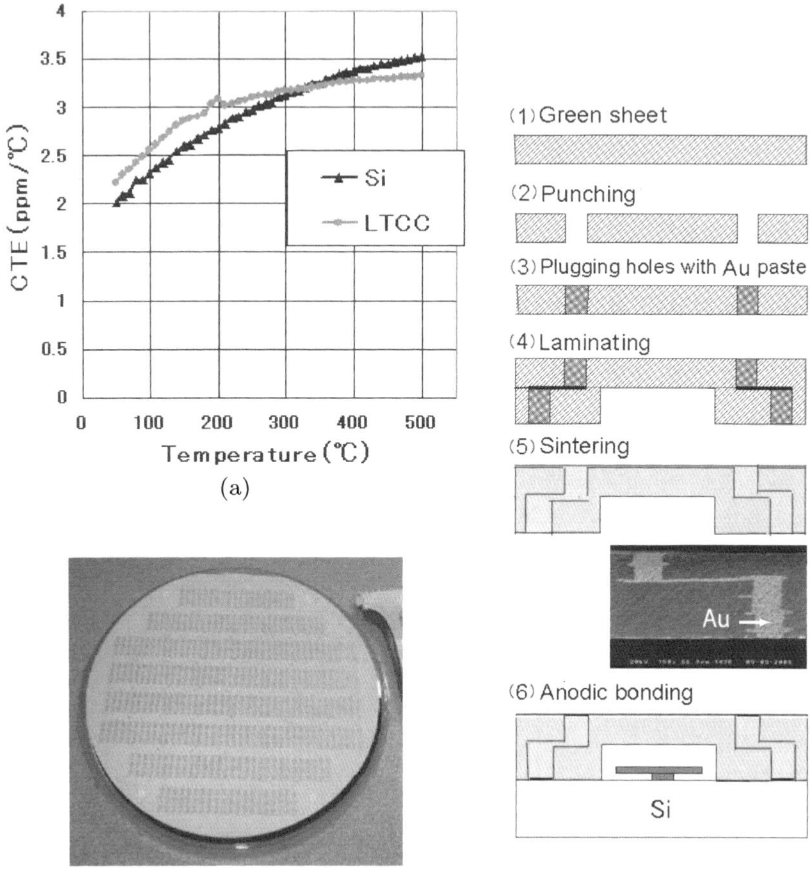

(b) (c)

Fig. 8.23. WLP using multilayer LTCC with feedthrough: (a) Coefficient of thermal expansion (CTE) of the LTCC compared with Si, (b) Photograph after wafer bonding, (c) WLP process using LTCC.

shows the bonding between Cu and Sn. The Sn is a low-melting temperature metal and the melt diffuses into the Cu, leaving an intermetallic compound [22]. This method is called transient liquid phase (TLP) bonding or solid liquid interdiffusion (SLID) bonding. The remelt temperature of Cu–Sn is 415°C, which is sufficiently higher than the reflow temperature for device mounting with a solder. The metal–metal bonding is carried out during the anodic bonding of the LTCC to Si at 400°C [23]. The oxides

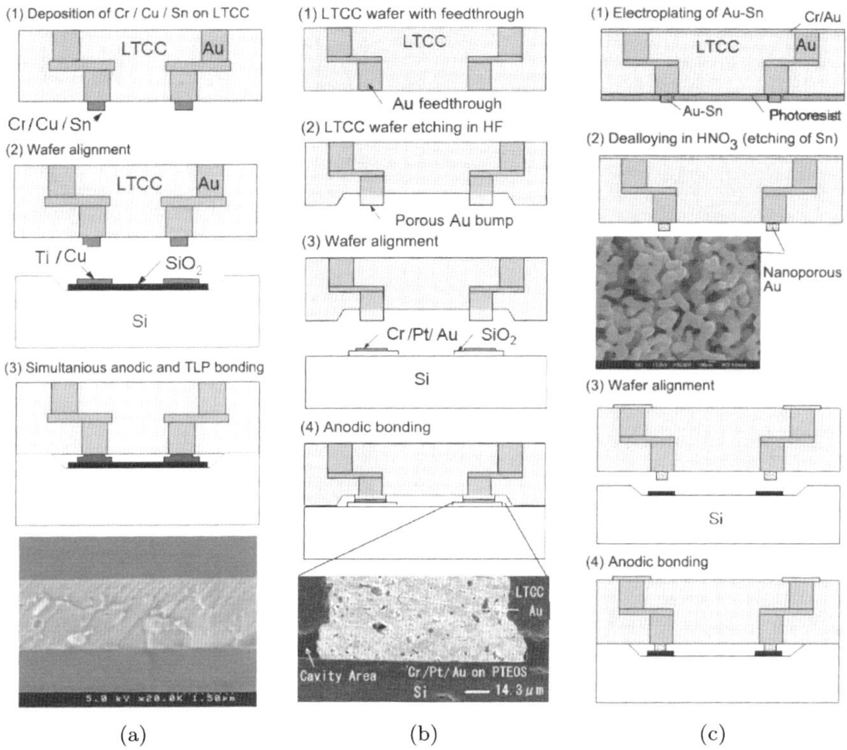

Fig. 8.24. Electrical interconnections between the Au via in the LTCC and the Si pad using three different methods: (a) Cu–Sn intermetallic compound, (b) Porous Au bump made by etching in HF, (c) Nanoporous Au by selective etching of Sn from Au–Sn.

of the Cu surface is reduced by formic acid (HCOOH) inside the bonding chamber.

The other methods to connect the Au vias in the LTCC and the pads on the Si are to use compliant porous Au bumps as shown in Figs. 8.24(b) and (c). The porous bump offers sponge-like function and forms electrical contacts during the anodic bonding. In the method of Fig. 8.24(b), the porous Au bump is spontaneously formed from a part of the Au via by wet-etching the LTCC wafer in HF solution, because glass-based filler in the Au paste is etched away and leaves pores [24]. The wet-etched LTCC wafer was anodically bonded to the MEMS wafer. The high yield of both hermetic sealing and electrical interconnection was confirmed.

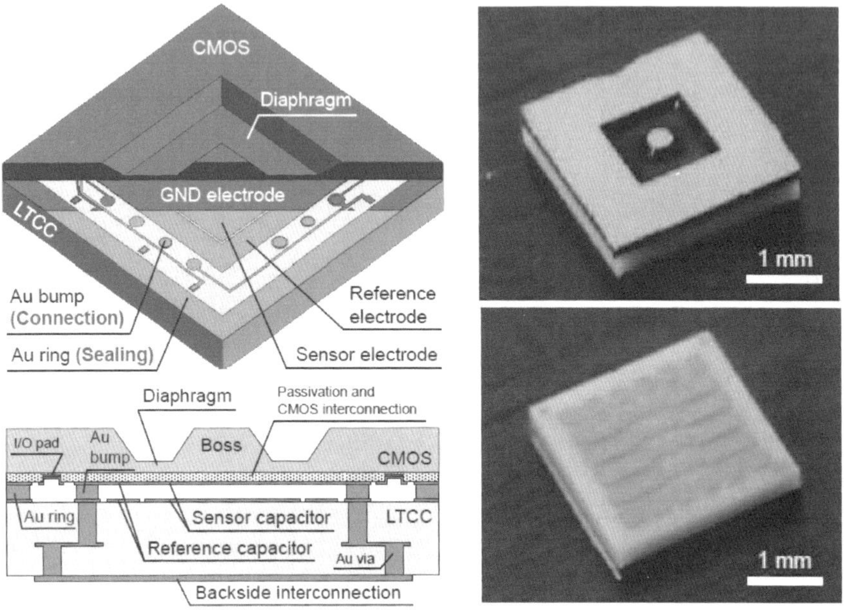

Fig. 8.25. Tactile sensor for network using the LTCC.

In the method in Fig. 8.24(c), nanoporous Au (NPG) is fabricated for the bumps [25]. AuSn alloy is electroplated on the Au via using photoresist as a mold. A 50:50 composition of AuSn is electroplated from the solution containing $KAuCl_2$ and $SnCl_2$. After electroplating, the AuSn structures are annealed at 70°C for 30 min. Then the NPG is formed by chemical dealloying process, in which Sn is etched away selectively by immersing it in concentrated HNO_3 solution (wt. 60%) for 3 days.

The tactile sensor for network was explained in Figs. 8.10–8.12. the advanced tactile sensor was fabricated by the WLP using the LTCC [26]. The structure and the photographs are shown in Fig. 8.25. The diaphragm for the capacitive sensing circuit are formed on the CMOS wafer. The fabrication process of the tactile sensor is shown in Fig. 8.26. Au is electroplated for the purpose of bumps on the CMOS LSI wafer (Fig. 8.26(a)). Planarization of the Au bump is performed mechanically (Fig. 8.26(b)). Rewiring and formation of the GND electrodes are made on the CMOS wafer (Fig. 8.26(c)). The capacitor electrode, Au layer for the bumps, and the seal ring are formed on the LTCC wafer (Fig. 8.26(d)). The CMOS wafer and LTCC wafer are bonded by Au–Au bond (Fig. 8.26(e)).

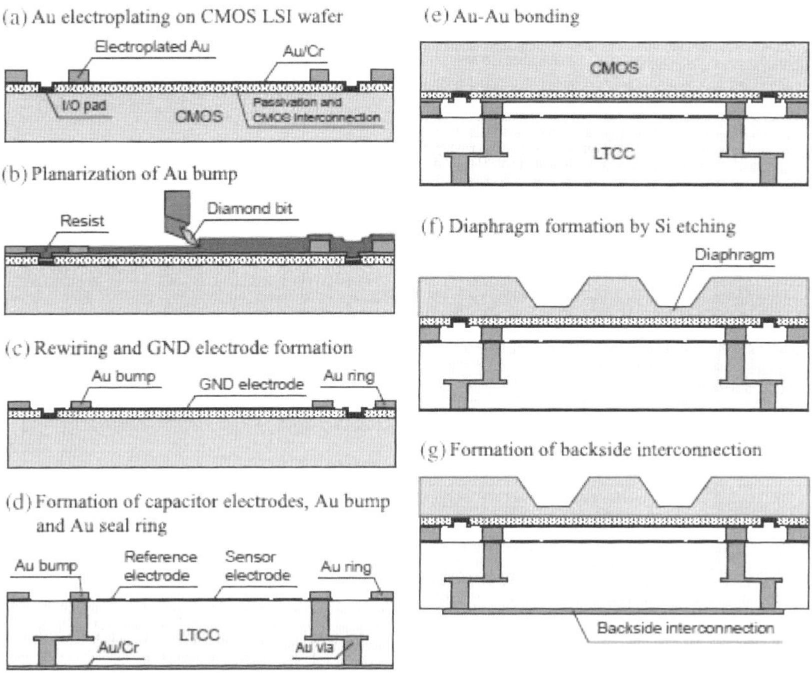

Fig. 8.26. Fabrication process of the tactile sensor using the LTCC: (a) Au electroplating on CMOS LSI wafer, (b) Planarization of Au bump, (c) Rewiring and GND electrode formation, (d) Formation of capacitor electrodes, Au bump and Au seal ring, (e) Au-Au bonding, (f) Diaphragm formation by Si etching, (g) Formation of backside interconnection.

Diaphragms are made by etching the backside of the CMOS wafer (Fig. 8.26(f)). Interconnections are made on the backside of the LTCC wafer (Fig. 8.26(g)). Finally, the bonded wafer is diced to form individual packaged tactile sensors.

8.4 Conclusions

Examples of heterogeneously integrated microsystems developed in Tohoku University have been briefly explained. The heterogeneous integration processes must fit the standard CMOS LSI. The integration processes presented here are based on a wafer-level device/film transfer at low temperature. The wafer-level transfer and the WLP have great advantages in contrast to the assembly-level integration and packaging. Not only we can reduce the chip size and the cost but also we can increase the number of the interconnections between the MEMS and the LSI (such as the massive parallel EB exposure system in Figs. 8.14–8.17) and reduce the stray capacitance and the stray inductance for high-performance capacitive sensors (such as the capacitive tactile force sensor in Figs. 8.10–8.12) and RF devices (such as SAW filters in Figs. 8.7 and 8.8 and Fig. 8.13). As compared with surface micro-machining, process incompatibility, especially damage to LSI, can be avoided and functional materials like diamond (as the BDD electrode array on LSI for electrochemical biosensing in Figs. 8.5 and 8.6) and PZT (as the PZT MEMS switch on LSI in Fig. 8.9) can be used in the wafer-level transfer. The difference in die size, which is a primal problem for cost-effective wafer-level integration, can be solved by the laser-assisted selective device transfer (as the tunable SAW filter in Figs. 8.7 and 8.8 and the multi SAW filters on a LSI chip in Fig. 8.13).

WLP is necessary for small size, low cost, high yield and high reliability. WLP using glass wafers with feedthrough (as in Figs. 8.19–8.22) and the advanced laminated LTCC wafers with feedthrough (as in Figs 8.23–8.26) are shown with applications.

Acknowledgments

The work was supported by "JSPS Core-to-Core Program", "Funding Program for World-Leading Innovative R&D on Science and Technology", and "Creation of Innovation Centers for Advanced Interdisciplinary Research Areas Program".

References

1. Esashi, M. and Tanaka, S. (2013). Heterogeneous integration by adhesive bonding, *Micro and Nano Systems Letters*, 1(3).
2. Esashi, M. (2008). Wafer level packaging of MEMS, *Journal of Micromechanics and Microengineering*, 18(7), 073001, 13pp.
3. Van Kessel, P. F., Hornbeck, L. J., Meier, R. E., and Douglass, M. R. (1988). A MEMS-based projection display, in *Proceedings of the IEEE*, 86(8), pp. 1687–1704.
4. Kirsten, M., Wenk, B., Ericson, F., Schweitz, J. A., Riethmüller, W., and Lange, P. (1995). Deposition of thick doped polysilicon films with low stress in an epitaxial reactor for surface micromachining application, *Thin Solid Films*, 259, pp. 181–187.
5. Seeger, J., Lim, M., and Nasiri, S. (2010). Development of high-performance, high-voltage consumer MEMS gyroscope, In *Technical Digest Solid-State Sensor, Actuator and Microsystems Workshop*, pp. 61–64, Hilton Head, USA.

6. Lapisa, A., Stemme, G., and Niklaus, F. (2011). Wafer-level heterogeneous integration for MOEMS, MEMS and NEMS, *IEEE Journal of Selected Topics in Quantum Electronics*, 17(3), pp. 629–644.
7. Guerre, R., Drechsler, U., Jubin, D., and Despont, M. (2008). Selective transfer technology for microdevice distribution, *Journal of Microelectromechanical Systems*, 17(1), pp. 157–165.
8. Samoto, T., Hirano, H., Somekawa, T., Hikichi, K., Fujita, M., Esashi, M., and Tanaka, S. (2013). Wafer-to-wafer transfer process of barium strontium titanate metal-insulator-metal structures by laser pre-irradiation and gold-gold bonding for frequency tuning applications, In *Technical Digest 17th Int. Conf. on Solid-State Sensors, actuators and Microsystems (Transducers 2013 & Eurosensors XXVII)*, pp. 171–174, Barcelona, Spain.
9. Hayasaka, T., Yoshida, S., Inoue, K. Y., Nakano, M., Matsue, T., Esashi, M., and Tanaka, S. (2015). Integration of boron-doped diamond microelectrode on CMOS-based amperometric sensor array by film transfer technology, *Journal of Microelectromechanical Systems*, 24(4), pp. 958–967.
10. Hirano, H., Samoto, T., Kimura, T., Inaba, M., Hashimoto, K., Matsumura, T., Hikichi, K., and Kadota, M. (2014). Bandwidth-tunable SAW filter based on wafer-level transfer-integration of $BaSrTiO_3$ film for wireless LAN system using TV white space, In *Proceedings of the IEEE Ultrasonics. Symposium Journal*, pp. 803–806, Chicago, USA.
11. Matsuo, K., Moriyama, M., Esashi, M., and Tanaka, S. (2012). Low-voltage PZT-actuated MEMS switch monolithically integrated with CMOS circuit, In *Technical Digest 25th IEEE International Conference on Microelectromechanical Systems (MEMS 2012)*, pp. 1153–1156, Paris, France (Jan. 29–Feb. 2, 2012).
12. Makihata, M., Muroyama, M., Nakano, Y., Tanaka, S., Nakayama, T., Yamaguchi, U., Yamada, H., Nonomura, Y., Funabashi, H., Hata, Y., and Esashi, M. (2013). A 1.7mm^3 MEMS-on-CMOS tactile sensor using human-inspired autonomous common bus communication, In *Technical Digest 17th International Conference. on Solid-State Sensors, actuators and Microsystems (Transducers 2013 & Eurosensors XXVII)*, pp. 2729–2732, Barcelona, Spain (June, 2013).
13. Hikichi, K., Seiyama, K., Ueda, M., Taniguchi, S., Hashimoto, K., Esashi, M., and Tanaka, S. (2014). Wafer-level selective transfer method for FBAR-LSI integration, In *Proc. 2014 IEEE International Frequency Control Symposium*, pp. 246–249, Taipei, Taiwan.
14. Koshida, N., Kojima, A., Ikegami, N., Yoshida, T., Suda, R., Yagi, M., Shirakashi, J., Yoshida, T., Miyaguchi, H., Muroyama, M., Nishino, H., Yoshida, S., Sugata, M., Totsu, K., and Esashi, M. (2015). Development of ballistic hot electron emitter and its applications to parallel processing; active-matrix massive direct-write lithography in vacuum and thin-films deposition in solutions, *Journal of Micro/Nanolithography MEMS MOEMS*, 14(3), p. 031215.
15. Esashi, M., Kojima, A., Ikegami, N., Miyaguchi, H., and Koshida, N. (2015). Development of massively parallel electron beam direct write lithography

using active-matrix nanocrystalline-silicon electron emitter arrays, *Microsystems and Nanoengineering*, 1, 15029.

16. Kudoh, T., Shoji, S., and Esashi, M. (1991). An integrated miniature capacitive pressure sensor, *Sensors and Actuators A*, 29, pp. 185–193.
17. Ura, N., Nakaichi, K., Minami, K., and Esashi, M. (1992). Vacuum packaging by anodic bonding, In *11th Sensor Symp. Late News*, p. 63, Tokyo, Japan.
18. Henmi, H., Shoji, S., Shoji, Y., Yoshimi, K., and Esashi, M. (1994). Vacuum packaging for microsensors by glass-silicon anodic bonding, *Sensors and Actuators A*, 43, pp. 243–248.
19. Miyashita, H., and Kitamura, Y. (2005). Micromachined capacitive diaphragm gauge, *Anelva Technical Report*, 11, pp. 37–41.
20. Nakamura, K., Takayanagi, F., Moro, Y., Sanpei, H., Onozawa, M., and Esashi, M. (2004). Development of RF MEMS switch, *Advantest Technical Report*, 22, pp. 9–16.
21. Tanaka, S., Matsuzaki, S., Mohri, M., Okada, A., Fukushi, H., and Esashi, M. (2011). Wafer-level hermetic packaging technology for MEMS using anodically-bondable LTCC wafer, In *Technical Digest 24th IEEE Int. Conf. on Micro Electro Mechanical Systems (MEMS 2011)*, pp. 376–379, Cancum, Mexico.
22. Bartels, F., Morris, Jr, J. W., Dalke, G., and Gust, W. (1994). Intermetallic phase formation in thin solid-liquid diffusion couples, *Journal of Electronic Materials*, 23(8), pp. 787–790.
23. Matsuzaki, S., Tanaka, S., and Esashi, M. (2010). Anodic bonding between LTCC substrate and Si substrate with electrical connections, In *Proceedings of the 27th Sensor Symposium*, p. 16, Matsue, Japan.
24. Tanaka, S., Mohri, M., Ogashiwa, T., Fukushi, H., Tanaka, K., Nakamura, D., Nishimori, T., and Esashi, M. (2012). Electrical interconnection in anodic bonding of silicon wafer to LTCC wafer using highly compliant porous bumps made from submicron gold particles, *Sensors and Actuators A*, 188, pp. 198–202.
25. Lin, Y.-C., Wang, W.-S., Chen, L. Y., Chen, M. W., Gessner, T., and Esashi, M. (2013). Nanoporous gold for MEMS packaging applications, *IEEJ Transactions on MEMS and Micromachines*, E-133(2), pp. 31–36.
26. Asano, S., Muroyama, M., Bertley, T., Kojima, T., Nakayama, T., Yamaguchi, U., Yamada, H., Nonomura, Y., Hata, Y., Funabashi, H., and Tanaka, S. (2015). Flipped CMOS-diaphragm capacitive tactile sensor surface mountable on flexible and stretchable bus line, In *Technical Digest 19th International of Conference. on Solid-State Sensors, Actuators and Microsystems (Transducers 2015)*, pp. 97–100, Anchorage, USA.

Masayoshi Esashi received B.E. and Dr.E. degrees in Electronic Engineering from Tohoku University in 1971 and 1976, respectively. He was Research Associate at Department of Electronic Engineering, Tohoku University from 1976 to 1981, and Associate Professor from 1981 to 1990. He was Professor at Department of Mechatronics and Precision Engineering, Tohoku University from 1990 to 2013, and he is currently Professor at Microsystem Integration Center, Tohoku University. His research interests include MEMS, integrated sensors. He was awarded Purple Ribbon Award (2006), IEEE Andrew S. Grove Award (2015), IEEE Jun-ichi Nishizawa medal (2016), etc.

Shuji Tanaka received B.E., M.E. and Dr.E. degrees in Mechanical Engineering from The University of Tokyo in 1994, 1996, and 1999, respectively. He was Research Associate at Department of Mechatronics and Precision Engineering, Tohoku University from 1999 to 2001, Assistant Professor from 2001 to 2003, and Associate Professor at Department of Nanomechanics from 2003 to 2013. He is currently Professor at Department of Bioengineering and Robotics. He was awarded The Young Scientists' Prize, The Commendation for Science and Technology by the Minister of Education, Culture, Sports, Science and Technology (2009), German Innovation Award, Gottfried Wagener Prize (2012), and other 7 prizes. His research interests include heterointegration, MEMS packaging, acoustic devices, RF MEMS, and Power MEMS.

Chapter 9

Packaging of Membrane-Based Polymer Microfluidic Systems

Yu-Chuan Su

9.1 Introduction

Accomplished by micro-fabrication technologies, innovative ideas, and inter-disciplinary cooperation, the field of microelectromechanical systems (MEMS) has been growing at a promising pace over the last few decades. MEMS generally refer to miniature devices and systems that have integrated electrical and mechanical components for actuation and/or sensing applications. MEMS have numerous applications in fluidics. After years of intensive research and development driven by the prospects of vast market demand, microfluidics, the subset of MEMS dealing with fluids, is being wisely applied to a variety of chemical analysis and synthesis processes. Polymeric materials usually play important roles in microfluidic systems for applications in life sciences. The advantage of polymer-based MEMS includes broad range choices of material properties, low raw material cost, and availability for mass production. As a result, many research groups and industrial enterprises have been heavily investigating polymer-based MEMS, including disposable devices. One of the major manufacturing issues in the fabrication of plastic microsystems is the assembly and packaging processes of components made of different materials. A common example is the sealing process for microfluidic channels, which are normally fabricated by etching or molding the channel shapes into substrates without the top cap. A second substrate is required to be bonded on top of the micro-channels without clogging or changing their geometrical shapes. A polymeric membrane is a selective barrier, which allows specific substances to pass through but stops others. Such substances may be molecules, ions or other small particles. In this chapter, packaging and integration schemes of functional membranes into polymer microfluidic systems are presented. These include spin casting and solvent welding, adhesive bonding, plasma enhanced bonding, localized bonding, plug insertion, micro-encapsulation and 3D printing. Porous, elastic, impermeable, semi-permeable, breathable, bio-compatible, and bio-degradable membranes can be integrated into microfluidic systems, which realize the functions of actuation, pumping, filtering, encapsulation, and drug delivery.

9.2 Polymer Micro-Fabrication

Polymers are often the least expensive substrate materials when compared to metals, ceramics, and composites. Polymeric materials, especially synthetic polymers, offer a diverse combination of properties that assists the achievement of a variety of applications in modern life. On the other hand, it is more complicated to design polymer based devices because the properties of polymers are significantly affected by changes in temperature and chemical environment within the normal range of usage. Depending on applications and economic considerations, polymeric microstructures can be fabricated by either mass production or rapid prototyping approaches. The availability of mass production processes that can be extended to the microscale means that polymeric devices can be extremely inexpensive to produce in volume. This allows for disposable devices, which aim for one-time usage and therefore eliminate long-term issues such as drift, clogging, and sterilization. On the other hand, there are a variety of rapid prototyping processes, which helps significantly reduce product development time and cost, available for research and development purposes. Most of these processes are batch processes and capable of duplicating a whole wafer at a time. While these methods are not conventional micro-fabrication processes, most of them require standard silicon or glass micro-fabrication in the beginning to make the mold. This section presents an overview of current polymer micro-fabrication techniques.

9.2.1 *Material Properties*

In strict terms, a polymer is a substance composed of molecules which have long sequences of one or more species of atoms or groups of atoms linked to each other by primary, usually covalent, bonds. This long sequence nature sets polymers apart from other materials and gives rise to their characteristic properties. According to their molecular structures and corresponding material properties, polymers can be classified into three groups: thermoplastics, elastomers, and thermosets.

9.2.1.1 *Thermoplastics*

Thermoplastics, often referred to just as plastics, are linear or branched polymers which can be melted repeatedly upon the application of heat. They can be molded into virtually any shapes and now constitute by far the largest proportion of the polymers used in industry. Generally, thermoplastics do not crystallize easily upon cooling to the solid state because this requires considerable ordering of the highly coiled and entangled macromolecules present in the liquid state. Those which do crystallize invariably do not form perfectly crystalline materials but instead are semi-crystalline with both crystalline and amorphous regions.

9.2.1.2 *Thermosets*

In addition to the normal covalent bonds that join the atoms together along the polymer chains, thermosets also have crosslinks that join the chains together. Thermosets normally are rigid materials and network polymers in which chain motion is greatly restricted by a high degree of crosslinking. Because of the insufficient chain mobility, they are amorphous and usually do not exhibit any elasticity behavior, and even their glass transition may be suppressed. They are intractable once formed and degrade rather than melt upon the application of heat. Generally, thermosets have stronger chemical resistance and better performance in high temperature because of crosslinks.

9.2.1.3 *Elastomers*

Elastomers are crosslinked rubbery polymers that can be stretched easily to high extensions and which rapidly recover their original dimensions when the applied stress is released. This extremely important and useful property is a reflection of their molecular structure in which the network is of low crosslink density. The rubbery polymer chains become extended upon deformation but are prevented from permanent flow by the crosslinks, and driven by entropy, spring back to their original positions on removal of the stress. As for thermosets, they are intractable once formed and degrade rather than melt upon the application of heat.

9.2.2 *Bonding*

In order to assemble and package polymeric microstructures and microdevices, chemical bonds between polymer–polymer or polymer–metal/ceramic/composite interface need to be formed.

9.2.2.1 *Adhesive and solvent sealing*

Adhesives join materials by adhering to the surfaces, whereas solvents melt the surfaces such that they intermingle, and form a bond when the solvents evaporate. They are performed in room temperature by bringing two surfaces into intimate contact to achieve strong bonding.

9.2.2.2 *Ultrasonic welding*

In ultrasonic welding, vibrations produced by a horn create intermolecular and boundary friction at the bonding interface. The polymer begins to soften locally and, as the dampening factor increases, the reaction accelerates itself, causing a greater proportion of the vibration energy to be translated into heat. That raises the temperature of the polymer until their melting point is reached. After the vibrations are halted, the pressure is maintained until a molecular bond is created and the bonding is sufficiently cooled. This bonding approach is only applicable to thermoplastics.

9.2.2.3 *Laser welding*

Laser can also be used to provide the required heat to locally melt polymers in the bonding interface. However, it is a serial process and takes longer time to complete the bonding of large area.

9.2.2.4 *Dielectric welding*

The principle of dielectric welding is that some polymeric materials become hot when subjected to a high frequency, alternating electric field. The generated heat will melt polymers locally in the bonding interface and bonds are formed after cooled.

9.2.2.5 *Hot resistance welding*

The surfaces of the two parts to be bonded are melted and then bonded together with the heat generated by resistance when electrical current is passed through it. On the microscale, this can be achieved by employing micro-heaters with designed geometry to control the overall heating profiles.

9.3 Spin Casting and Solvent Welding

The first packaging and integration scheme introduced in this chapter is spin casting and solvent welding. Figure 9.1 shows the schematic diagram of an osmotic micro-actuator, which utilizes osmosis to produce mechanical actuation without consuming any electrical energy [1]. The actuation speed is controlled by the characteristics of the semi-permeable diaphragm, such as area and permeability, and a constant volume expansion output is expected when the osmotic actuator is operated under saturated state. Dry osmotic driving agent is packed in a chamber. An actuation diaphragm is constructed on top and a semi-permeable diaphragm is constructed on the bottom of the chamber. Water diffuses into the chamber due to osmosis to reduce the concentration difference. The opposite osmotic solute diffusion process is blocked by the semi-permeable diaphragm. This results in a net flow of water across the semi-permeable diaphragm into the chamber that contains osmotic driving agent.

The manufacturing process starts with the fabrication of the silicon mold by using the deep reactive ion etching process. Cylindrical columns of 500–2000 μm in diameter and 200–1000 μm in height are designed and fabricated on silicon wafers. A micro-hot embossing process is followed to make micro-chambers on the surface of cellulose acetate sheet. Cellulose acetate is chosen as the material to construct the micro-actuator because of its semi-permeable property. A constant pressure of 0.2 MPa is applied throughout

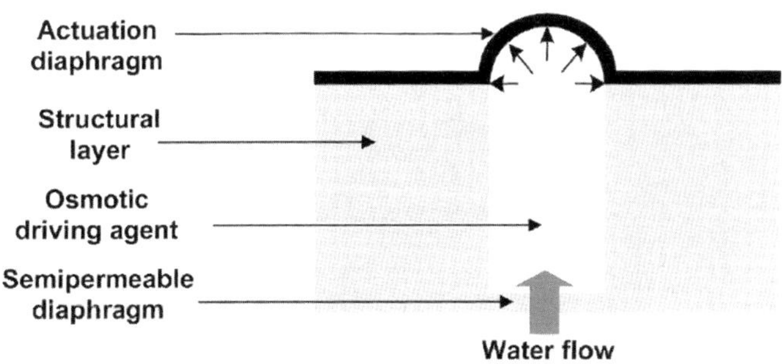

Fig. 9.1. Schematic of osmotic actuation [1].

the hot embossing process, including the imprinting stage and the cooling stage, to obtain good replications. The standard process takes about 30–50 min at an elevated temperature of 130°C followed by a 15-min cooling period. If the depth of the chamber is properly designed, this hot embossing process can fabricate semi-permeable diaphragm of 150–400 μm in thickness at the backside of the chamber. If the desirable thickness of this backside semi-permeable diaphragm is less than 150 μm, it is difficult to accomplish the requirement by this approach. A thin cellulose acetate diaphragm of desirable thickness can be prefabricated and be solvent-welded on the backside of a cellulose acetate substrate with mechanically-drilled through holes.

Figure 9.2 shows the SEM micrograph of a fabricated micro-actuator and Fig. 9.3 shows the asymmetric semi-permeable membrane, which is composed of a thin dense skin layer supported by a thicker porous substrate layer. This asymmetric membrane combines the advantages of high selectivity of a dense membrane, high stiffness of a thick membrane, and high permeation of both porous and thin membranes. As such, high actuation rates, which correspond to the delivery rates of the drug-delivery system, can be achieved. The polymer solution used to form the asymmetric membrane is prepared by dissolving the polymer of choice, 15% cellulose acetate, in a mixture of two solvents, 33% ethanol and 52% acetone. The solvents are chosen to have different boiling points and so that the more volatile solvent (acetone) is a good solvent for the polymer and a

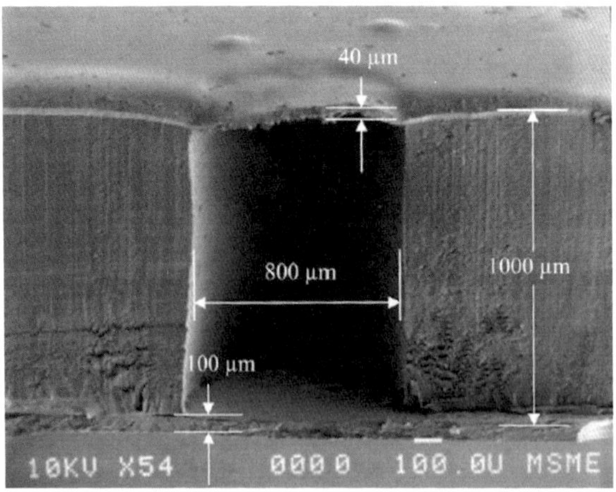

Fig. 9.2. SEM micrograph of a fabricated osmotic micro-actuator [1].

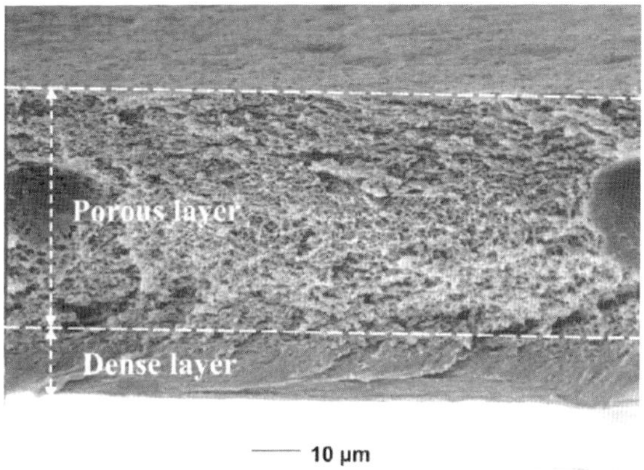

Fig. 9.3. SEM micrograph of an asymmetric semi-permeable membrane [2].

less volatile solvent (ethanol) is not a solvent for the polymer. When this solvent mixture evaporates, the progressive composition shift (from solvent to non-solvent) results in abrupt precipitation of the polymer, which can be exploited to control porosity of the membrane. First, the polymer solution is slowly coated on a supporting substrate and then dried in room-temperature air for 5 s. Afterward, it is immersed in a water quench bath also at room temperature for 20 min and then dried in room-temperature air for at least 12 h. The fabricated white-color membrane has a smooth surface on one side and a rough surface on the other side because of its porous structure. It is found that ethyl acetate, which dissolves cellulose acetate partially, is suitable as the solvent to bond cellulose acetate films.

At the end, an impermeable diaphragm consisting of Saran F-310 solvent-soluble barrier polymer (vinylidene chloride and acrylonitrile copolymer) is employed by spin-on coating as the actuation diaphragm. The powder of Saran F-310 resin is first dissolved in the solvent of 65% methyl ethyl ketone and 35% toluene. Experimentally, up to 20% in weight of F-310 powder is added to the solvent and thc mixture is kept at 23°C for 1 h until the F-310 powder is completely dissolved. Figure 9.4 shows the SEM micrograph of a micro-actuator after the actuation caused by osmosis. A dome shaped structure is formed. Figure 9.5 shows the micro-actuator that is cut in half after actuation. Delamination of actuation diaphragm at the boundary of chamber is clearly observed. It appears that the effective normal force generated by the hydrostatic pressure difference across the

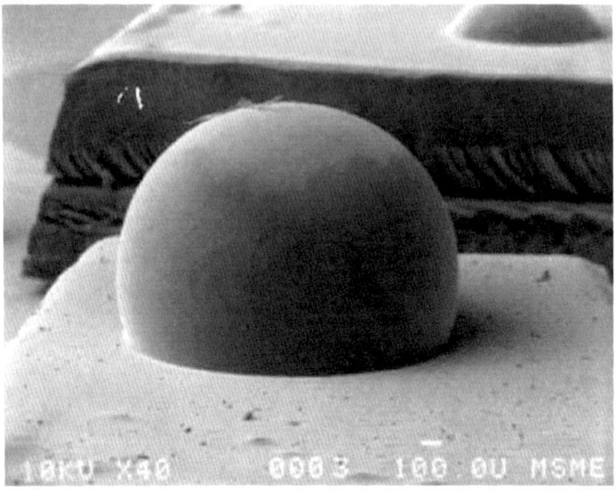

Fig. 9.4. SEM micrograph of an osmotic micro-actuator after immersing into water [1].

Fig. 9.5. SEM micrograph of the cross section of a micro-actuator after actuation [1].

actuation diaphragm will eventually be larger than the bonding strength to cause the separation.

9.4 Plasma Enhanced Bonding

Due to its simple processing, high elasticity, optical transparency, and biocompatibility, polydimethylsiloxane (PDMS) is widely used for microfluidic devices. To permanently bond a PDMS chip to a glass slide or another

PDMS chip, a plasma treatment is utilized to change the surface properties of PDMS and glass. The oxidation of the surface layer increases the concentration of hydroxyl groups and this leads to the formation of strong intermolecular bonds. The silanol groups on the surfaces condense with each other when brought into conformal contact. For both PDMS and glass, the reactions yield Si–O–Si bonds after loss of a water molecule. These covalent bonds form the basis of a tight irreversible seal. Typically, such seals can withstand 2 to 3.5 atm of pressure difference and are practically inseparable.

Figure 9.6 shows the schematic diagram of a micro drug-delivery system [2]. The system is composed of two major parts: an osmotic micro-actuator at the bottom and a PDMS cover at the top, including a drug storage reservoir, a delivery micro-channel and a delivery port. Induced by the concentration difference across the semi-permeable membrane at the bottom of the osmotic micro-actuator, water from aqueous environment is drawn through the semi-permeable membrane into the chamber filled with osmotic driving agent to power the drug-delivery system. By virtue of the incompressibility of water, the top actuating membrane, which is hundreds of times more flexible than the bottom semi-permeable membrane, expands to drive and release the liquid drug stored in the reservoir. To fabricate the top PDMS cover, a layer of 100 μm thick negative photoresist (MicroChem SU-8 100) is spin-coated and patterned on top of a clean silicon wafer to create a mold for duplicating microfluidic components in the following

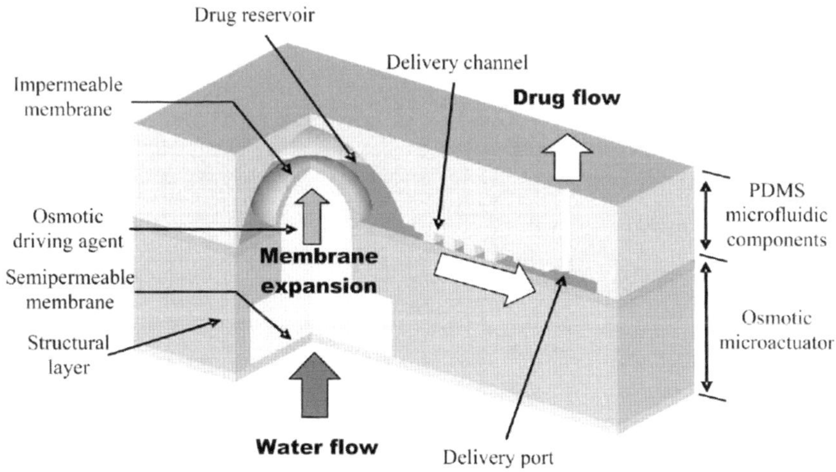

Fig. 9.6. A schematic diagram of micro drug-delivery system [2].

polymer casting process. Placing a droplet of SU-8 on top of the designated spot naturally forms the dome-shaped mold of drug reservoir. After the SU-8 mold is fully cured under UV light, it is placed in a desiccator under vacuum for 2 h with a vial containing a few drops of tridecafluoro-1,1,2,2-tetrahydrooctyl-1-trichlorosilane to silanize the surfaces. A mixture of 10:1 PDMS prepolymer and curing agent (Dow-Corning Sylgard 184) is stirred thoroughly and then degassed under vacuum. The prepolymer and curing agent mixture is then poured onto the mold, degassed, and cured for 2 h at 75°C. After thoroughly curved, this PDMS replica is peeled off from the mold. Because PDMS is permeable to vapor, a 15 μm thick barrier film (DuPont Mylar M45) is added on the replica to prevent drug in the reservoir from diffusing and evaporating to outside environment. However, native PDMS has low surface energy and cannot bond strongly with Mylar. This problem is solved by treating both PDMS and Mylar surfaces with oxygen plasma before bonding. Figure 9.7 shows the SEM micrograph of the bonding interface between PDMS and Mylar.

The major challenge in fabricating the drug-delivery system lies in the integration of PDMS microfluidic components and osmotic micro-actuators. Because of its low surface energy, native PDMS surfaces do not generate good bonding strength with most adhesives. To promote the adhesion, researchers have unveiled that oxygen plasma treatment can chemically modify PDMS surfaces and the modified PDMS surfaces can irreversibly

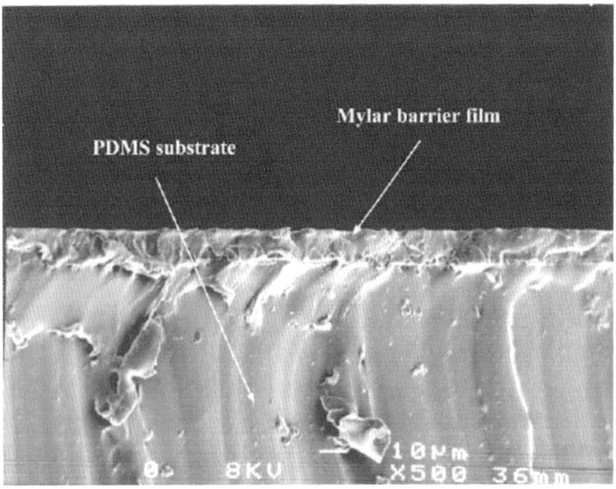

Fig. 9.7. Cross sectional SEM micrograph of the bonding interface between PDMS substrate and Mylar barrier film [2].

bond with another treated PDMS film, glass, silicon, and other polymers under room temperature. The top layer of the osmotic micro-actuator is a flexible and impermeable membrane made of vinylidene chloride and acrylonitrile copolymer (Dow Saran F-310). Unfortunately, it does not form a strong bond with oxygen plasma-treated PDMS surface. To solve the problem, an intermediate bonding layer is introduced to facilitate the integration. After intensive searching and testing, it is found that styrene–isoprene–styrene (S–I–S) copolymer (KRATON D-1193) has the right mechanical and chemical properties for the bonding purpose. It is dissolved in toluene and spin-coated on top of the Saran membrane to form a 30-μm thick layer. Both KRATON and PDMS are then exposed to oxygen plasma to activate the bonding surfaces using a parallel-plate plasma etching system (Technics PE-IIA). Strong bonding was achieved once two surfaces were brought to conformal contact. The next process challenge is to store liquid drug into the reservoir and microfluidic channel. It has been reported that treated PDMS surface becomes hydrophilic and remains active for bonding if it is placed under liquid immediately after the plasma treatment. After dispensing a certain amount of aqueous drugs into the reservoir, the aqueous fluid fills the hydrophilic reservoir and micro-channel spontaneously. Two bonding surfaces are immediately brought together and any extra liquid drug in the gap between two bonding surfaces eventually evaporates and strong bonding is achieved once two surfaces are in contact. The cross-sectional view of the assembled micro drug-delivery system is shown in Fig. 9.8 where the thickness of the PDMS layer is 1.5 mm. To further prevent liquid from diffusing and evaporating to outside environment, the four sides of the assembly are coated with a loop of barrier polymer to block the paths of potential loss.

9.5 Localized Bonding Processes

Most existing bonding techniques in micro-fabrication focus on the bonding of semiconductor, metal, and ceramic materials and require high processing temperatures. These schemes are not applicable for plastic materials, which are unstable at high temperature. It is important and necessary to develop reliable and repeatable bonding processes that do not alter the properties and performance of plastic materials for polymer-based MEMS devices. Thermoplastic polymers can serve as intermediate bonding layers and they are generally safe and environmentally preferred since no harmful solvent is involved during the bonding process. For example, in the

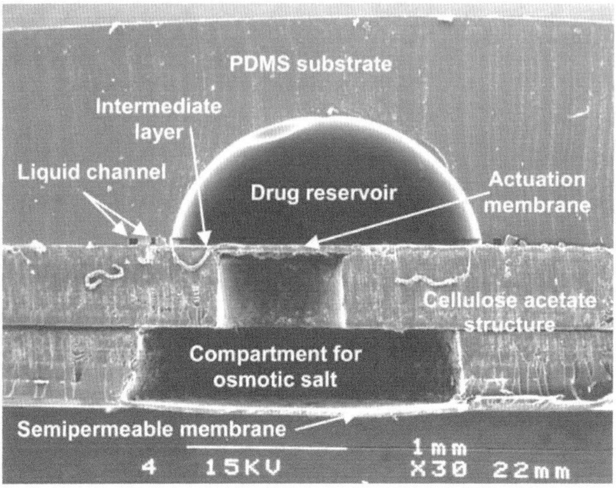

Fig. 9.8. Cross-sectional SEM micrograph of the assembled micro drug-delivery system [2].

lamination process, thin thermoplastic layers are rolled onto the substrates with heated rollers to melt the polymers and seal micro-channels without the requirement of a long curing period. In addition to heated rollers, laser, ultrasonic vibration and ac electromagnetic field can also be employed to supply the required energy to melt thermoplastics for bonding. However, since the heat is applied globally, the melted polymers tend to block tiny microfluidic channels, and the processing temperature is still too high for temperature-sensitive materials. Typically, biological materials will not survive in temperature higher than 45°C for a period of more than 2 min. This low-temperature requirement makes the bonding process extremely challenging.

All bonding processes should operate under two basic conditions. First, the two bonding surfaces must have intimate contact. Second, proper processing temperatures that are determined by the bonding material systems are required to provide the activation energy for bonding. In practice, it is also desirable that the lowest possible processing temperature should be employed to reduce the detrimental effect of thermal mismatch and to prevent degradation of plastic materials. The elastic properties of thermoplastic materials reduce the stress caused by the bonding process and the force required to keep intimate contact as compared with other common micro-fabrication materials, such as silicon and glass. Localized heating provides the energy for bonding and is achieved by using micro-heaters

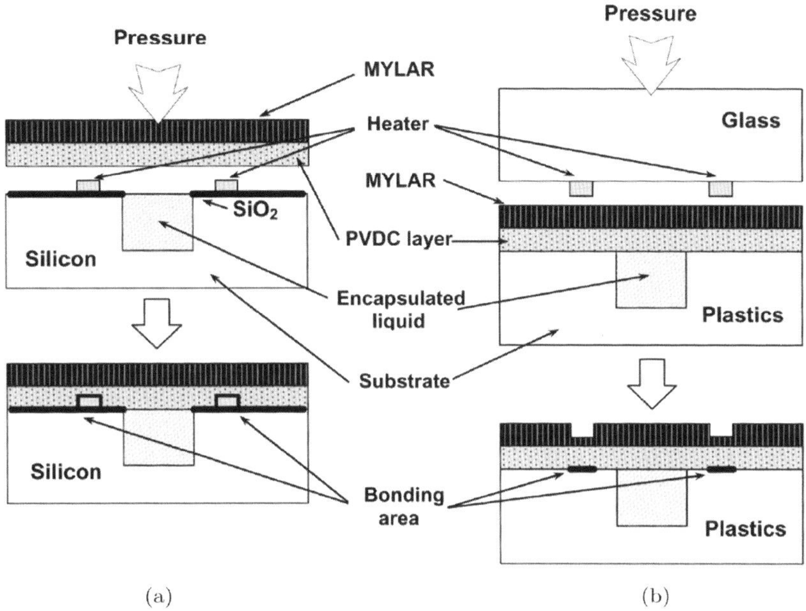

Fig. 9.9. Schematic of bonding processes. (a) Built-in heater configuration. (b) Reusable heater configuration [3].

instead of global heating furnaces. These micro-heaters are constructed in a way that heating is restricted in a small region that is surrounded by insulation materials. The effectiveness of localized heating depends on the selection of materials and the design of the geometrical shape of the overall structure. Two bonding schemes are demonstrated as shown in Fig. 9.9: (a) bonding by built-in type heaters and (b) bonding by external, reusable type heaters [3]. Both silicon and glass substrates have been used in the localized plastic bonding processes. For silicon substrates, a silicon dioxide layer of $1\,\mu$m thick is grown for electrical and thermal insulation before heaters are fabricated. If glass substrates are used, no oxide insulator layer is required. An aluminum thin film is then deposited and patterned as heaters on the surface of substrates. When an electrical current is applied, the temperature rises to activate the bonding process and plastic materials are localized melted and bonded with the substrates.

In the prototype demonstration, a Mylar film coated with a thermoplastic poly-vinylidene chloride (PVDC) copolymer layer on one surface is used as the plastic material to be bonded with silicon, glass, or another plastic substrate. The scheme with built-in micro-heaters is suitable for bonding thick plastic materials with various types of substrates including silicon, glass, and plastics because micro-heater can easily provide the activation energy locally for bonding. The scheme with external, reusable heater is best suitable for bonding of thin plastic films with substrates by using external, reusable heaters to provide required bonding energy instead of built-in heaters. Heat is transferred through plastic thin film and a heated zone is generated locally in the bonding interface where the bond is formed. In this case, it is preferable that the plastic thin films have two layers such that the top layer should have a higher melting temperature to prevent possible bonding with the external, reusable heaters such that heaters can be removed after the completion of bonding process and be reused repeatedly. Both bonding processes have been demonstrated in 0.25 s of heating and under about 0.4 MPa of contact pressure.

To test the quality of the bonding processes, several experiments have been designed. In the first experiment, bonded systems are put into a vacuum chamber and observed under an optical microscope. It is found that the top of the encapsulated chamber expands to form a dome shape as soon as the vacuum chamber is turned on due to the 1 atm pressure gradient across the membrane. Figure 9.10(a) is a micrograph taken under SEM showing the result of Mylar bonded on PMMA. The dome shape can be observed as the proof that a good seal is achieved, and there is no leakage of air via the bonding interface. In addition, the plastics-to-plastics bond is forcefully broken to examine the bonding interface. After breaking the Mylar-to-PMMA bond, the bonding interfaces are examined under SEM as shown in Figs. 9.10(b) and 9.10(c). It is observed that the bonding result is uniform, and part of the Mylar film is attached to the bonding substrate. These indicate that very strong bonding is achieved.

9.6 Plug Insertion

In addition to stacking layer by layer using permanent bonding, functional membranes can be packaged and integrated into systems by temporary embedment as well. As illustrated in Fig. 9.11(a), the preprogrammed osmotic pumping chip is composed of valves and switches with layers of functional polymeric materials, which are micro-fabricated utilizing milling, laser machining, molding and assembled by multilayer bonding processes [4]. These components are cylinder-shaped plugs that can be embedded into the microfluidic chip. A PDMS diaphragm powered by osmosis is employed to provide the desired volumetric displacement for microfluidic

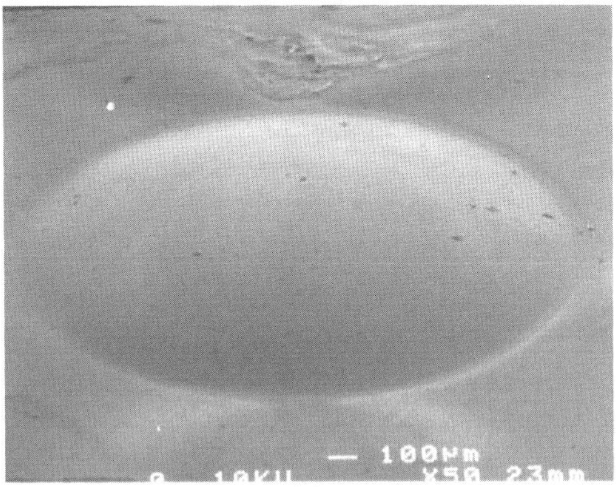

(a)

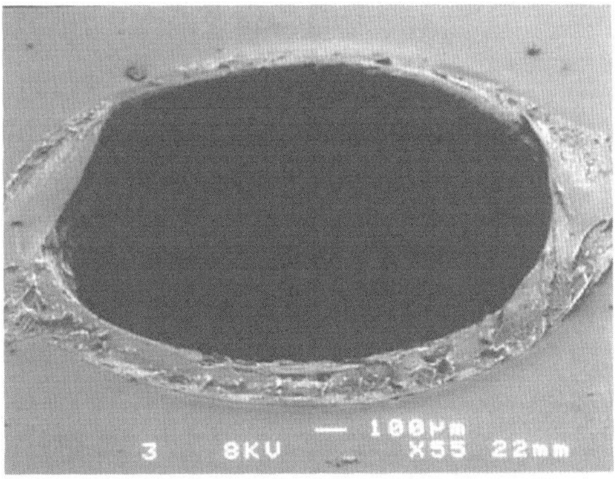

(b)

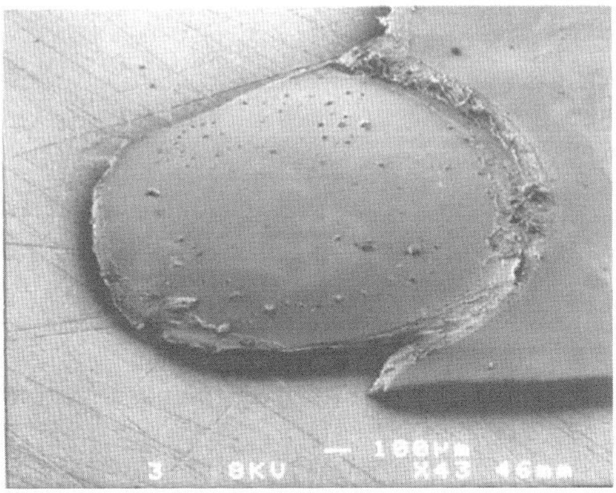

(c)

Fig. 9.10. SEM micrographs of (a) Dome shape formed under vacuum and a broken bonding interface of (b) PMMA substrate and (c) Mylar film [3].

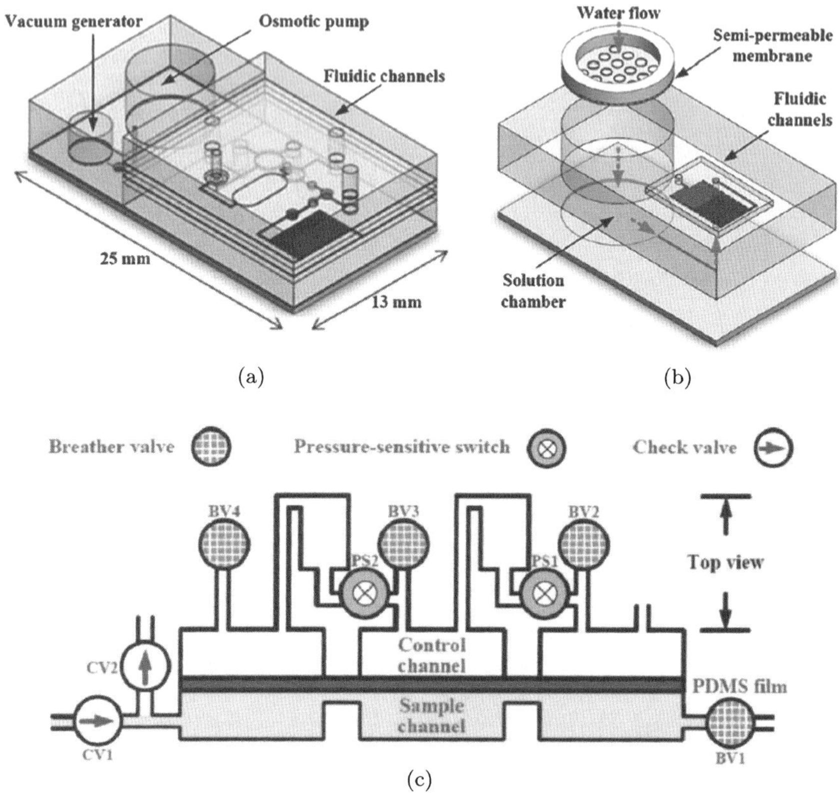

Fig. 9.11. (a) 3D model of preprogrammed osmotic pumping on a microfluidic chip, (b) Schematic illustration of the osmotic pump and (c) operating principle [4].

pumping. Regulated by integrated time switches, the diaphragm is actuated intermittently to realize time-varying pumping patterns. Meanwhile, vacuum generators with dedicated check valves and breather valves are employed to achieve the priming of liquid samples.

The main structure of the chip (i.e. a substrate with embedded diaphragm and microfluidic channels) is made of PDMS and shaped by casting and bonding. For control and sample channels that have minimum feature sizes down to tens of micrometers, molds fabricated with patterned photoresist films on silicon wafers are employed to duplicate the structures. In case that features with sizes up to millimeters are needed, PMMA molds fabricated using a computer numerical control (CNC) milling machine (MDX-15, Roland, with a tool of 300 μm in diameter) are employed. A mixture of 10:1 PDMS pre-polymer and curing agent (Sylgard 184,

Dow-Corning) is stirred thoroughly and then degassed under vacuum to remove trapped air bubbles. Afterward, the PDMS mixture is poured into the molds, degassed, cured for at least 1 h at 85°C and then peeled off from the molds. To irreversibly seal PDMS layers together, the surfaces of PDMS structures are modified with a hand-held corona treater (BD-20AC, Electro-Technic Products), which ionizes the surrounding air and creates localized plasma to activate the surfaces for irreversible bonding. The corona-treated surfaces are then pressed together and left undisturbed for at least 1 h at 85°C for the bonding to take effect. Once the main structure is ready, separately fabricated pumps, valves and switches are then plugged into the integrated microfluidic chip.

In the prototype demonstration, vapor-permeable thermoplastic polyurethane (Dintex FT-1880, Ding Zing Chemical Products) is used as the semi-permeable material for osmosis. The material is received in the form of 15 μm thick sheets, and can be shaped into structures with desired geometries by a thermal forming process, which heats the sheets up to around 220°C with the application of a pressure up to 5 MPa. As shown in Fig. 9.11(b), the semi-permeable membrane is mounted on a PMMA supporting structure, which has through holes for water to pass and is used to facilitate the sealing of underneath solution chamber. Figure 9.12 illustrates the fabrication of osmotic pumps in detail. The circular PMMA plates are manufactured using a CO_2 laser machine (LaserPro Venus, GCC) with a power of 1 W and a scanning rate of 0.2 m/s, while the through holes are formed with short pulses of laser. A hot melt film (Provecta FS2650-PE, Ding Zing Chemical Products) of 25 μm thick is employed to assemble semi-permeable membrane and PMMA substrates together. For the bonding to take effect, the assembly is heated up to 115°C, pressed together and left undisturbed when it is cooled down to room temperature. Meanwhile, breather valves are fabricated with a similar process as illustrated in Fig. 9.13. A supported hydrophobic PTFE membrane filter (J020A013A, Advantec MFS) is pasted on a PMMA substrate with through holes (300 μm in diameter) and a hot melt film on its top surface. Once the bonding has taken effect, cylindrical breather valves (800 μm in diameter and 1 mm in height) are cut from the PMMA substrate by CO_2 laser machining. With regard to the pressure-sensitive switch, a plastic mini-cylinder (1 mm in diameter) functioning as the pressure controller is inserted into the structure with an airtight seal around it. Figure 9.14 shows the photograph of a fabricated and assembled microfluidic chip for preprogrammed osmotic pumping. The chip is designed to have an overall

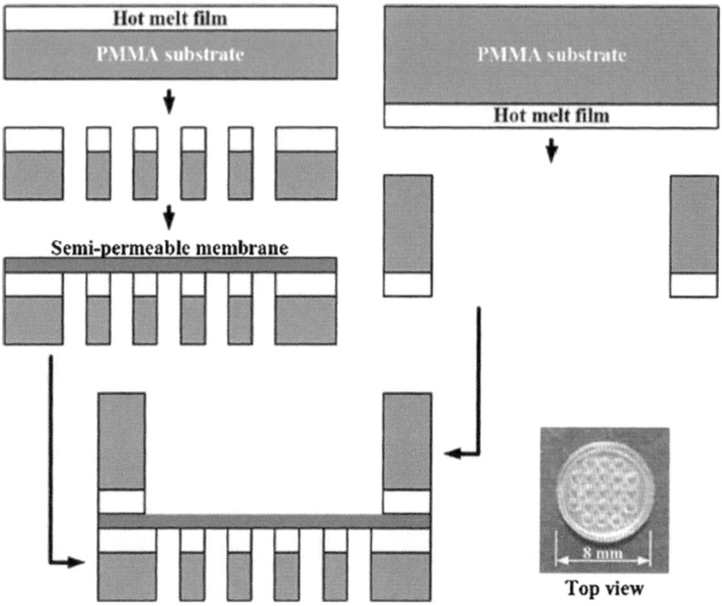

Fig. 9.12. Schematic illustration of the fabrication process forosmotic pumps [4].

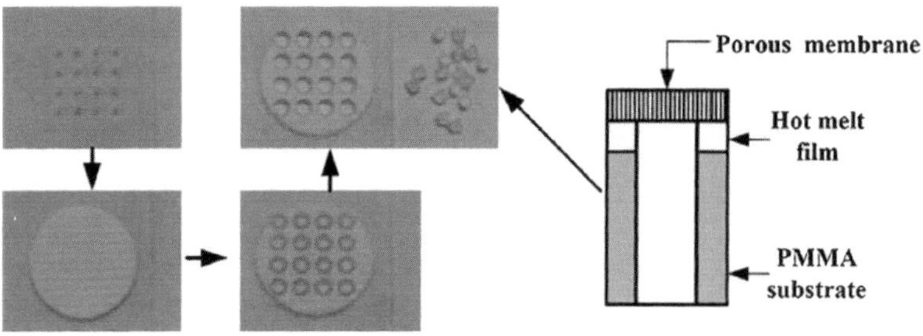

Fig. 9.13. Schematic illustration of the fabrication process for breather valves [4].

pumping volume of 1 μl, and dimensions of 25 mm in length, 13 mm in width, and 5 mm in thickness.

9.7 Micro-Encapsulation

Micro-encapsulation is the packaging technology of substances with thin polymeric membranes that forms tiny particles called micro-capsules. The polymeric coating acts as a protective film, isolating the core and avoiding

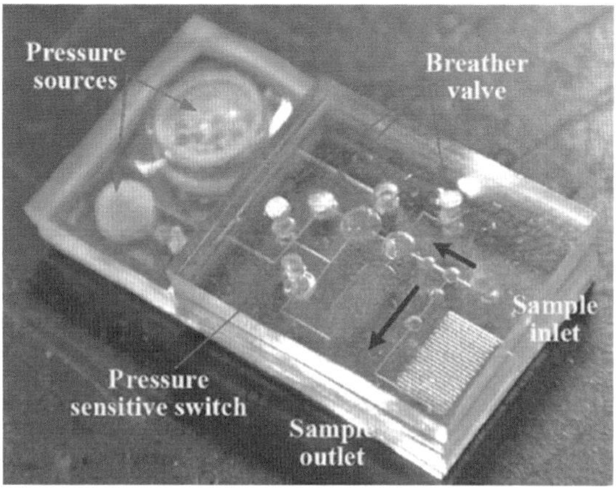

Fig. 9.14. Photograph of a fabricated chip for preprogrammed osmotic pumping [4].

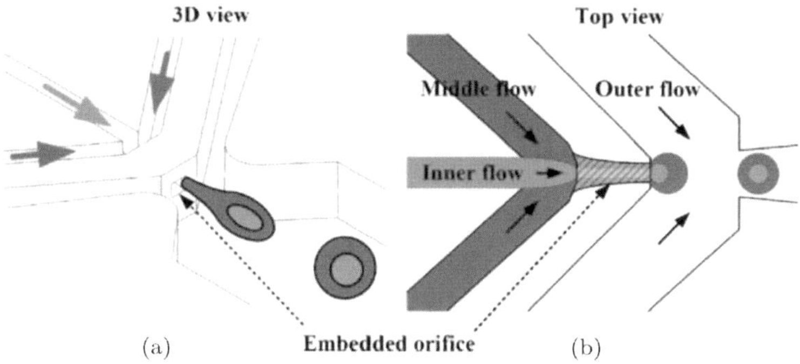

Fig. 9.15. Schematic illustration of 3D double emulsification devices [5]: (a) 3D view and (b) top view.

the effect of its inadequate exposure. This membrane dissolves itself through a specific stimulus, releasing the core in the ideal place or at the ideal time. Micro-encapsulation has numerous applications in areas such as the pharmaceutical, agricultural, medical, and food industries. Because of the strong application potential, such as the encapsulation of drugs and living cells, considerable attention has been attracted to the development of controllable and reliable fabrication schemes for double emulsions.

9.7.1 *Controlled Double Emulsification Using 3D Channels*

Figure 9.15 illustrates the geometries of a novel double-emulsification device for micro-capsule formation [5]. The device is made up of two adjacent focusing junctions with one embedded orifice connecting in between. The employed inner, middle, and outer fluids are driven by three independently controlled syringe pumps operating at varying flow-rates. Before entering the embedded orifice, which is located in the sidewall separating the two adjacent focusing junctions, the inner flow is intersected and focused by the middle flow at the upstream junction. Meanwhile, the embedded orifice is divergent and radius-edged around its entrance to ensure smooth transition. While flowing inside the embedded orifice, the inner flow is either shaped into a thread or broken into a series of droplets by the surrounding middle flow, depending mainly on the flow-rates and viscosities of the two fluids, and the geometries of the orifice as well. Once coming out of the embedded orifice, ideally the coaxial flow should be intersected and broken (by the outer flow) into drops with smaller droplets inside. The employment of an embedded orifice restricts the resulting coaxial flow away from the surface of downstream channel. Otherwise, it is very likely that the middle fluid would attach on the channel sidewall, which results in the distortion of flow pattern and eventually the failure of double emulsification. For an on-chip focusing and emulsification scheme, the employment of an embedded orifice is anticipated to reduce the undesired surface distraction dramatically.

The formation of biodegradable micro-capsules is realized by generating water-in-organic-solvent-in-water double emulsions first, followed by the extraction of solvent which results in the solidification of microstructures. Organic solvents which biodegradable materials can dissolve in are employed as the middle fluid. For example, ethyl acetate is chosen in this work to dissolve biodegradable materials including poly(L-lactic acid), trilaurin, and phosphocholine. In addition, materials such as surface-treated nanoparticles can also be dispersed in the middle fluid to render the resulting micro-capsules responsive to external stimulation. The solubility of ethyl acetate in water is roughly 8.3 g in 100 mL at room temperature. So the extraction can be achieved simply by rinsing the resulting emulsions with DI-water. Compared to water, organic solvents usually have much higher wettability on hydrophobic PDMS surfaces. While flowing inside a native PDMS channel, it is most likely that the middle organic solution would attach itself to the channel surfaces. The utilization of an embedded orifice could potentially prevent the attachment, but it is quite possible that some unexpected disturbance during the operation (especially the transient response in the starting period) might still lead to the undesired attachment and halt the emulsification process immediately. To better eliminate

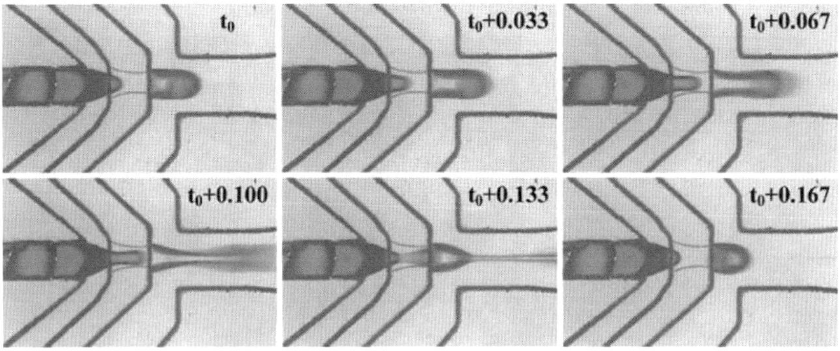

Fig. 9.16. Schematic illustration of 3D double emulsification devices [6].

the undesired surface distraction, the downstream PDMS channel surfaces are selectively modified into hydrophilic. Furthermore, various agents and surfactants are dispersed in both the middle and outer fluids in order to stabilize the interfaces. By doing this, the wettability of water on the downstream channel surface would be significantly enhanced. Although the undesired attachment could still occur accidently, it could be resolved simply by accelerating the outer water flow to drive the middle flow away from the PDMS sidewalls.

In the case shown in Fig. 9.16, the inner and middle flows share the same breakup frequency, while the time lag between the two breakups is around 0.22 s. Both breakups occur around the entrances of converging channels, where the flows are accelerated locally. The velocity gradient along the flow direction stretched the highly viscous middle flow into a long thread before finally broke it into an organic solution drop with just one water droplet inside. Meanwhile, the breakup frequency can be adjusted by varying the fluid flow-rates, and multiple droplets can be packed into one single drop. Figure 9.17 illustrates the measured relationship between the outer fluid flow-rate ($Q_{\rm o}$), inner fluid flow-rate ($Q_{\rm i}$) and the resulting emulsion dimensions. The flow-rates of the inner and middle fluids were set at 0.02 and 0.07 ml/hr, respectively, while $Q_{\rm o}$ was varied from 0.5 to 4 ml/hr, which resulted in double emulsions of 270 to 180 μm in drop diameter. In general, the drop diameter ($D_{\rm m}$) decreases in case that a higher $Q_{\rm o}$ is employed, while the droplet diameter ($D_{\rm i}$) and breakup period decrease as well. Meanwhile, $D_{\rm i}$ increases when $Q_{\rm i}$ rises, as also illustrated in Fig. 9.17. Once the water-in-organic-solvent-in-water structure was formed, ethyl acetate (the solvent) was extracted from the organic solution and the

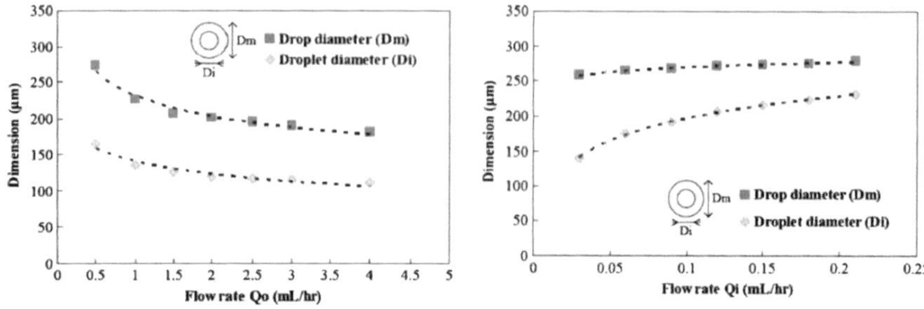

Fig. 9.17. Relationship between fluid flow-rates and the emulsion dimensions [6].

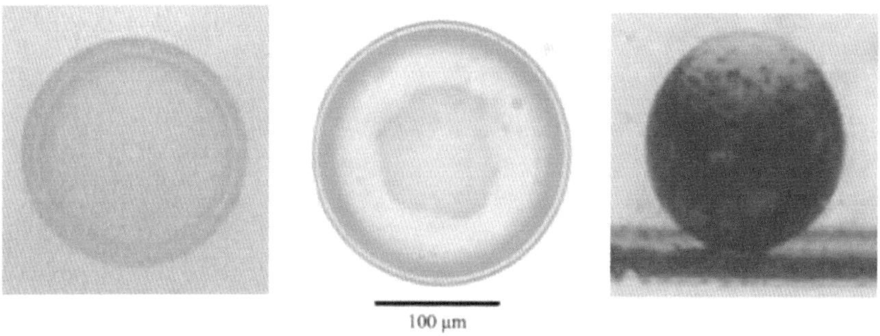

Fig. 9.18. Photographs of (a) afabricated trilaurin micro-capsule, (b) a phosphocholine micro-vesicle and (c) a magnetic poly(L-lactic acid) micro-capsule [6].

dissolved biodegradable material solidified into solid capsules afterward. In the prototype demonstration, it was also demonstrated that γ-Fe_2O_3 nanoparticles can be simultaneously embedded onto the micro-capsules to render them responsive to external magnetic stimulation. Figure 9.18 shows the photographs of fabricated trilaurin, phosphocholine, and poly(L-lactic acid) capsules with diameters of 163 μm, 181 μm, and 170 μm, respectively. The fabricated biodegradable micro-capsules could function as controlled delivery systems, which are desired for a variety of biological and pharmaceutical applications.

9.7.2 *On-Demand Micro-Encapsulation*

To address the need for real-time controllability on micro-encapsulation, an integrated microfluidic system that could form semi-permeable micro-capsules with desired geometries and compositions on demand is presented. Emulsion droplets functioning as templates and reactors are employed

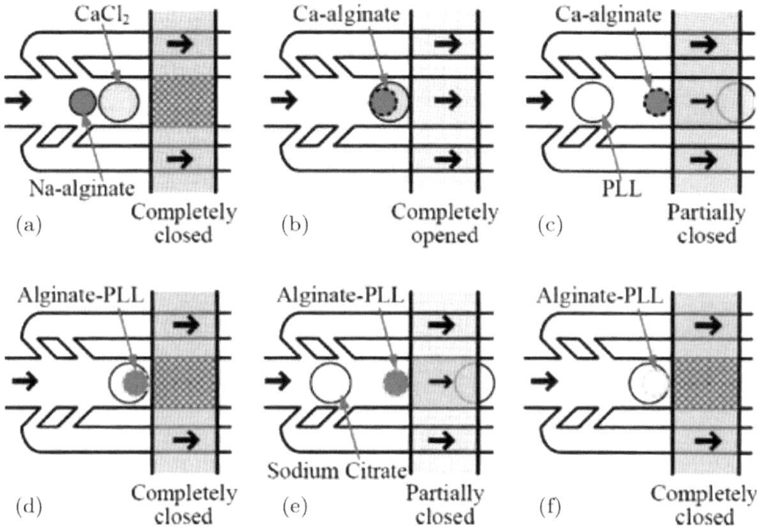

Fig. 9.19. Schematic illustration of an on-demand micro-encapsulation scheme [7].

to realize the synthesis process. Devices with pneumatically actuated diaphragm valves constructed on top of specially designed fluidic-channels are utilized to control the droplets and, consequently, the encapsulation process. Figure 9.19 shows a schematic illustration of an on-demand micro-encapsulation scheme [7]. The complete collapse of a diaphragm on the central flow-path stops the droplets and results in the merger into a double-emulsion structure. Two droplets containing aqueous Na-alginate and $CaCl_2$ solutions individually react on the interface to form a solid Ca-alginate capsule. The diaphragm valve is then partially opened, which allows outer un-reacted $CaCl_2$ solution to pass while the solid capsule remains trapped. The trapped capsule is rinsed by an incoming buffer droplet and then re-suspended into another incoming droplet containing poly-L-lysine (PLL), while the valve turns completely closed. Re-suspension of the capsules into PLL solution results in the formation of an alginate-PLL capsule due to the poly-ion complex. Once an alginate-PLL capsule is formed, it is rinsed and then re-suspended into a droplet of sodium-citrate solution to dissolve unreacted Ca-alginate gel inside the capsule. At the end, an alginate-PLL capsule is formed with its core in liquid state and released for sequential processing. It has been demonstrated that multi-step reactions could be performed on droplet-in-droplet interfaces to synthesize alginate-PLL capsules.

9.8 Future Trends and Summary

3D printing is known as additive manufacturing, defined by the ASTM as the process of joining materials to make objects from 3D model data, usually layer upon layer, as opposed to subtractive manufacturing methodologies (ASTM F2792). The major benefits of 3D printing include rapid production, low cost, increase design freedom, and innovation. Currently, there are many different techniques for 3D printing available on both industrial and commercial markets. Among them, stereolithography (SLA) is most relevant to the packaging and integration of membrane-based microfluidic systems. Figure 9.20(a) illustrates the working scheme of an osmotic orthodontic actuator [8]. It is mainly a hallow structure filled with supersaturated solution and mounted on the lingual side of teeth. At each end of the actuator, there is an osmotic pump consisting of a semi-permeable membrane sandwiched by honeycomb structures, as shown in Fig. 9.20(b). The actuator is manufactured/packaged using digital light processing (DLP) stereo-lithography and shrink wrapped by a 15 μm-thin breathable film to build a 3D and completely sealed interface. Compared to the packaging

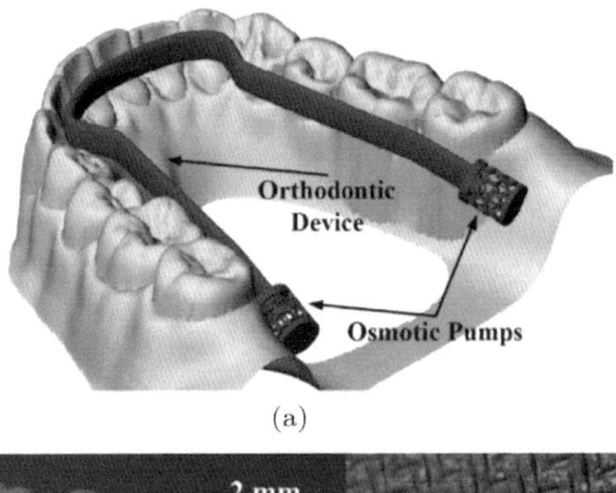

(a)

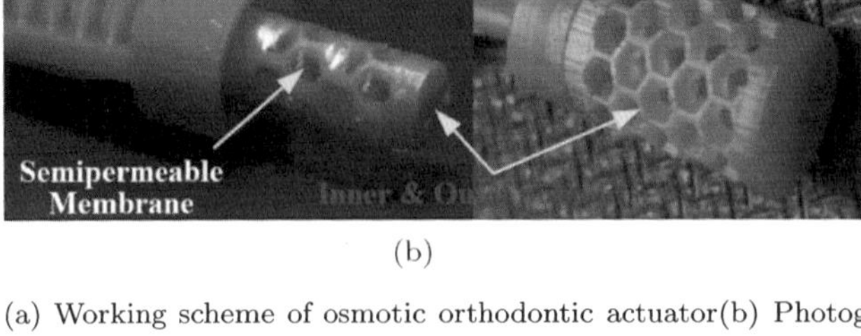

(b)

Fig. 9.20. (a) Working scheme of osmotic orthodontic actuator(b) Photographs of an osmotic pump.

and integration schemes presented earlier in this chapter, 3D printing is undoubtedly more accurate and consist since potential human errors are eliminated. The adoption of 3D printing for the fabrication of microfluidics can dramatically lower the technical barrier and accelerate the creation of sophisticated devices and systems.

References

1. Su, Y. C., Lin, L., and Pisano, A. P. (2002). A water-powered osmotic microactuator, *Journal of Microelectromechanical Systems*, 11(6), pp. 736–742.
2. Su, Y. C. and Lin, L. (2004). A water-powered micro drug delivery system, *Journal of Microelectromechanical Systems*, 13(1), pp. 75–82.
3. Su, Y. C. and Lin, L. (2005). Localized bonding processes for assembly and packaging of polymeric MEMS, *IEEE transactions on advanced packaging*, 28(4), pp. 635–642.
4. Peng, P. J. and Su, Y. C. (2015). Preprogrammed, rate-switchable osmotic pumping on a chip, *Microfluidics and Nanofluidics*, 18(2), pp. 267–277.
5. Chang, F. C. and Su, Y. C. (2008). Controlled double emulsification utilizing 3D PDMS microchannels, *Journal of Micromechanics and Microengineering*, 18(6), 065018.
6. Liao, C. Y. and Su, Y. C. (2010). Formation of biodegradable microcapsules utilizing 3D, selectively surface-modified PDMS microfluidic devices, *Biomedical Microdevices* 12(1), pp. 125–133.
7. Chang, S. C. and Su, Y. C. (2011). On-demand micro-encapsulation utilizing on-chip synthesis of semi-permeable alginate-PLL capsules, *Microfluidics and Nanofluidics* 10(6), pp. 1165–1174.
8. Hsu, H., Liu, L. Y., and Su, Y. C. (2017). 3D printed, programmable osmotic actuators for orthodontic application, in *Proceedings of the. 30th IEEE International. Conference. on Micro Electro Mechanical Systems, MEMS*, pp. 301–304.

Yu-Chuan Su received the B.S. and M.S. degrees in Power Mechanical Engineering from the National Tsing Hua University, Taiwan, in 1993 and 1995, respectively. In 2003, he received the Ph.D. degree in Mechanical Engineering from the University of California, Berkeley, with an emphasis on MEMS technology. Since 2004, he joined the National Tsing Hua University, Taiwan, and is currently a professor at the department of Engineering and System Science. His research interests are in design and fabrication of polymer-based nano- and microfluidic systems for biomedical and energy related applications, including 3D printing, droplet-based microfluidics, energy harvesting, power MEMS and controlled drug delivery. He is a member of IEEE, ASME, and Phi Tau Phi.

Chapter 10

Wafer-Level Solder Bonding by Using Localized Induction Heating

Hsueh-An Yang, Chiung-Wen Lin and Weileun Fang

10.1 Introduction

10.1.1 *Advanced Semiconductor Package*

In recent years, the system in package (SiP) has been applied in personal electronic devices to provide highly complex system integration. Figure 10.1 shows different package types in SiP technology. The package developing trend started from horizontal integration such as multichip module (MCM) [1, 2] to vertical integration likes multichip package (MCP) technology [3, 4]. Figure 10.2 shows a typical MCM package architecture, which enables the side-by-side integration of several chips and discrete devices. Thus, all of components can be assembled in a thinner package. However, this package needs a high-density substrate for component integration. The power consumption, substrate cost, and footprint of package size will be a complicated issue for consumer electronic devices. For example, the MCM graphic processing unit (GPU) module in Fig. 10.2 has a footprint of 37.5 mm × 37.5 mm, which consists of RAM and a processor [5].

Although the MCM package can provide a multichip integration, with more easy chips assembly process and better thermal management package solution, the package footprint is still larger. The developing 3D package architecture would shrink the package footprint further. This vertical integration solution provides not only smaller package footprint but also shorter interconnection. Stack dies, package on package (PoP) [7, 8], and package in package (PiP) technology [9, 10] are widely adopted 3D package architectures. In applications, the stack dies technology is applied in memory stick production where memory dies are stacked, connected by

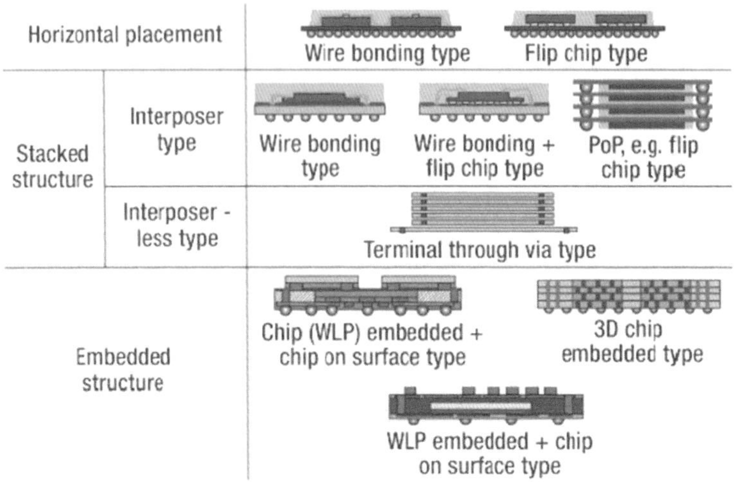

Fig. 10.1. SiP Package architectures (*Source*: iNEMI 2009 Industry Roadmap [6]).

Fig. 10.2. Multichip package (MCP) architecture (*Source*: AMD) [5].

wire bonding, and covered by molding to achieve a high-density footprint. Figure 10.3 shows 20 stacked NAND flash memory chips with multilayer wire bonding, which is only 1.4 mm thick [11]. Figure 10.4 shows a typical cross section of PoP package. The PoP technology involves stacking two individual packages on a redistributed substrate. The bond wire or solder

Fig. 10.3. MCP with 20 stacked NAND flash memory chips (*Source*: Hynix Semiconductor [11]).

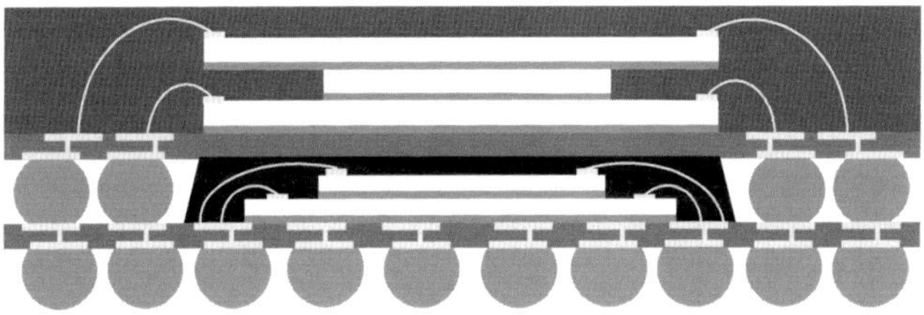

Fig. 10.4. Typical PoP (package on package) architecture (*Source*: Siliconware Precision Industries Co., Ltd. [12]).

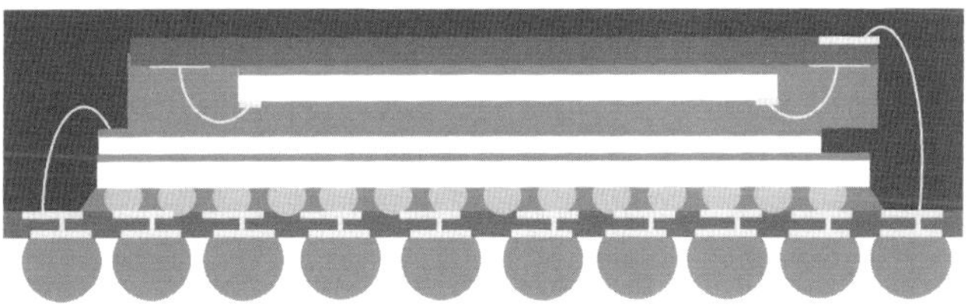

Fig. 10.5. Typical PiP (package in package) architecture (*Source*: ASE Group [13]).

ball can be used for interconnection. This package provides a simple, reliable, but thicker package solution. Compared to PoP package, the package height can be further reduced in PiP technology. Figure 10.5 shows an individual package stacked on a flip-chip, substrate for package integration.

Those 3D package technologies are based on the existing assembly process flow but vertical stacking process is needed for footprint shrinkage.

The current assembly process steps comprise wafer thinning, wafer sawing, die bonding, wire bonding, molding, and singulation. The outsourced semiconductor assembly and test (OSAT) companies can use current assembly tools to provide those package solutions, without additional investment in tools or resources for developing new technologies. Moreover, lower cost and shorter development time to market can be offered by these approaches.

Besides, the existing interconnection technologies still have some disadvantages when applied in 3D package to address the abovementioned obstacles. Though wire bonding can provide a simple, reliable process and cost-efficient way for multilayer die stacking, the top die will have much longer wire length than the bottom one. Thus, the chip at the top will have much higher power consumption than the chip at the bottom. On the other hand, the wire-bonding pads must be arranged around the chips so as to limit the quantity of I/O ports. The wire bonding technology cannot satisfy the higher quantity of I/O ports in IC chips. In this regard, developing through silicon via (TSV) interconnect technology to provide a finer pitch for interconnect with shorter TSV wire length to lower power consumption becomes critical for advanced semiconductor packages [14, 15].

TSV technology has some advantages such as shorter interconnect, fine pitch, and compatibility with wafer foundry process. However, the current package substrates like land grid array (LGA) capability still cannot provide fine pitch to meet the foundry process. The costs of substrate and assembly are still high and thus become a bottleneck for this technology development. The silicon interposer will be a solution to bridge both process capabilities shown in Fig. 10.6. There are two advantages of this silicon interposer technology. First, it can be acted as a redistribution layer to fan-out fine pitch I/O ports to substrate. The second is using silicon interposer as a buffer layer to balance coefficient of thermal expansion (CTE) mismatch issue between substrate and thinner IC chips.

10.1.2 *MEMS Package*

Currently, MEMS products have been widely adopted nowadays either in industry or consumer applications. In general, chip size is one of the major concerns in consumer electronics (CE). Moreover, the number of sensors in wearable devices has continuously increased to create the challenges of chip size and space arrangement. Figure 10.7 shows different MEMS sensor package types. The package development trend still starts from horizontal to vertical integration. STMicroelectronics has demonstrated a

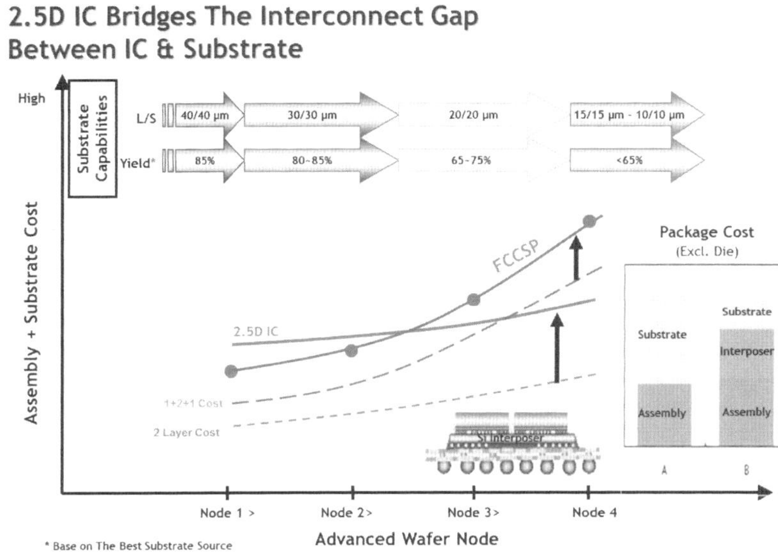

Fig. 10.6. Developing trend for silicon interposer (*Source*: ASE Group [13]).

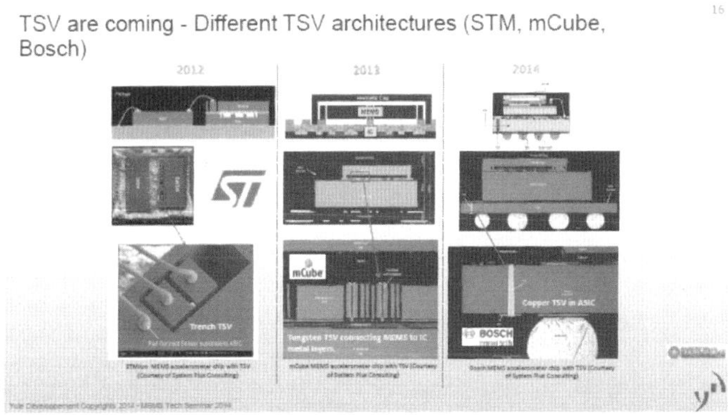

Fig. 10.7. MEMS sensor package architectures (*Source*: Yole Development [16]).

MEMS accelerometer where ASIC assembled nearby MEMS chips within wire bonding in 2012. Bosch has used stack MEMS chip on ASIC within TSV for interconnection, and final package by wafer-level bumping process to shrinkage package size.

Invensense has demonstrated a CMOS MEMS process platform. The MEMS wafer is directly bonded to ASIC wafer as shown in Fig. 10.8, where to thin down the thickness of MEMS wafer by lapping is required

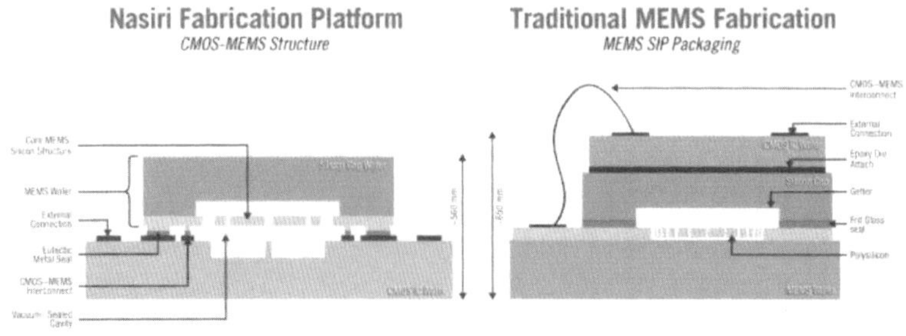

Fig. 10.8. CMOS MEMS package architecture. (*Source*: Invensense).

to accommodate the overall space requirements [17]. As a result, several additional design and process issues arose after the thinning of the MEMS substrate. First, it is not straightforward to handle the MEMS device on a thin substrate, at either wafer or chip level. Second, thinning of the MEMS substrate will induce unexpected deformation as the packaged product is mounted on a printed circuit board (PCB) or operated during temperature change in a normal operation. The deformation of substrate may cause increase of the sensing gap between electrodes that physical excitation is needed to detect pressure, gravity, or linear acceleration. Thus, the performance of the sensors is affected and extra calibration for sensors may also be required.

The monolithic integration of CMOS circuit and MEMS device in the same wafer is a promising solution for the aforementioned size and stiffness considerations. The thickness of ASIC die could contribute an additional stiffness to support the MEMS structure after wafer thinning. Moreover, as compared with the conventional dual-chip connection between ASIC and MEMS, the parasitic capacitance from wire bonding can be eliminated by using the monolithic integration of CMOS and MEMS devices. In general, the approaches to achieve the monolithic integration of MEMS and CMOS processes on the same wafer can be categorized as "MEMS first" (fabricating MEMS structures prior to CMOS processes), the "MEMS middle" (fabricating MEMS structures in between CMOS process steps), and the "MEMS last" (fabricating MEMS devices after the completion of CMOS processes). Among these three approaches, the "MEMS last" is the preferred one since it has the following advantages:

1. MEMS processes could be achieved flexibly outside the CMOS foundry.
2. The contamination of process facilities at CMOS foundry is prevented.

10.2 Potential Issues for Vertical Chip Stacking Integration

Wafer thinning process, thin wafer handling, and thin chips assembly are very critical processes for the stacking-type package solutions. In Fig. 10.3, 20 memory are stacked on chips in one package, and the final package height is only 1.4 mm. The thickness of each chip would be thinner than 70 mm. However, the thin wafer usually has a serious warpage issue. The die warpage will cause bump creak, bump cold joint, and delamination between chip to chip or chip to substrate. There are two potential risks to assemble thin chips [18, 19]:

1. The programmable temperature reflow oven can control the soldering reflow profile. This traditional reflow process provides a global heating environment, high throughput, and low cost for chips assembly. Since the thicker chips have more robust mechanical support, smaller chip warpage, and lower thermal stress during temperature change in reflow oven, the solder joint will be more reliable during thicker chips assembly. Thus, the warpage in thin chips would be difficult to be assembled by this standard reflow process.

   The thermal compression bonding is a workable solution to assemble thin chips for multichip stacking package, which controls chip warpage by heating the chip's solder locally from the bonder head. This local heating technique will reduce CTE mismatch issue between substrate and thinner chip. Figure 10.9 shows the process flow for the thermal compression bonding [20].
2. The disadvantage of the thermal compression bonding is process throughput. This bonding process only assembles one chip at a time. It takes a long time to reach multilayer stacking. Also, multireflow temperature cycle of this assembly procedure will affect the solder bump reliability of the bottom chips.

Assembly cost is another concern for high density, thinner package development. Figure 10.10 shows the trend of assembly cost for 3D package. For lower cost solution, flip-chip (FC) is preferred as the pitch/wafer thickness is higher. As the wafer thickness is reduced, TSV technique will be advantageous with higher cost.

To incorporate MEMS device in multidie stacking package, fragile microstructure must be well protected. Wafer bonding technology had been widely implemented to protect MEMS device from damage during

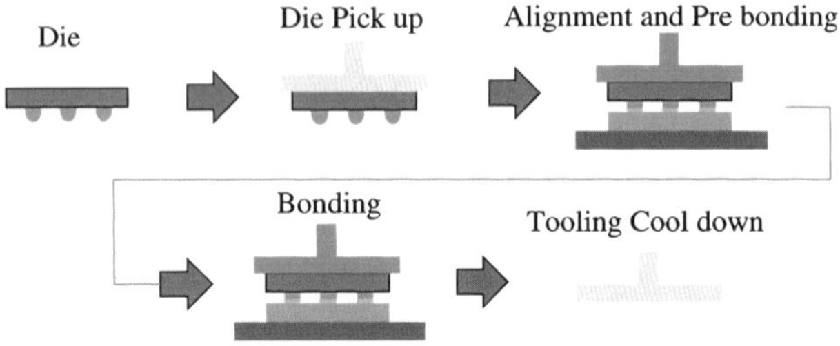

Fig. 10.9. A typical thermal compressive bonding process flow.

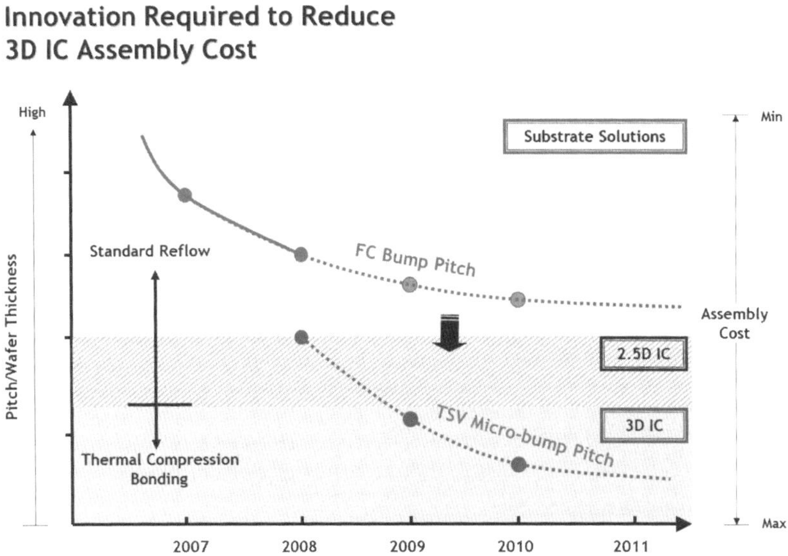

Fig. 10.10. The trend of assembly cost for 3D package (*Source*: ASE Group [13]).

assembly. To incorporate wafer bonding with "MEMS last" CMOS-MEMS wafer, one major concern is the thermal budget of the CMOS device underneath, especially for the interconnection aluminum (Al) film. In other words, the temperature of wafer bonding process must be lower than 660°C, which is the melting temperature of Al [21, 22]. Thus, there are two concerns for the bonding of CMOS–MEMS wafer:

1. To achieve the bonding quality at temperature well below 450°C in CMOS area.
2. To reduce the process time for bonding to prevent the irreversible affect to CMOS at temperature approach 450°C.

However, most of the existing bonding techniques are conducted at high temperature. The temperature will be 1000°C for the conventional fusion bonding and at 300–500°C [23, 24] for anodic bonding. These high-temperature processes may induce thermal problems and damage the microstructures or the integrated circuits. As reported in References 25 and 26, a localized heating approach by an embedded heater was presented to prevent thermal problems during bonding. As a result, wafer bonding that can "localize" the heat generated at bonding region could fulfill the thermal budget requirement. Moreover, the localized wafer bonding technique with fast bonding time is even an additional benefit to reduce the unwanted heat accumulation around CMOS devices.

10.3 Localized Induction Heating Concept and Fabrication Process

Induction heating is a well-known technology for producing a temperature rise in specific materials very quickly [27, 28]. This technology has been extensively employed to heat electrically conductive materials in applications such as melting, soldering, and hardening of metals. As shown in Fig. 10.11, the induction heating system consists of a high-frequency power supply, an induction coil, and the testing vehicles. In this system, two testing vehicles were designed to demonstrate this concept. Figure 10.11(a) shows the stack dies connected by a copper pillar bump using localized induction heating. The other application is wafer-level MEMS package

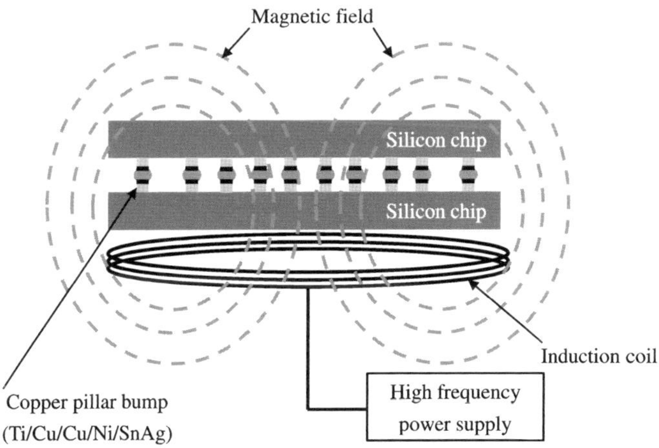

Fig. 10.11. (a) is the concept for the stack dies connect by copper pillar bump using localized induction heating.

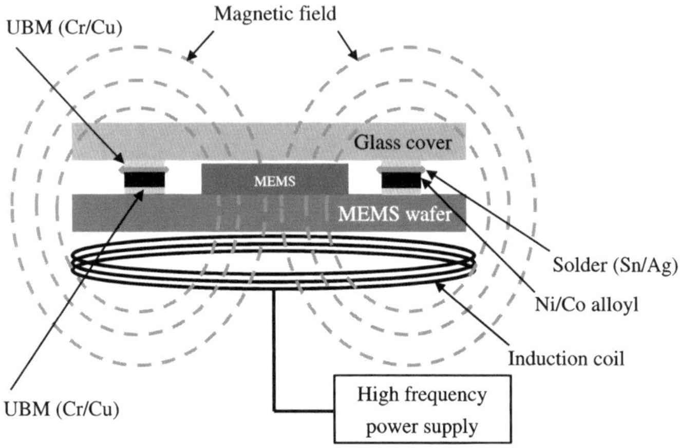

Fig. 10.11. (b) is the concept for the MEMS wafer level capping by localized induction heating.

shown in Fig. 10.11(b). A high-frequency power supply drives the induction coil, generating a magnetic field. Meanwhile, an eddy current is generated in the testing vehicle as it is placed inside the magnetic field. The heat will be generated by the ferromagnetic material, nickel, due to the eddy current loss and hysteresis loss [29]. According to the characteristics of the heating mechanism, the packaged chips can be free of contact from the coil and locally heated in specific regions. Using the induction heating technology, a higher temperature will be generated at the metal parts, especially in the ferromagnetic nickel. These ferromagnetic materials can be exploited as a localized heating element for the composite chips. Wafer-level packaging (WLP) can be realized if the whole wafer can be placed inside the magnetic field. The corresponding fabrication process has been established by conducting the presented bonding technique for silicon substrates. Compared to traditional solder reflow process, this process can be adopted without any changes to the current wafer bumping process. In addition, fabrication processes used to encapsulate the MUMPs devices by Si and glass covers can also be accomplished by the presented bonding method.

10.3.1 *Si–Si Bonding Processes*

UBM of Cr/Cu with 20 nm Cr as the adhesive layer and 20 nm Cu as the seed layer were deposited on a 4-inch single-side polished silicon wafer, as shown in Fig. 10.12(a). A Ni/Co alloy was electroplated on top of the seed layer to act as the spacer. The solutions for the electroplating of the Ni/Co

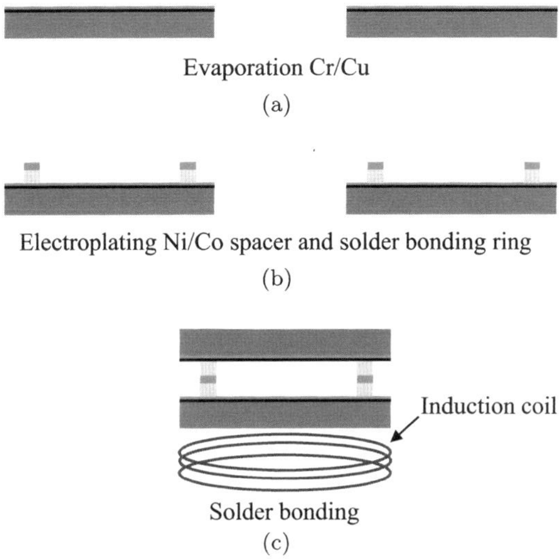

Fig. 10.12. The concept and fabrication process for the present study.

are listed in Table 10.1; in addition, the deposition rate was 0.1–0.2 μm/min and the current density was 1–5 ASD (ampere/dm^2). Since the Ni/Co alloy is a ferromagnetic material, it has very large permeability [27]. Thus, a much higher temperature was generated by the Ni/Co spacer for a given driving current during induction heating. The Sn/Pb alloy solder was electroplated on top of the spacer, as shown in Fig. 10.12(b). The electroplating solutions of the solder are listed in Table 10.1; the deposition rate and the current density were 0.1–0.5 μm/min and 1–4 ASD, respectively. The composition of the Sn/Pb solder was Sn–Pb (63/37), and its melting point was 183°C. The spacer and solder provided more than 50 μm of space in providing large range for motion of the MEMS devices. As shown in Fig. 10.11(b), the induction coil driven with high-frequency ac current generated a magnetic field. After the two silicon wafers were aligned, the magnetic field was employed to heat the thin film Ni/Co magnetic spacer, as illustrated in Fig. 10.12. In this study, the process was performed at atmospheric pressure (1 atm). The induction heating temperature is significantly influenced by the magnetic field, the material property (permeability) and the area of the magnetic layer [22]. Hence, a large area of Ni/Co alloy and a higher driving frequency were required for microscale induction heating. The driving frequency and current of the magnetic coil were 100 kHz and 25 A, respectively. Moreover, the width of the electroplated Ni/Co bonding

Table 10.1. The plating solution for the electroplating of (a) Ni/Co and (b) Pb/Sn solder.

(a) Electroplating of Ni/Co	
Nickel sulfamate, $Ni(NH_2SO_4)_2 \cdot 4H_2O$	450–550 ml/l
Cobalt sulfamate, $Co(NH_2SO_4)_2$	15–25 ml/l
Nickel chloride, $NiCl_2 \cdot 6H_2O$	2–7 g/l
Boric acid, H_3BO_3	30–50 g/l
Stress reducer	2–10 ml/l
(b) Electroplating of Pb/Sn solder	
Tin	12–18 g/l
Lead	1.5–2.5 g/l
pH value	3.0–5.0
Specific weight	1.1–1.2

ring is about 600 μm. It took only 15 s for the magnetic spacer to reach 200°C using induction heating where the temperature will cause the solder on top of the spacer began to reflow. Thus, the two silicon wafers were bonded together by the solder. The SEM photograph in Fig. 10.13(a) shows the typical result of silicon substrates after bonding. The sample was purposely broken so that the space provided by the spacer and solder could be clearly observed. The close-up SEM photograph in Fig. 10.13(b) shows a typical cross section of the bonded silicon–silicon test sample as indicated by the dashed box in Fig. 10.13(a). There are three different layers, including two Ni/Co spacer layers and one Sn/Pb solder layer, between the silicon substrates. According to the characteristics of the solder reflow, no interface has been observed in the bonded Sn/Pb layers.

10.3.2 *Die-Stacking Solder Bonding Process*

A typical copper pillar bump was defined by thick photoresist or dry film. In this study, a UBM film Ti/Cu was deposited on an 8″ wafer by a sputtering system. Then, selective electroplating of copper post, barrier layer nickel, and solder cap on the UBM film was performed. The thickness of each layer is 100 nm Ti, 500 nm Cu, 5 μm thick plating Cu post, 2 μm thick barrier plating Ni, and 3 μm thick plating Sn/Ag alloy as a solder material. Finally, the photoresist had been removed by stripper and then etching of UBM was achieved by metal wet etching.

Two wafers are aligned precisely in wafer bonding system. Aligned wafers are placed on a high-frequency generator, and an AC magnetic field

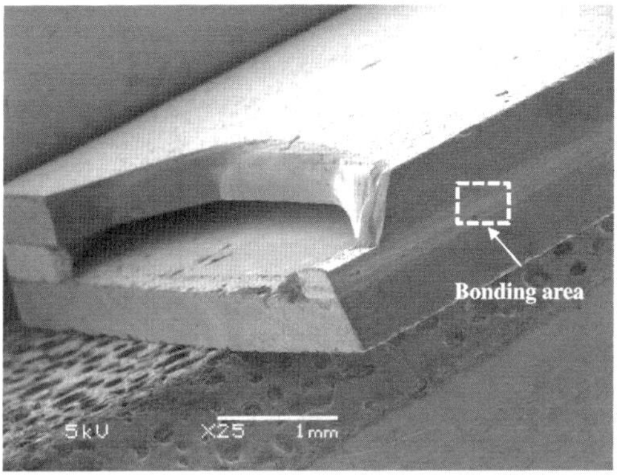

(a)

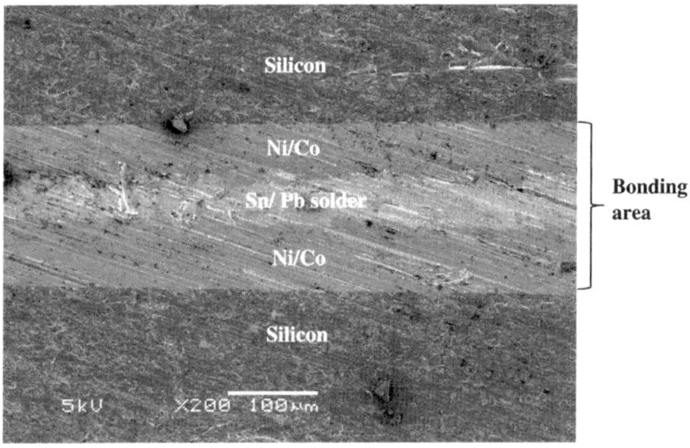

(b)

Fig. 10.13. The SEM photographs of (a) the silicon substrates after bonding and (b) the close-up of the bonded specimen in the dashed box region.

was applied. The eddy current will be induced on the defined nickel part of the copper pillar bump. After reaching reflow temperature, the corresponding copper pillar bump will be connected by soldering. The totally process time took just around 1 min. Figure 10.14 shows the cross section of the copper pillar bump.

10.3.3 *Integration of MUMPs and Bonding Processes*

It will be straightforward to integrate the presented localized bonding technique with other fabrication processes. Moreover, it is possible to employ this technique to bond silicon and glass. This part established a process,

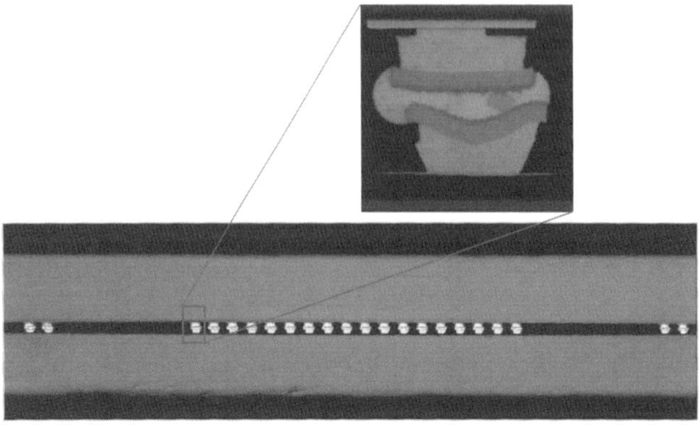

Fig. 10.14. Localized induction heating for copper pillar bump.

depicted in Fig. 10.15, to demonstrate the feasibility of integration of 1-poly surface micro-machined structures with the presented packaging techniques. As shown in Fig. 10.15(a), the fabrication started with a 2-poly surface micro-machined process. The first poly-silicon (poly0) acted as the conducting/grounding layer and the second polysilicon (poly1) acted as the structural layer. As illustrated in Fig. 10.15(b), the Cr and Cu films were then evaporated and patterned to perform as the electroplating adhesive layer and the seed layer, respectively. It was necessary to properly cover the conducting routing formed by the poly0 layer using the Cr/Cu films. As shown in Fig. 10.15(c), a thick photoresist was spun on and patterned using photolithography. The patterned photoresist was exploited as a mold to electroplate the thick Ni/Co alloy spacer. Since the poly0 layer was properly covered by the Cr/Cu films, the step height of the conducting routing could be sealed by the electroplated Ni/Co spacer. After a 10-μm-thick solder was electroplated on top of the Ni/Co alloy, the sacrificial layer was etched away to release the polysilicon surface structures. As shown in Figs. 10.15(d) and (e), the silicon substrate with surface micro-machined devices was covered with a glass cap after the localized bonding process. Hence, the poly0 conducting routing acted as an interconnection. The SEM photograph in Fig. 10.16(a) shows the electroplated Ni/Co spacer on the substrate, and a 4×4 micro-torsional-mirror array inside an $850\,\mu$m $\times$ $550\,\mu$m region. The spacer was $50\,\mu$m in height, allowing for a large space in the out-of-plane direction for the torsional mirrors. The closeup photograph in Fig. 10.16(b) shows the conducting wires and the mirror bonding pads. The conducting routing acted as interconnection to connect the pads

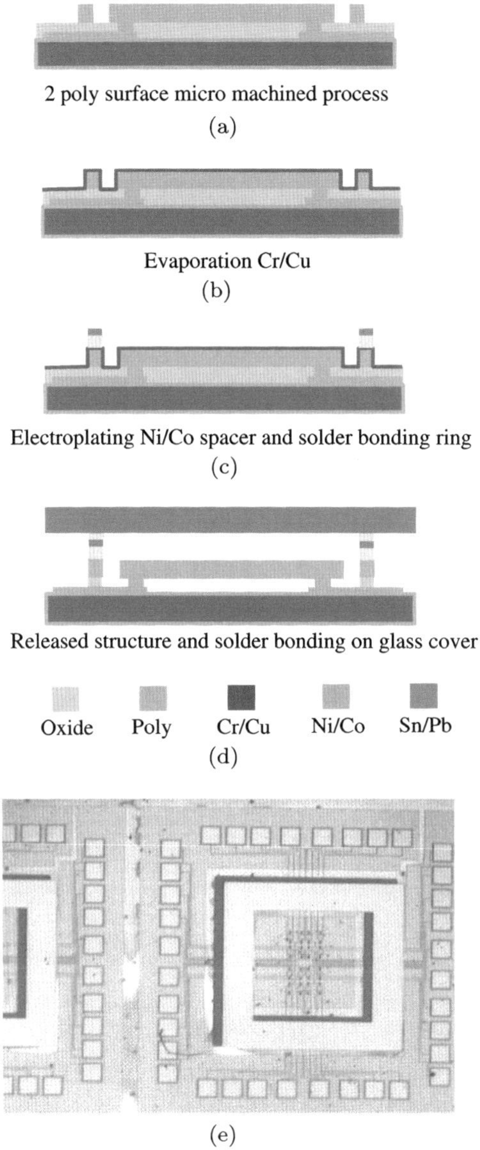

Fig. 10.15. Fabrication process to integrate the surface micro-machined devices with the present bonding method.

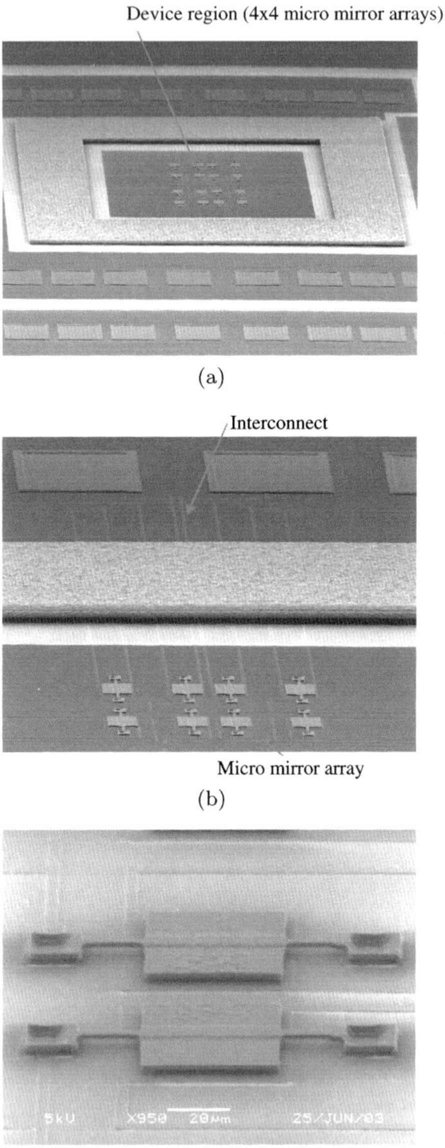

(c)

Fig. 10.16. The SEM photographs of the integration of surface micro-machined devices and the spacer: (a) a 4 × 4 mirror array inside the 50 μm height Ni/Co spacer, (b) conducting routing and bonding pads of mirrors, and (c) the microscopic section of the bonding region.

and the mirrors after encapsulation of the mirror array. In addition, the SEM photograph in Fig. 10.16(c) shows the microscanning mirrors. It is found that the 0.5 μm step height poly0 conducting routing was perfectly sealed by the Ni/Co spacer.

10.4 Testing

According to the basic principles of induction heating, heating efficiency should significantly decay in a microscale thin film [20]. This study has conducted four different tests, including temperature distribution, bonding strength, leakage, and feedthrough to evaluate the performance and feasibility of the bonding process.

10.4.1 *Temperature Distribution*

The measurement setup for the localized heating process is shown in Fig. 10.17. Briefly, the substrate was placed on top of the induction coil. Because the whole wafer can be placed inside the magnetic field, the WLP

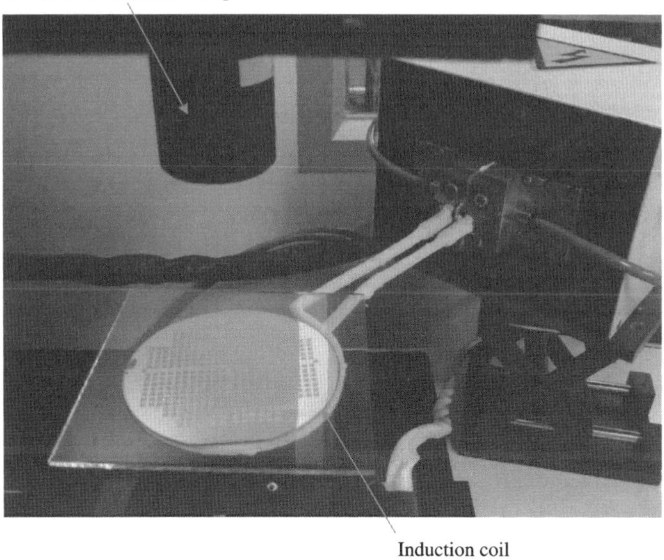

Fig. 10.17. The experiment setup in this study for induction heating. The sample was placed on top of a magnetic coil. The temperature distribution of the sample was inspected by an infrared microscope.

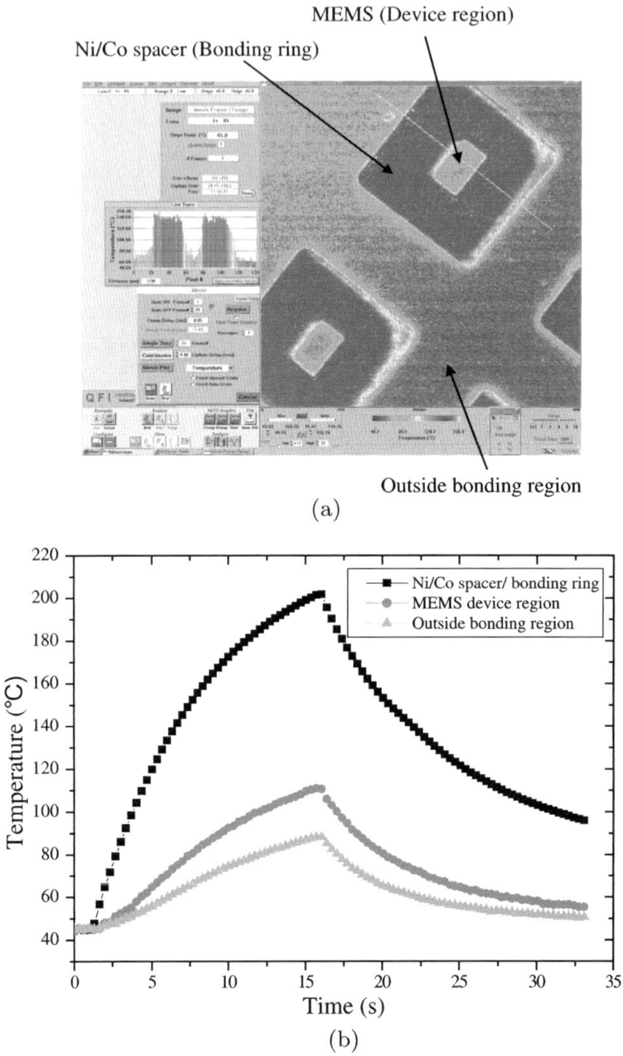

Fig. 10.18. The measured temperature of the test sample during induction heating: (a) spatial distribution and (b) temporal distribution.

can be realized. An infrared thermal microscope was used to monitor the temperature distribution during heating. The spatial resolution ranges from 3 to 60 μm and the temperature resolution of the infrared thermal microscope is 0.1°C. Figure 10.18(a) shows typical measurement results. The measured temperature distribution of the sample was displayed by colors. The temperature distribution of the specimen during bonding can be characterized as the following three regions: device region, bonding region, and outside bonding region. Figure 10.18(b) shows the quantitative temperature distribution measured from these three regions during the heat-up and cool-down processes. Moreover, Fig. 10.18(b) also shows the variation of the specimen temperature with the heating and cooling time. It took 15 s for the bonding region to exceed a temperature of 200°C, so as to enable the solder to reflow. Under such a circumstance, the Si–Si or the Si–glass will be bonded by the solder. Since the bonding process was completed within a few seconds, the temperature rise of the device region due to thermal conduction was not significant. Accordingly, the temperature of the device region surrounded by the spacer was only 110°C, so that the thin film devices inside this region would not be damaged. In addition, the maximum temperature outside the bonding region was only approximately 90°C.

10.4.2 *Bond Strength and Leakage/Interconnection Test*

The mechanical strength of the bonded specimen was measured using a tensile test. As shown in Fig. 10.19, the size of the test sample was 10 mm × 10 mm. The bonded area as indicated in the figure was 12 mm^2. During the test, the sample was fixed between two socles by epoxy and then loaded by a mechanical force, as shown in Fig. 10.19. The bonding strength measured

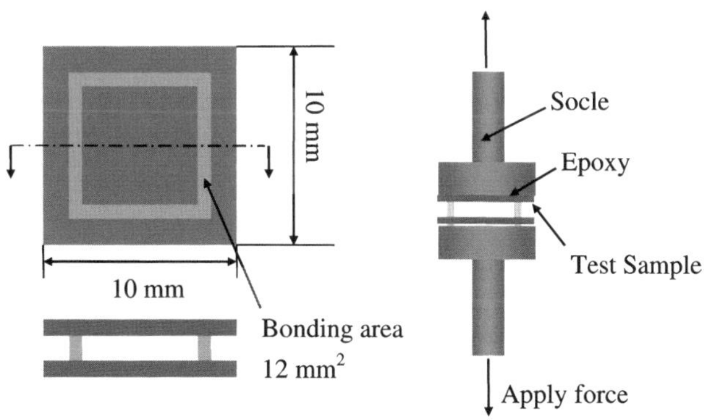

Fig. 10.19. The experiment setup for the evaluation of bond strength. The sample was under a tensile test.

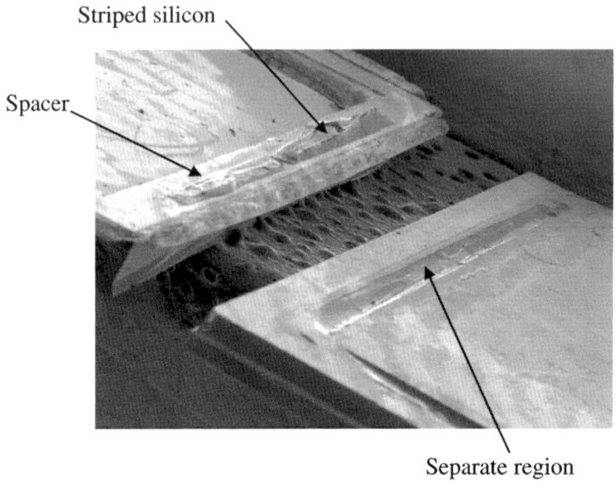

Fig. 10.20. The SEM photograph of the specimen and its bonding region after stripped away.

for eight different tests ranged from 10.6 to 18.3 MPa. After the tensile test, the broken sample was examined under SEM, as shown in Fig. 10.20. It was found that the sample was not separated at the solder bonding region. Moreover, part of the damaged region was at the silicon substrate as indicated in Fig. 10.20.

A bonded sample was also immersed in color ink to conduct the leakage test of liquid. Figure 10.21 shows that no leakage occurred for the 1 h test. The test results show that this bonding approach can be exploited to protect the micro-machined structures during the dicing process. The results also indicate that the step height of the conducting wires was perfectly covered by the electroplated Ni/Co alloy.

In addition, the rough surface of the electroplated Ni/Co was fully filled with the solder after reflow. A helium leakage test is required to further demonstrate the hermeticity of the packaged device. Figure 10.22(a) shows the driving test of the surface micro-machined mirror after bonding. The driving ac signal was imported from the bond pad via the feedthrough cable to the mirror surrounded by spacer. The typical measured frequency response of the mirror driven by a 150 V voltage is shown in Fig. 10.22(b). The test results demonstrate that the feedthrough wire successfully transported the driving voltage to the mirror isolated by the bonding ring.

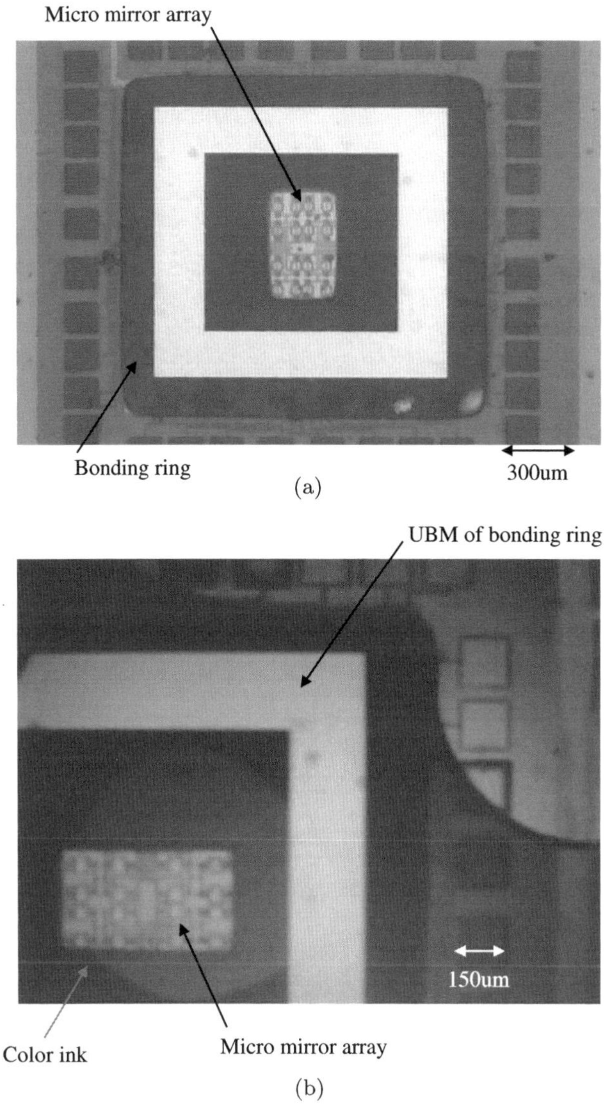

Fig. 10.21. The liquid leakage test for the chip after the 1 h test using colored ink.

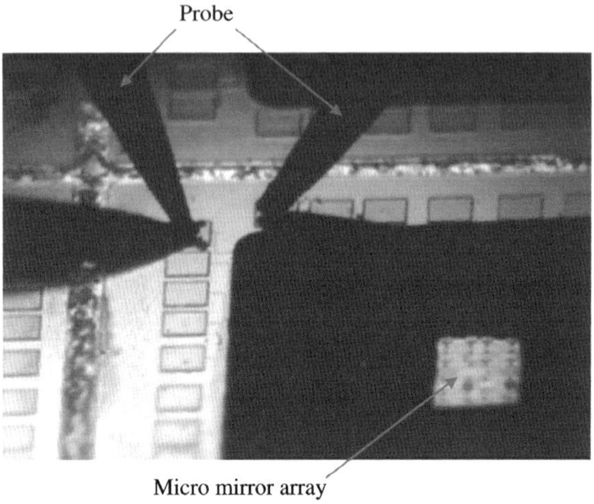

(a)

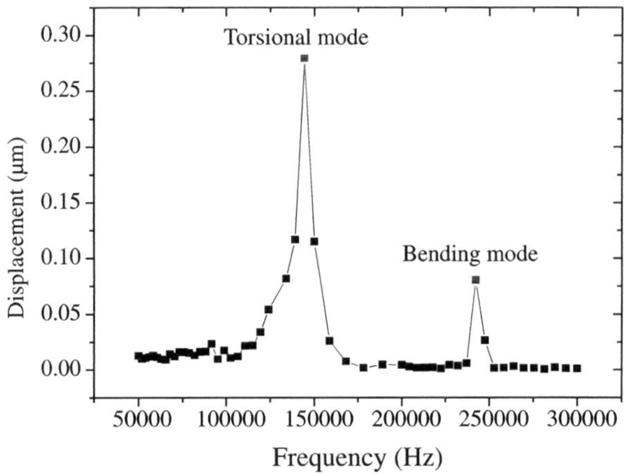

(b)

Fig. 10.22. (a) The driving test of the surface micro-machined mirror after bonding and (b) the typical measured frequency response.

10.5 Conclusion

The localized induction heating and solder reflow bonding assisted WLP has been successfully demonstrated in this chapter. The experimental results show that it took less than 1 min to complete the bonding process. Implementation of this localized heating technique had been successfully demonstrated in flip-chip technology and MEMS wafer capping. Flip-chip assembly of silicon substrate with copper pillar bump had been achieved successfully. With the further integration of TSV inside silicon substrate, this will be a promising technique to achieve higher density interconnection with enhanced reliability since heat is localized where nickel is located. As for MEMS wafer capping, the MEMS devices experienced a temperature of only 110°C during bonding, low enough so that the thin film materials would not be damaged. Moreover, the bond strength between silicon and silicon wafer was found to be higher than 18 MPa. In summary, four advantages are available through this approach. First, the magnetic induction coil will only heat up magnetic materials, such as nickel and cobalt, to a very high temperature within several seconds. Hence, localized heating is realized. Second, the low-temperature electroplating process can integrate with various fabrication processes, such as MUMPs and CMOS. For the advanced IC package process such as wafer bumping or WLCSP, the process of material composition does not change. Third, the electroplated nickel and cobalt alloy can also perform as a spacer for MEMS devices. Fourth, the step height of the feedthrough wire is sealed by the electroplated film, and the flatness and roughness of the electroplated surface are recovered by the solder reflow. Thus, the package for preventing water leakage could be achieved.

References

1. Hagge, J. and Wagner, R. (1992). High-yield assembly of multichip modules through known-good IC's and effective test strategies, in *Proceedings of the IEEE*, pp. 1965–1994.
2. Siu, W. (1992). MCM and monolithic VLSI: perspectives on dependencies, integration, performance, and economics, in *Proceedings of the. IEEE Multi-Chip Module Conference' 92, IEEE Press*, Piscataway, N.J.
3. Tummala, R. R. (1992). Multichip packaging — A tutorial, in *Proceedings of the IEEE*, 80, pp. 1924–1941.
4. Oh, B. H., Lee, E. K., Loo, H. Y., and Oh, P. T. (2006). Package routing and thermal analysis on a multi-chip package, in *International. Conference. on Electronic Materials and Packaging*, pp. 1–5.
5. www.amd.com
6. 2009 iNEMI Industry Roadmap, Component and Subsystem Technologies Section.
7. Yoshida, A., Taniguchi, J., Murata, K., Kada, M., Yamamoto, Y., Takagi, Y., Notomi, T., and Fujita, A. (2006). A study on package stacking process for package-on-package (PoP), ECTC, San Diego.
8. Kada, M. and Smith, L. (2000). Advancements in stacked chip scale packaging (S-CSP), provides system-in-a-package functionality for wireless and

handheld applications, *Pan Pacific Microelectronics Symposium Conference*, Maui, Hawaii.

9. Karnezos, M. (2004). Package-in-Package (PiP): A new 3-D Module for Wireless Applications, MEPTEC Symposium on SIPs or SOCs? The Multi-million Dollar Question, in *Proceedings, Santa Clara*, CA.
10. Bantog, E *et al.* (2006). Wire bond, flip chip and chip scale package solution to high silicon integration, in *Proceedings of the 56th Electronic Components and Technology Conference*, San Diego, CA, pp. 1386–1391.
11. http://www.skhynix.com
12. http://www.spil.com.tw
13. http://www.aseglobal.com
14. Lau, J. H. (2011). TSV interposer: The most cost-effective integrator for 3D IC integration, *ASME Interpack*, pp. 53–64.
15. Rao, V. S. (2009). TSV interposer fabrication for 3D IC packaging, *IEEE 11th Electronics Packaging Technology Conference.*
16. http://www.yole.fr
17. https://www.invensense.com
18. Chino, M. (2002). Reliability Improvement of 4-stacked FBGA Package, *International Conference on Electronics Packaging*, pp. 339–344.
19. Yoshida, A. and Ishibashi, K. (2002). An extremely thin, BGA format chip-scale package and its board level reliability, *Electronic Components and Technology Conference*, pp. 1335–1340.
20. http://www.toray-eng.com
21. Sood, S., Farrens, S., Pinker, R., Xie, J., and Catabay, W. (2010). AI-Ge eutectic wafer bonding and bond characterization for CMOS compatible wafer packaging, *218th Electrochemical Society Meeting*, 33(4), pp. 93–101.
22. Crnogorac, F., Birringer, R., Dauskardt, R. and Pease, F. (2009). Aluminum-Germanium eutectic bonding for 3D integration, *IEEE International Conference on 3D System Integration*, pp. 1–5.
23. Fung, C. D., Cheung, P. W., Ko, W. H., and Fleming, D. G. (eds) (1985). *Micromachining and Micro Packaging of Transducers* (Amsterdam: Elsevier).
24. Kovacs, G. T. A. (2000). *Micromachined Transducers Sourcebook* (New York: McGraw-Hall).
25. Cheng, Y. T., Hsu, W. T., Najafi, K., Nguyen, C. T. C., and Lin, L. (2002). Vacuum packaging technology using localized aluminum/silicon-to-glass bonding, *Journal of. Microelectromechanical. Systems.*, 11, pp. 556–565.
26. Cao, A., Chiao, M., and Lin, L. (2002). Selective and localized wafer bonding using induction heating, *Technical Digest of Solid-State Sensors and Actuators Workshop* (*Hilton Head Island, NC, June 2000*), pp. 153–156.
27. Stansel, N. R. (1949). *Induction Heating*, New York: McGraw-Hill.
28. Guy, A. G. (1976). *Essentials of Materials Science*, New York: McGraw-Hill.
29. Hagemaier, D. J. (1990). *Fundamentals of Eddy Current Testing*, Colombus, OH: The American Society for Nondestructive Testing, Inc.

Hsueh-An Yang was born in Taipei, Taiwan, R.O.C., in 1975. He received the Ph.D. degree from National Tsing Hua University, Hsinchu, Taiwan, in 2006. His thesis was focused on the analysis and applications of electromagnetic eddy current for MEMS.

Since 2006, he has been with the Advanced Semiconductor Engineering Group, Kaohsiung, Taiwan, which is the world's biggest chip packager and assemble house, as a department manager. He was in charge of various projects such as 2.5D/3D package, TSV process development, wafer bumping, WLCSP, Fan-out WLCSP, and MEMS package. In 2010, he joined TSMC as a technology manager for MEMS WLCSP process development. He is now working at Opus Microsystem, Taipei, Taiwan, and involved in MEMS actuator design, process and package development for pico projector application.

Chiung-Wen Lin received his Ph.D. degree from National Tsing Hua University (Taiwan) in 2010. His research interests including MEMS 3D integration, micromachining technology and MEMS sensor design. After receiving his Ph.D. degree, he joined Sitronix Technology Corp. (Taiwan) where he was a senior MEMS designer for MEMS accelerometer. Two years later, the sensor department was spun off as Sensortek Technology Corp. (Taiwan) after successful production of the MEMS accelerometer. In 2013, he joined Richtek Technology (wholly owned subsidiary of MediaTek) as Principal MEMS designer for MEMS gyroscope. He is now with Pixart Imaging Inc. as a manager for sensor technology development.

Weileun Fang received his Ph.D. degree from Carnegie Mellon University (Pittsburgh, PA) in 1995. He joined the National Tsing Hua University (Taiwan) in 1996, where he is now a Chair Professor. His research interests include MEMS with emphasis on micro fabrication/packaging technologies, CMOS MEMS, micro transducers, and characterization of thin film mechanical properties. Prof. Fang is now the Chief Editor of *Journal of Micromechanics and Microengineering.* He also served as the TPC of IEEE MEMS conference, the ETPC of Transducers conference, and the TPC chair of IEEE Sensors Conference. He has become the ISC member of Transducers from 2009, and is the General Chair of Transducers 2017. Moreover, he also serves as a technical consultant for many MEMS companies in Taiwan. Dr. Fang is a member and Fellow of the IEEE, and the IOP (Institute of Physics, UK).

Chapter 11

Localized Sealing Schemes for MEMS Packaging

Y. T. Cheng, Y. C. Su and Liwei Lin

11.1 Introduction

Microelectronics packaging can protect chips from harsh environment for long lifetime and also facilitate communications for better system performances. Heterogeneity is the main feature of microelectromechanical systems (MEMS) chips as introduced in previous chapters. In MEMS packaging, hermetic or vacuum sealing plays a critical role to provide special environments for micromechanical components to achieve high quality factor, Q, for resonator-based systems and/or good reliability [1, 2]. As such, complicated MEMS packaging structures and processes are inevitable to accomplish the packaging requirements and the result is high packaging cost.

In the past decades, varieties of MEMS packaging techniques have been developed for micro-encapsulation, which has been recognized as the main embodiment of the zero-level packaging approach Basically, micro-encapsulation can be realized by two kinds of packaging approach: (1) capping a MEMS substrate by bonding another silicon or glass wafer micromachined with micro-cavities; (2) sealing the substrate directly by thin film deposition. Low temperature and short processing time are desirable specifications in the developments of these micro-encapsulation technologies for less thermal budget and high throughput. Meanwhile, micro-capping is usually carried out after the CMOS and MEMS device fabrication process, which should take into the consideration of the process compatibility and effects, such as thermal stress, interface degradation, outgassing contamination, etc. To relieve the process predicament, localized sealing schemes including wafer level permeable hole seal [1, 3, 15], localized heating and encapsulation [16–29], localized reactive bonding and capping [30–33], etc.

have been developed and will be introduced in this chapter. Unlike the conventional electronic packaging practice, these local encapsulation techniques have disclosed a possible path to diminish the stress and compatibility issues resulting from the choice of packaging material and process on as-fabricated MEMS devices due to the special characteristics of "localization".

11.2 Principle of Localized Processing

11.2.1 *Localized Sealing*

"Localized Processing" in MEMS micro-encapsulation can refer to either sealing locally or bonding locally. "Sealing locally" means that a process to hermetically seal local areas for housing the MEMS devices with a controlled environment is formed after the sacrificial layer in the MEMS process is removed as shown in Fig. 11.1 [16]. The fabrication of the microcaps typically comprises three major process steps in sequence: the deposition of a capping layer to encapsulate MEMS devices without removing the sacrificial materials; formation of the permeable holes followed by micromechanical

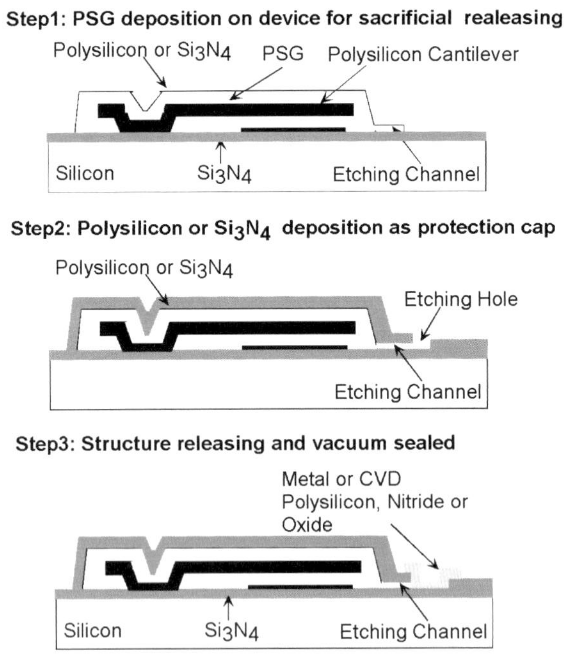

Fig. 11.1. Schematic process flow concept of "sealing locally" [16].

structure release using chemical etching to remove the sacrificial materials; and local sealing of the permeable holes to form the microcaps. The capping layer can be CVD-deposited polysilicon, silicon nitride, electroplated Ni etc., and its thickness can be up to tens of micrometers for mechanical robustness. The permeable holes are opened in the capping layer to provide the path for etchant to etch away the sacrificial materials and to release the encapsulated MEMS structures. The holes are later sealed by depositing another dielectric or metal layer The deposition condition will determine the capping environment inside the shells at the end. Low processing cost can be achieved by taking advantage of wafer-level processing.

In contrast, "bonding locally" can be regarded as post-MEMS packaging, which is done after all MEMS device fabrication including the removal of the sacrificial materials for the release of MEMS structures [16–20, 31–37]. To accomplish the micro-encapsulation package, devices are protected by bonding another micro-machined silicon or glass substrate as illustrated in Fig 11.2 [16]. Low packaging cost can be expected due to the wafer-level process. Because devices and protection caps are fabricated on two different wafers, the capping scheme can have more flexibility in terms of the MEMS fabrication processes. Nevertheless, successfully assembling the substrates together to achieve the final encapsulation requires a good bonding technique, which must provide not only a strong and reliable bond between the two substrates but also minimal processing side effects such as mechanical stresses. Sufficient thermal or mechanical energy is needed to trigger the chemical reaction to form a strong bond. As a result, bonding at high temperature is inevitable, especially for a short processing time to reach the same bonding quality as compared with a similar process conducted at a low bonding temperature. In order to mitigate the thermal effects, localized bonding techniques have been developed in the recent two

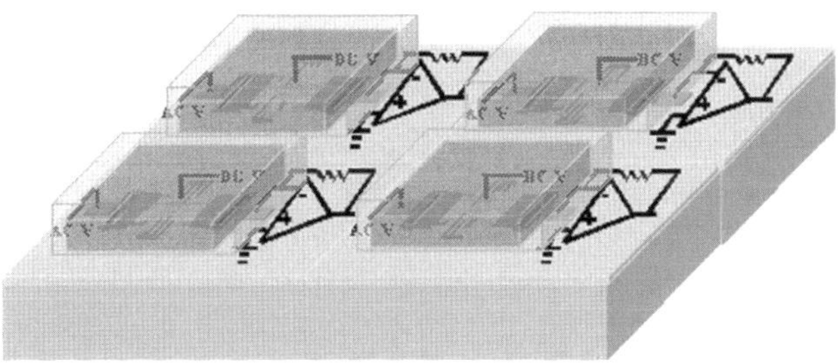

Fig. 11.2. Post-process packaging using wafer bonding techniques to assemble the device substrate with the micro-machined cap substrate [16].

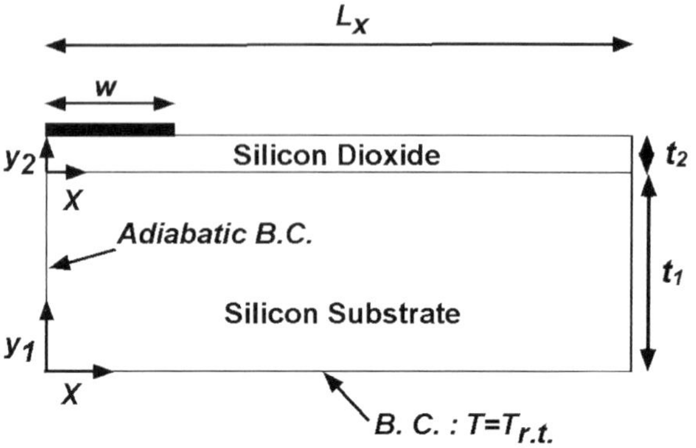

Fig. 11.3. The schematic 2D thermal model, geometry and boundary conditions (B.C.).

decades for micro-encapsulation process to prevent the MEMS devices and circuits from high temperature processing damages.

11.2.2 *Localized Heating and Bonding*

Considering a die structure where a micro-heating element is devised on top of a layer of silicon dioxide deposited on a silicon substrate as shown in Fig. 11.3, the temperature distribution around the heating zone can be calculated by solving the two dimensional (2D) governing heat conduction equations as follows:

$$\alpha_1 \left(\frac{\partial^2 T_1}{\partial x^2} + \frac{\partial^2 T_1}{\partial y_1^2} \right) = 0 \tag{11.1}$$

$$\alpha_2 \left(\frac{\partial^2 T_2}{\partial x^2} + \frac{\partial^2 T_2}{\partial y_2^2} \right) = 0 \tag{11.2}$$

These equations are solved under steady state with constant heat flux and adiabatic boundary conditions, so the corresponding temperature profiles, T_1 and T_2, are derived as follows:

$$T_1(x, y_1) = \sum_0^{\infty} A_n(n) \cos\left(\frac{n\pi}{L_x} x\right) \sinh(Q_n y_1) + T_{r.t.} \tag{11.3}$$

$$T_2(x, y_2) = \sum_0^{\infty} \cos\left(\frac{n\pi}{L_x} x\right) [B_n \cosh(Q_n y_2) + C_n \sinh(Q_n y_2)] + T_{r.t.} \tag{11.4}$$

$$A_n = \frac{qR_n}{k_2 Q_n \left[\sinh(Q_n t_1)\sinh(Q_n t_2) + \frac{k_1}{k_2}\cosh(Q_n t_1)\cosh(Q_n t_2)\right]} \tag{11.5}$$

$$B_n = A_n \sinh(Q_n t_1) \tag{11.6}$$

$$C_n = \frac{k_1}{k_2} \cosh(Q_n t_1) A_n \tag{11.7}$$

$$R_n = \frac{w}{L_x}(n = 0) = \frac{2}{n\pi} \sin\left(\frac{n\pi w}{L_x}\right) \tag{11.8}$$

$$Q_n = \frac{n\pi}{L_x} \tag{11.9}$$

where $T_1, T_2, \alpha_1\alpha_2, t_1, t_2, k_1$, and k_2 are the temperature, thermal diffusivity, thickness and thermal conductivity of silicon substrate and silicon dioxide respectively; $T_{r.t.}$ is the bottom temperature of the substrate; q is the heat flux density at the die-surface; w is the width of the heating element, and L_x represents the die size. Because the thermal conductivity of silicon is about 100 times higher than that of silicon dioxide, the temperature at the point right below the heating element and at the interface of silicon dioxide and silicon substrate, i.e. T_2 (0,0), or T_1 $(0, d_1)$, can be approximated as follows:

$$T_2(0,0) = T_1(0, t_1) \approx \frac{q}{k_1}\left(\frac{wt_1}{L_x}\right) + T_{r.t.} \cong \frac{Q}{k_1}\left[\frac{t_1}{L_x}\right] + T_{r.t.} \tag{11.10}$$

where Q is the total power input per unit length onto a micro-heater. As a result, the temperature of silicon substrate, T_1 $(0, t_1)$, can be kept at low or close to room temperature as long as the thickness of silicon dioxide is much smaller than the die size and a large heat sink is placed underneath the silicon substrate to constrain $T_{r.t.}$ to the ambient temperature.

Based on the analysis, localized heating and bonding can be easily accomplished by applying a micro-heating source on the aforementioned structure to trigger bonding reactions for the assembly of two substrates together. The source can be generated by resistive heating, inductive heating, lasing, etc. Figure 11.4 shows a 2D heat transfer simulation of a hermetic package using Ansys Software. A polysilicon resistive micro-heater is devised to provide thermal energy for the aluminum/silicon to glass bonding in the package. As the prediction of the prior theoretical analysis, most of the heating region is locally confined and the silicon substrate can be kept at low temperature when the bottom of the substrate is constrained at room temperature by closely contacting a good cooling chuck. The temperature of aluminum/(silicon) solder is close to the micro-heater temperature even though there is heat conduction loss along the electrical

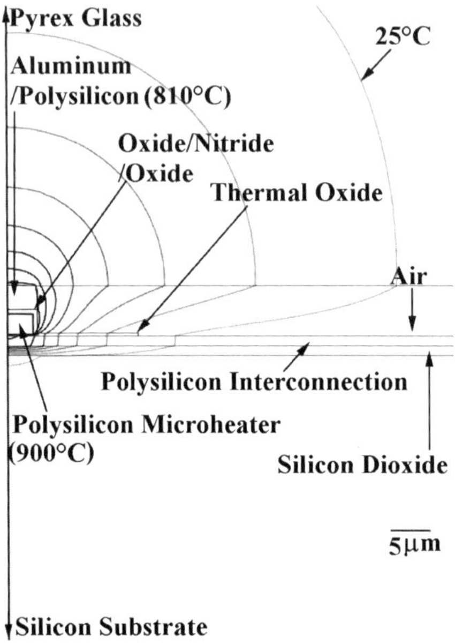

Fig. 11.4. A 2D heat transfer simulation of the localized heating and bonding package using Ansys FEA (Finite Element Analysis) [17].

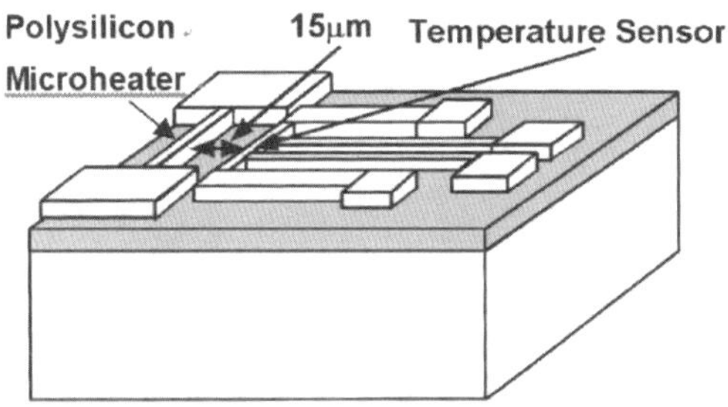

Fig. 11.5. The surrounding temperature measurement of a polysilicon micro-heater using line-shaped polysilicon temperature sensor.

insulation layer. The localized resistive heating concept has been verified experimentally. Figure 11.5 shows a tested polysilicon four-point resistance measurement structure as a temperature sensor 15 μm away from a 5 or 7 μm wide line-shape polysilicon micro-heater. According to the linear temperature dependence of polysilicon resistance [38], line-shaped polysilicon was used as the temperature sensor for measuring the temperature

Table 11.1. The resistivity change of the polysilicon temperature sensor.

Input Current (5 μm wide Polysilicon Micro-heater)	0 mA (~25°C)*	20 mA (~340°C)*	30 mA (~1300°C)*	
Sensor Resistivity (10^{-4}, Ω-cm)	9.4 ± 0.3	9.7 ± 0.3	9.7 ± 0.3	
Input Current (7 μm wide Polysilicon Micro-heater)	0 mA (~25°C)*	20 mA (~340°C)*	41 mA (~980°C)*	46 mA (~1300°C)*
Sensor Resistivity (10^{-4}, Ω-cm)	9.9 ± 0.4	9.5 ± 0.4	9.2 ± 0.4	9.3 ± 0.4

*The heater temperature is estimated based on a 1D electrothermal model [16].

surrounding a micro-heater. No drastic resistivity change from the temperature sensor was measured while the micro-heater was heated up over 1000°C as listed in Table 11.1. It was noted that every 100°C temperature rise on the temperature sensor should transform to a resistivity change up to 12% of the polysilicon temperature sensor. These results further assured that the heating region is confined locally at no more than 15 μm from the heating source.

11.2.3 *Selective Heating and Bonding*

Instead of resistive heating, laser heating is another way to realize localized heating [19–29]. Following the principle of laser welding, localized laser heating and bonding are typically utilized to assemble two substrates, which are transparent and absorbing the employed laser irradiation, respectively. Since the bonding takes place only in the regions absorbing laser irradiation, the laser heating and bonding technique can be referred to one kind of selective bonding processes. Different from the localized resistive heating, the laser or inductive heating generate a local hot spot for bonding during the transient responses. In principle, localized laser heating can be analyzed by solving the governing equations as follows:

$$\rho C \frac{\partial T}{\partial t} = k\left(\frac{\partial^2 T}{\partial r^2} + \frac{1}{r}\frac{\partial T}{\partial r} + \frac{\partial^2 T}{\partial z^2} + I(r,t,z)\right) \tag{11.11}$$

where ρ, C, t, and k are mass density, specific heat temperature, thermal diffusivity, time and thermal conductivity of silicon substrate and silicon dioxide respectively, r and z are the coordinate along the radius and vertical directions, and $I(r,z,t)$ represents the laser power intensity distribution.

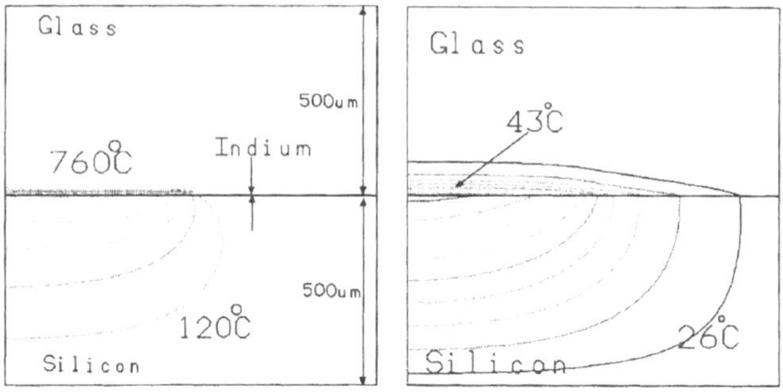

(a) (b)

Fig. 11.6. Simulated isothermal results regarding the transient response of nanosecond laser heating and bonding: (a) after 1 μs and (b) after 1 ms [19].

Thus, via the control of laser pulse duration and repetition, and power, the bonding temperature and region can be selectively defined according to the heat absorption and conduction properties of the two substrates. Prior studies have utilized silicon substrate as a heating element for direct bonding and employed an intermediate layer, such Au, In, BCB, glass frits, etc. as the bonding agent to assemble silicon and glass substrates together. In addition, metal masking can be also used as a photomask to define the bonding region.

Figure 11.6 showed two simulated isothermal results regarding the transient response of nanosecond laser heating and bonding where indium is used as the intermediate layer for glass to silicon bonding [19]. In this case, laser power is 22 mJ with a beam size of 1 mm and partly irradiated on the indium film. It was estimated that the initial temperature at the bonding interface is about 2500°C and then become 760°C and 43°C after 1 μs and 1 ms, respectively, due to the rapid thermal heat dissipation along the silicon substrate to environment. The results disclosed two critical points: the high temperature region could be confined locally and only last for microseconds. Unlike the prior localized resistive heating, which is used for chemical bonding at a steady state, localized laser heating will not be suitable for diffusion bonding.

Inductive heating and bonding can also be regarded as a selective heating and bonding technique [20]. An induction coil is driven by a high frequency AC current to generate a time-variant magnetic field where a conductive substrate can be heated up due to the formation of eddy current. More heat can be generated using ferromagnetic materials such as nickel due to the eddy current losses and hysteresis losses. Since silicon is a semiconductor material and the inductive coil does not require to contact the silicon substrate, the usage of ferromagnetic materials can be exploited for the realization of localized heating in the MEMS chips. The details regarding the inductive heating and bonding can be found in Chapter 12.

11.2.4 *Reactive Bonding*

In fact, the thermal energy to trigger the chemical bonding process not only comes from the external heating sources as mentioned before but also originates from its exothermic reaction [30–33]. The enthalpy change of bonding reactions like the exothermic reaction of a multilayer foil, Ni/Al, triggered by spark or laser impulse can result in a great amount of heat to facilitate more atomic diffusion and compound formation. The reactive reaction system can be categorized into three energetic systems, which are small, medium, and high with the corresponding formation heat smaller than 40 kJ/mol-atom, in between 40 and 80 kJ/mol-atom, and above 80 kJ/mol-atom, respectively [30, 33]. To make the concept more practical for real packaging applications, it is necessary that the heat generated by atomic diffusion is faster than that removed by thermal diffusion and the reaction will become self-propagating as shown in Fig. 11.7. Moreover, the multilayer foils can act as a localized heating source to melt solder for micro-encapsulation in a second or less. The MEMS susceptible to magnetic field can also be encapsulated using this method.

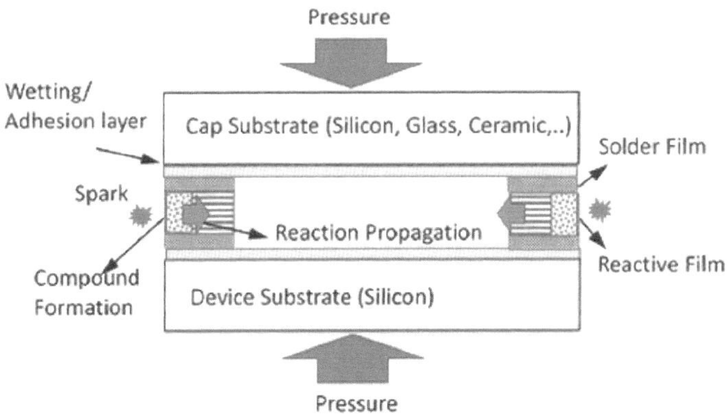

Fig. 11.7. Scheme of micro-encapsulation formed by reactive heating and soldering, where the reactive film is triggered by a spark to start a self-propagating exothermic reaction and the solder absorbing the released heat will form a bond with the two substrates.

11.3 Localized Sealing Techniques

11.3.1 *Wafer-Level Permeable Hole Sealing Techniques*

The capping concept was firstly proposed by Mastrangelo *et al.* [5] to fabricate vacuum encapsulated light source. Owing to the characteristics of thin film process in wafer level, varieties of MEMS devices such as microresonators, filters, pressure sensors, etc., have been developed based on the technique. It is required to have permeable holes to release the encapsulated device structures in the technique. As a result, their working environments were determined by the method to seal its permeable holes. Figure 11.8 shows a broken vacuum-encapsulated micro-resonator where a $\sim 2\,\mu$m thick low stress LPCVD (Low Pressure Chemical Vapor Deposition) silicon nitride film was deposited to form the micro-shell [1]. Measurement results showed that the quality factor of the encapsulated 24 KHz comb-driven resonator is about 2200 corresponding to a pressure of $\sim$300 mTorr, which is the processing pressure for another layer of $1\,\mu$m thick nitride deposition to seal the permeable holes in the shell. The nitride micro-shell would also raise two design issues: structural deposition and collapse. The resonant frequency of the resonator will shift due to the deposition of a thin nitride layer, i.e. 1500 Å, during the sealing process. Resulting from the pressure difference between ambient and the inner cavity, there was a 2–$4\,\mu$m concave deflection at the center of the $\sim 3.5\,\mu$m-thick and 300–$400\,\mu$m in length nitride shells. Recent progress in the development of the encapsulated technique has focused on the improvement of the

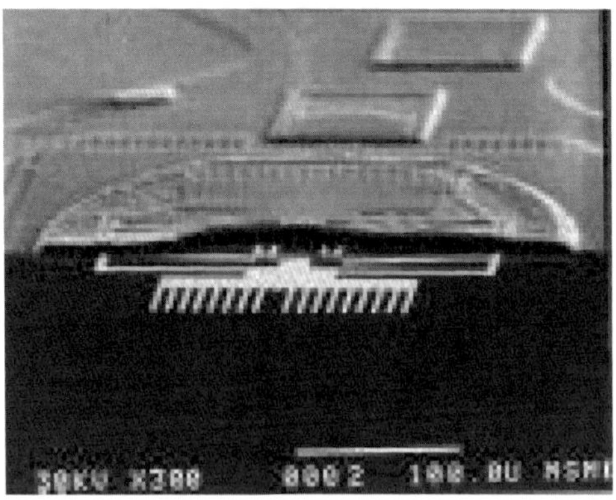

Fig. 11.8. SEM image of a broken microencapsulated micro-resonator fabricated using the wafer-level permeable hole sealing techniques.[1]

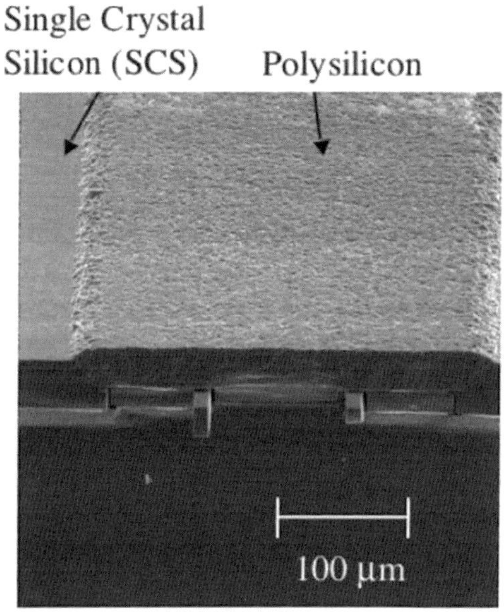

Fig. 11.9. SEM view on the cross section after deposition of encapsulation, before CMP. The step height between polysilicon and single crystal silicon resulting from an oxide step and different deposition rates for single and polycrystalline silicon [11].

micro-shell rigidity and the employment of different capping and sealing materials to resolve the issues.

Figure 11.9 shows cross sectional SEM images of encapsulated silicon resonators where a 20 μm thick polysilicon layer is deposited to form a microcap in a pressure less than 1 Pa [11]. The resonator was first fabricated on a silicon-on-insulator (SOI) substrate followed by similar wafer-level permeable hole sealing process. Thickening the capping structure can make the structure strong enough to survive in the following packaging steps including CMOS frontend and backend fabrication, wafer dicing, die handling, and injection molding of plastic, etc. In this technique, the thick polysilicon deposition was performed at high temperature (950°C) in a low-pressure reactor with only hydrogen which not only results in epitaxial silicon deposition on certain areas of the device substrate as shown in the figure but also makes the environment inside the encapsulation more clean and stable. The epitaxial single crystal silicon can also facilitate the fabrication of integrated circuitry beside the resonator after the micromechanical device fabrication. As aforementioned, the environment inside the microcap was controlled by the sealing process and the residual gas remaining in the cavity was hydrogen. Since silicon is permeable to hydrogen molecules, which diffuse out of the cavity during annealing in a nonhydrogen ambient, this technique has been further modified by the deposition of a 2 μm polycrystalline 3C silicon carbide (poly-SiC) film on the thick polysilicon layer as a poly-SiC composite layer to hermetically seal MEMS devices [14]. Via the non-permeable characteristic of the poly-SiC layer, the pressure inside the packaging can be controlled from 447 Pa to 15.5 kPa while it was thermally annealed at 400°C in an ambient with different partial pressure of the hydrogen. Besides, owing to the structure deposition with the poly-SiC thin film during the sealing process and the characteristics of high chemical resistivity, low stiction and friction, and high wear resistance, the poly-SiC encapsulation can offer stable packaging for the resonators with low frequency change and no hysteresis of the frequency change over tested time and temperature range.

Instead of high temperature LPCVD silicon deposition, electroplating nickel as the capping material could provide better process flexibility. Figure 11.10(a) shows a hermetically encapsulated Pirani gauge with polysilicon packaging feedthroughs where a 40 μm thick electroplated Ni film was used as the protection cap rigid enough to sustain the pressure difference between the ambient and the cavity inside [10]. The technique utilized PR (photoresist) and polysilicon as sacrificial materials, respectively, which were both etched in TMAH and dried in supercritical CO_2. To accomplish the hermetic encapsulation, thin film processing, localized welding, sputtering/evaporating, and solder bumping can be applied for the sealing process that requires a special design in the fluidic access port, i.e.

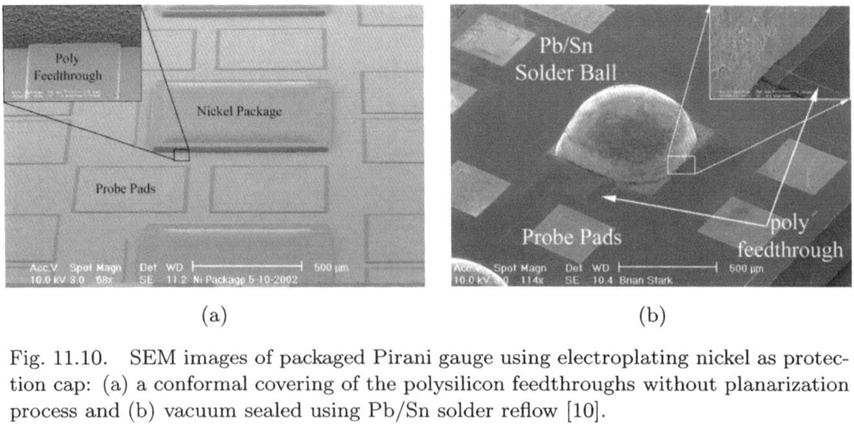

(a) (b)

Fig. 11.10. SEM images of packaged Pirani gauge using electroplating nickel as protection cap: (a) a conformal covering of the polysilicon feedthroughs without planarization process and (b) vacuum sealed using Pb/Sn solder reflow [10].

permeable holes, for the MEMS structure release process in the technique. For instance, as shown in Fig. 11.10(b), Pb/Sn solder balls are used to encapsulate the package. After the release of the Pirani gauge inside the Ni cap, 63Sn/37Pb solder paste was stenciled over the package using a custom stainless steel microstencil, and then the sealing was performed by heating the device substrate over the soldering temperature in a vacuum chamber to wet the solder over the metal package to seal the access ports.

Line-of-sight deposition of amorphous SiC using ion beam sputtering deposition is a method to prevent the deposition of sealing material in the CVD process onto the MEMS structure. By tilting the device substrate for "off-normal" deposition in the ion beam sputtering system, the permeable holes can be sealed by amorphous SiC film without visible down-hole deposition as shown in Fig. 11.11 [12]. Experimental results show that the SiC film was accompanied with compressive intrinsic stresses, which was

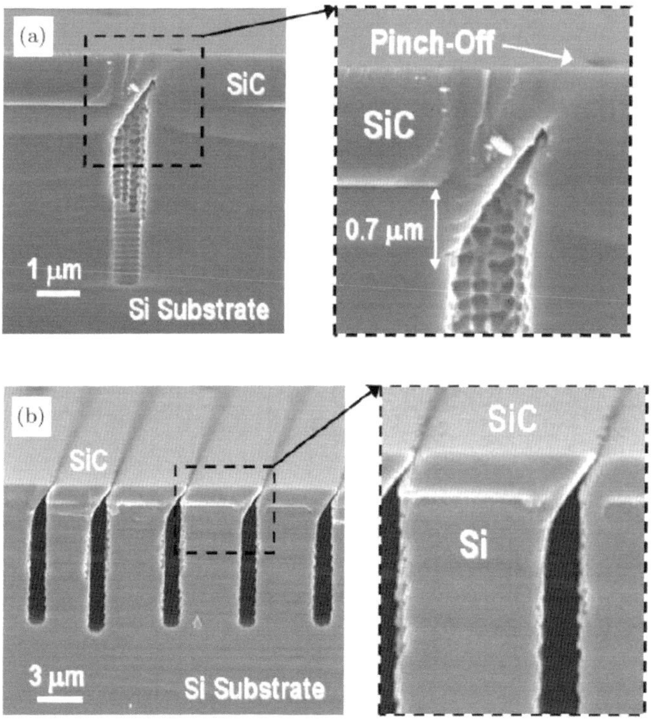

Fig. 11.11. SEM images of the amorphous SiC sealing layer topography for an angle of incidence of 50° light-of-sight deposition: (a) etched holes and (b) etched trenches [12].

dependent on the incidence angle of sputtered atoms, about 1250 MPa for an angle of incidence of 50° in this case. Another low temperature sealing method is the utilization of compressed wire-bonded gold bumps [15]. The bumps were placed close to the access ports with partial overlap. Sealing process is then accomplished by pressing the bump into the ports under vacuum. As a result, a bump with a diameter of 90 μm and in height of 50 μm were flattened to cover the port and the cavity connected to the port can keep a vacuum working environment for the MEMS devices. The corresponding leak rate for the sealed cavity is less than 2.9×10^{-14} mbarL/s indicating no measurable external leak. Owing to the low thermal budget, cost effective manufacture, and no structure deposition with sealing material, this vacuum sealing method also showed its potential for hermetic MEMS packaging application.

11.3.2 *Localized Heating and Bonding Encapsulation*

Localized heating and bonding is a way to facilitate all kinds of chemical bonding formation for MEMS packaging without having any possible thermal side effects. Figure 11.12 shows a SEM micrograph of a vacuum encapsulated resonator surrounded by a 30 μm wide microheater with aluminum/silicon bonding layer on top [16, 18]. A Pyrex glass cap with a 10 μm deep recess was placed on top with an applied pressure of ~0.2 MPa under a 25 mtorr vacuum environment, and the heater was heated using 3.4 watts input power. The results showed that a strong glass–/silicon–glass bond

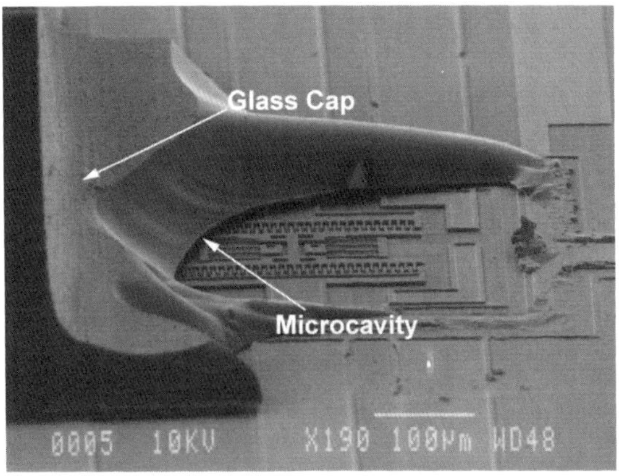

Fig. 11.12. SEM microphoto of encapsulated μ-resonators after the glass cap is forcefully broken [16].

is formed between the Pyrex glass cap and device substrate. According to the broken interface of glass and silicon substrate, the bonding strength of glass/silicon-glass is higher than the failure strength of Pyrex glass or silicon which is above 10MPa [39] strong enough for micropackaging. Long-term testing of the vacuum-encapsulated resonator with a quality factor of 9600 has been demonstrated with stable operation for more than a year. This encapsulation technique has successfully accomplished the scheme of post-MEMS zero-level packaging, which requires controllability of the cavity pressure, low temperature processing at the wafer level, excellent bonding strength, low fabrication cost and high reliability.

The resistive heater can not only be incorporated directly in the package but also be devised in the bonding tool for reusable function as schematically shown in Fig. 11.13 [21]. Figure 11.14 shows the hermetic sealed PMMA substrate with about 0.18 μL of water using a mylar film heated by a reusable-type heater cavity [21]. In that case, the external, reusable heater aligned with the cavity on the plastic substrates was used for bonding the plastic film to the plastic substrate with the insertion of a mylar film coated with a thermoplastic polyvinylidene chloride (PVDC) copolymer layer on one surface A higher heating power was required to ensure the penetration

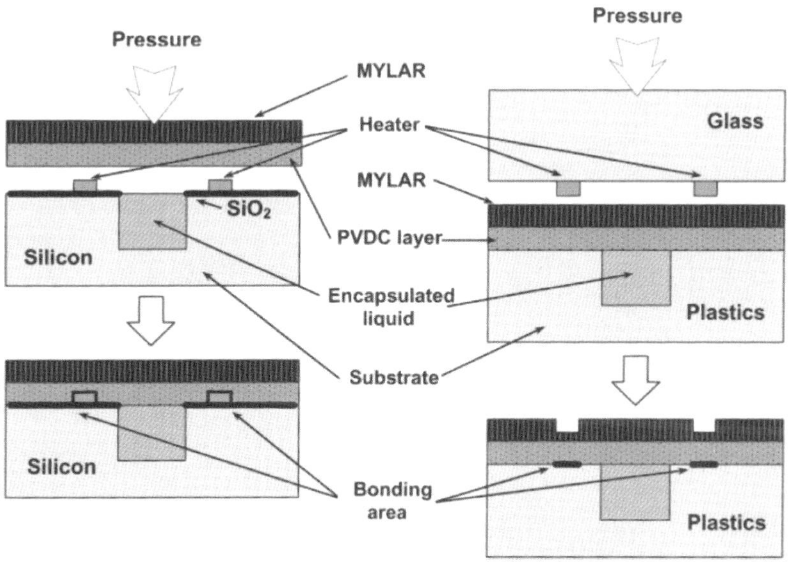

Fig. 11.13. Bonding process scheme: (a) Built-in heater configuration and (b) reusable heater configuration [21].

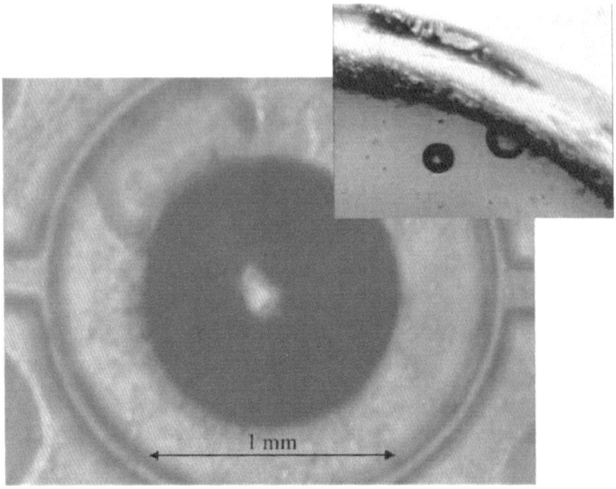

Fig. 11.14. Optical micrograph of the encapsulated water using the localized resistive heated polymer bonding [21].

of heat through the thin film to the bonding interface. Owing to the localized heating characteristics the temperature at the bonding interface could be kept around 150°C, whereas the substrate temperature away from the heaters was under 40°C as long as the aluminum heater was controlled at around 200°C. Loss of water inside the sealed cavity to the environment is only from the evaporation and diffusion through the top mylar membrane and the substrate.

Localized heating and bonding can also be accomplished by the employment of laser. Figure 11.15(a) shows a glass-LCP (liquid crystal polymer)-Si bonded sample by localized laser heating which were immersed in a color solution for 24 h indicating that a good hermetic sealing is accomplished [29]. Figure 11.15(b) shows the corresponding cross-sectional view of the glass-LCP-Si bonding interface. No voids and other observable defects were found due to good bond quality. In-situ temperature monitoring experiment has been performed to verify the localized laser heating and bonding. Figure 11.16 shows the measured temperature during the bonding process of a kovar lid to a leadless chip carrier (LCC) package as shown in the figure using localized laser heating and glass frit bonding process [26]. Three thermocouples were devised on three positions as described in the figure caption: two on the bonding ring for bonding temperature monitoring and one in the center of the package. The results showed that the temperature in the center part was always kept below 140°C which is substantially

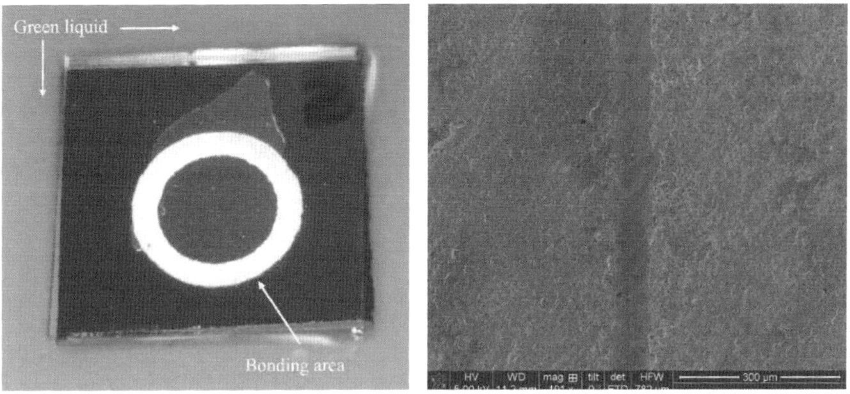

(a) (b)

Fig. 11.15. Optical micrograph of the bonded sample using localized laser glass-LCP-silicon bonding technique: (a) after immersion for 24 h in a color liquid indicating no leakage was found, and (b) SEM micrograph of the bonding interfaces [29].

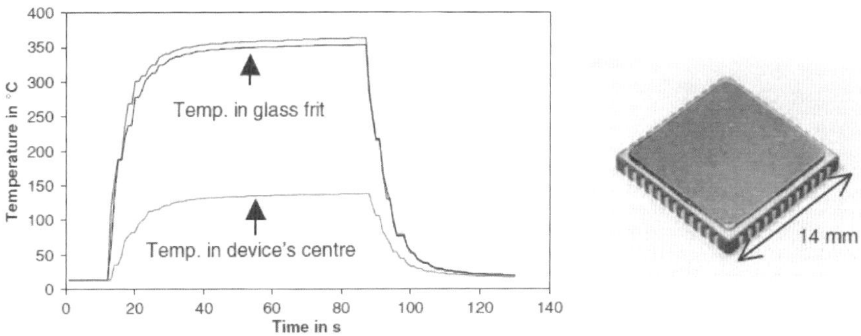

Fig. 11.16. Temperature measurement results plotted as a function of time during the laser bonding process (laser power 75 W, time 75 s). Blue and black traces in the bonding region, Green one on the center of the LCC substrate [26].

lower than the bonding temperature of 375°C for the packaging process revealing the truly localized heating characteristic of the heating process. It is noted that the temperature inside the joining layer is dependent on laser power. A power of 75 W was continuously applied to heat the bonding layer up to 375°C. During the bonding process, the thermal equilibrium of heat energy removal by the copper block underneath the package and the laser energy being absorbed was established, so that the whole experimental setup and device temperature will not be continuously elevated. Therefore, as discussed previously, laser heating and bonding like the resistive heating case still requires an excellent heat sink in good thermal contact with the package to realize the local heating process.

11.3.3 *Localized Reactive Bonding and Capping*

Figure 11.17 shows the cross sectional SEM image and EDX analysis of the Al/Pd multilayers reactive bonding interface before and after the reaction.

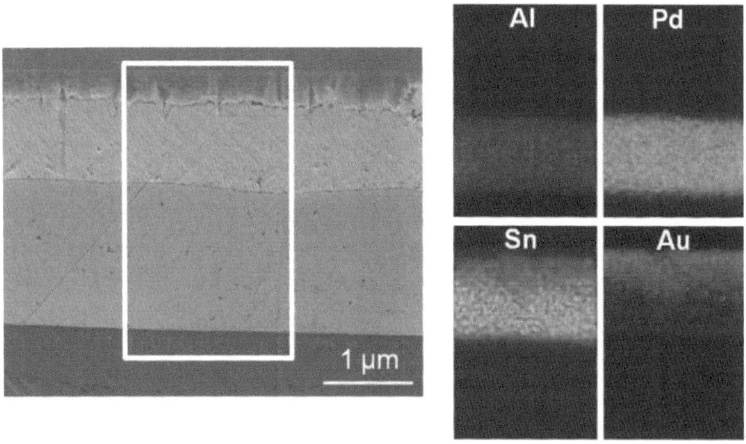

(a)

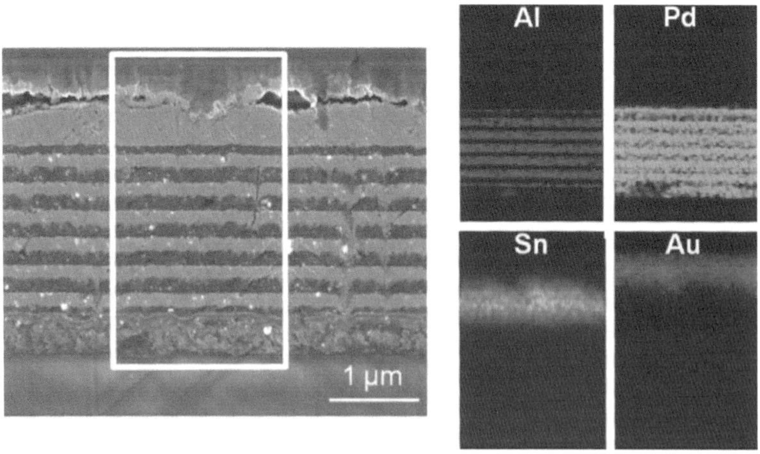

(b)

Fig. 11.17. Cross sectional SEM image and EDX analysis of the Al/Pd multilayers reactive bonding interface: (a) before and (b) after the reaction [32].

The multilayer of Al/Pd was sputtered and covered by the solder layer (Sn) and the wetting layer (Au) on a silicon substrate [32]. After the reactive bonding procedure, Al and Pd were completely intermixed to form one AlPd layer and a layer of Sn/Au was formed to bond the top Si cap with the bottom silicon substrate. Sn and Au are not diffusing into Al and Pd and vice versa. In this case, the exothermic reaction was initiated by a spark generated by a dc power input of 5 V and the released heat melted Sn to form the alloy of AuSn.

The utilization of the reactive heat for soldering two substrate together at room temperature has been successfully applied for hermetic packaging applications. For example, Ni/Al based reactive foil coated with solder material has been used for hermetically sealing an Au-plated stainless steel lid with an Au-plated stainless steel base with as low as 1×10^{-10} atm-cc/s leakage rate of helium gas [40]. Figure 11.18 shows the cross sectional view on a typical Si-to-Si joint using the Ni/Al based reactive foil with two pieces of free standing AuSn solder films [31]. In the process, solder thickness and joining pressure are two critical parameters for accomplishing good seal. If the solder is too thick, the exothermic heat will not be enough to sustain the melting and reflowing processes of the solder for good joint. If the solder is too thin, voids could form in the bonding interface due to the lack of material to fill-in all of the surface defects on the samples like the crack of the foil filled with the solder shown in the figure. The joining pressure

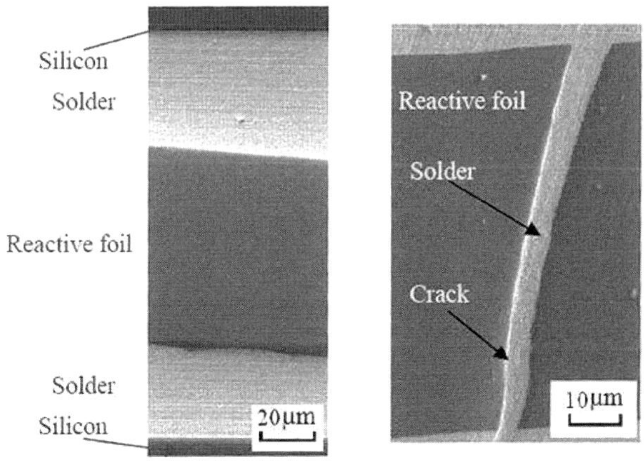

Fig. 11.18. SEM micrograph of two silicon wafers that were bonded using two pieces of free-standing AuSn solder and one Ni/Al reactive foil. There was solder filling in a crack in the reacted foil [31].

loaded on the package during the process is to ensure solder reflowing and wetting for good joint formation, whereas the pressure cannot be so high that excessive solder movement out of the joint area could result in voids in the sealing interface.

The concept of localized reactive heating and bonding has also been applied for wafer-level MEMS packaging applications. The reactive nanoscale multilayered films such as Ni/Al, Al/Pd, etc. can be alternatively deposited at wafer level with the packaging process schematically shown in Fig. 11.7. Meanwhile, the exothermic heat of the reactive film, i.e. Al/Pd in this case, could facilitate the bond formation of Ti/Au–AuAlPd alloy-Al or Ta/TaN/Cu–CuAlPd alloy-Al to hermetically seal the packages comprising of Si–Si, Si–glass as well as Si–ceramic substrates at room temperature [33]. Very strong bond interfaces (maximum shear strengths of 235 MPa) can be achieved without using any additional solder material. The measured helium gas leak rates were below 1×10^{-8} m/bar/s for the bonding rings with a minimal width of 50 μm and the rates could remain constant over 180 days.

11.3.4 *Other Localized Sealing Techniques*

Heat is not the only source to trigger bonding reactions for hermetic microencapsulation. Instead of heating, ultrasonic energy, mechanical compression, UV light, etc. can also be the sources to facilitate the bonding formation [41]. Figure 11.19 shows a hermetically glass-encapsulated capacitive accelerometer using UV curable adhesive. The unique characteristic

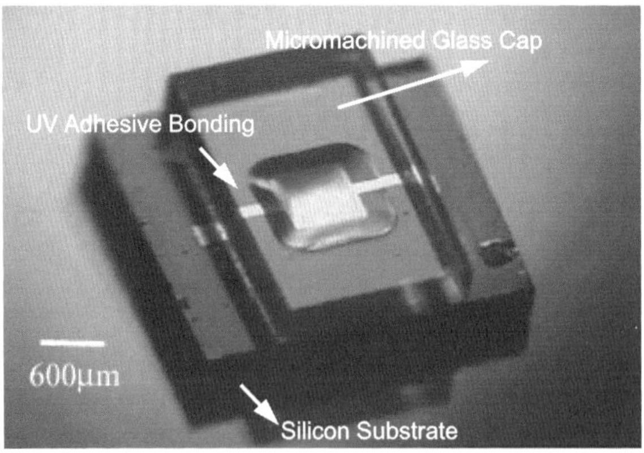

Fig. 11.19. A packaged capacitive accelerometer using UV curable adhesive [41].

of this process is that the bonding interface is formed by the UV-cured adhesive spin-coated onto the glass cap substrate. UV light exposure without heating makes the process more practical for packaging temperature-sensitive devices and the adhesive bond was only locally formed in the pre-defined contact region between the glass cap and device substrate. As a result, low packaging cost with high throughput and high design flexibility can be expected.

Cold welding of overlapping metal rings could create metal-to-metal bonding and sealing. Figure 11.20 shows cross sectional SEM images with (a) Au sealing ring structures that are wedged into each other and (b) enlarged view of the interface of two counter fitting sealing ring structures [42]. The bonding could provide good wafer-to-wafer hermetically sealing at room temperature via the two counter fitting sealing ring structures with slight overlap wedged into each other and the fill of underfill material in the gap of two bonded substrates like Si-to-Si, glass, or SOI. Gross leak testing results showed that the main failure mechanism were localized fractures of the wafers underneath the gold sealing rings due to the excessive shear or compressive stresses resulting from insufficient alignment or too large applied pressure during the bonding process. The cold welding process can also be realized using ultrasonic energy. Multiplication and migration of dislocations in metal resulting from the absorption of ultrasonic energy could effectively reduce shear stress generated in the plastic deformation of metals [43]. As a result, easy slip mechanism takes place in the crystal lattice of metal to cause yield strength reduction. Once metal

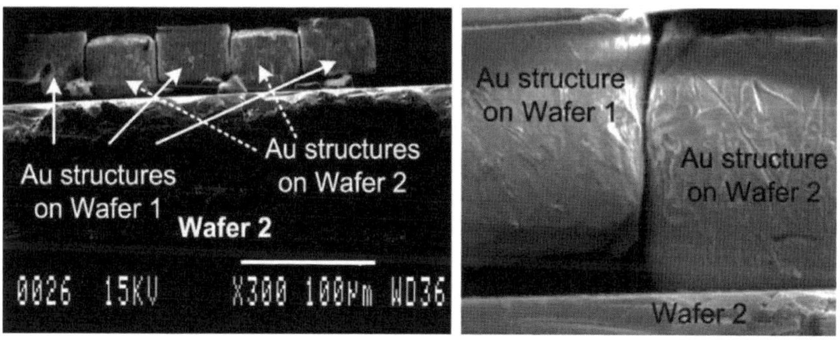

(a) (b)

Fig. 11.20. shows cross sectional SEM micrographs: (a) Au sealing ring structures were wedged into each other and (b) an enlarged view on the interface of two counter fitting sealing ring structures [42].

was sheared and flowed under ultrasonic loading, most of contaminants, such as oxides and dirt, would put aside into debris areas to expose clean areas for direct contact and diffusion for bonding formation. Prior investigation has demonstrated the feasibility of ultrasonic bonding including In-to-Au, Al-to-Al, and plastics-to-plastics for hermetic MEMS packaging applications [44].

11.4 Conclusion

Localized sealing schemes with corresponding process techniques have been thoroughly introduced in the chapter. All of the technologies can be accomplished at wafer level to protect MEMS devices with less detrimental process effects owing to the characteristic of "sealing localization". Since bonding structures, materials, process, and related tooling must be pertinent to the device characteristics, these technologies are still in the research stage only with the demonstration of localization. Further packaging process optimization and the reliability tests such as thermal cycling test, drop test, leakage test, etc. are still required for the commercial applications of hermetic MEMS encapsulation.

Acknowledgments

The works here were supported by the ATU Program funded by MOE in Taiwan and in part by the Novel Bioengineering and Technological Approaches to Solve Two Major Health Problems in Taiwan sponsored by the Taiwan Ministry of Science and Technology Academic Excellence Program under Grant Number: MOST 105-2633-B-009-003, and the projects of MOST 106-2221-E-009-181 and 106-3114-E-009-006.

References

1. Lin, L., Howe, R. T., and Pisano, A. P. (1998). Micromechanical filters for signal processing, *ASME/IEEE Journal of Microelectromechanical Systems*, 7(3), pp. 286–294.
2. Peng, Z., Palego, C., Hwang, J. C., Forehand, D. I., Goldsmith, C. L., Moody, C., Malczewski, A., Pillans, B. W., Daigler, R., and Papapolymerou, J. (2009). Impact of humidity on dielectric charging in RF MEMS capacitive switches. *IEEE Microwave and Wireless Components Letters*, 19(5), pp. 299–301.
3. Guckel, H. (1991). Surface micromachined pressure transducers, *Sensors and Actuators, A* 28(2), pp. 133–146.

4. Ikeda, K., Kuwayama, H., Kobayashi, T., Watanabe, T., Nishikawa, T., Oshida, T., and Harada, K. (1990). Three-dimensional micromachining of silicon pressure sensor integrating resonant strain gauge on diaphragm, *Sensors and Actuators, A* 23(1–3), 1001–1010.
5. Mastrangelo, C. H. and Muller, R. S. (1989). Vacuum-sealed silicon micromachined incandescent light source, *IEEE, IEDM*, pp. 503–506.
6. Bartek, M., Foerster, J. A., and Wolffenbuttel, R. F. (1990). Vacuum sealing on microcavities using metal evaporation, *Sensors and Actuators, A* 61(1), 364–368.
7. Elderstig, H. and Wallgren, P. (1995). Spin deposition of polymers over holes and cavities, *Sensors and Actuators, A* 46(1), pp. 95–97.
8. Lebouitz, K. S., Mazaheri, A., Howe, R., and Pisano, A. P. (1999). Vacuum encapsulation of resonant devices using permeable polysilicon, *IEEE International Conference on MEMS*, pp. 470–475.
9. Candler R. N., Park, W. T., Li, H., Yama, G., Partridge, A., Lutz, M., and Kenny, T. W. (2003). Single wafer encapsulation of MEMS devices, *ASME/IEEE Jorunal of Microelectromechanical Systems*, 26(3), pp. 227–232.
10. Stark, B. H. and Najafi, K. (2004). *ASME/IEEE J. Microelectromechanical Systems*, 13(2), pp. 147–157.
11. Candler, R. N., Hopcroft, M. A., Kim, B., Park, W. T., Melamud, R., Agarwal, M., Yama, G., Partridge, A., Lutz, M., and Kenny, T. W. (2006). Long-term and accelerated life testing of a novel single-wafer vacuum encapsulation for MEMS resonators, *ASME/IEEE Journal of Microelectromechanical Systems*, 15(6), pp. 1446–1456.
12. Jones, D. G., Azevedo, R. G., Chan, M. W., Pisano, A. P., and Wijesundara M. B. J. (2007). Low temperature ion beam sputter deposition of amorphous silicon carbide for wafer-level vacuum sealing, *IEEE International Conference on MEMS*, pp. 275–278.
13. Rico, H., Celis, J.-P., Baert, K., Van Hoof, C., and Witvrouw, A. (2007). A new generic surface micromachining module for MEMS hermetic packaging at temperatures below 200°C, *Microsystem Technology*, 13, pp. 451–1456.
14. Yoneoka, S., Roper, C. S., Candler, R. N., Chandorkar, S. A., Graham, A. B., Provine, J., Maboudian, R., Howe, R. T., and Kenny, T.-W. (2010). Characterization of encapsulated micromechanical resonators Sealed and coated with polycrystalline SiC, *ASME/IEEE Journal of Microelectromechanical Systems*, 19(2), pp. 357–366.
15. Antelius, M., Fischer, A. C., Roxhed, N., Stemme, G., and Niklaus, F. (2013). Wafer-level vacuum sealing by coining of wire bonded gold bumps, *ASME/IEEE Journal of Microelectromechanical Systems*, 22(6), pp. 1347–1353.
16. Cheng, Y. T. (2000). Localized Heating and Bonding for MEMS Package, Ph.D. Thesis.
17. Cheng, Y. T., Lin, L., and Najafi, K. (2001). A hermetic glass-silicon package formed using localized aluminum/silicon-glass bonding *IEEE/ASME Journal of Microelectromechanical Systems*, 10(3), pp. 392–399.

18. Cheng, Y. T., Hsu, W. T., Najafi, K., Nguyen, C. T., and Lin, L. (2002). Vacuum packaging technology using localized aluminum/silicon-toglass bonding, *IEEE/ASME Journal of Microelectromechanical Systems,* 11(5), pp. 556–565.
19. Luo, C. and Liwei, L. (2002). The application of nanosecond-pulsed laser welding technology in MEMS packaging with a shadow mask, *Sensors and Actuators A* 97, pp. 398–404.
20. Yang, H.-A., Wu, M., and Fang, W. (2005). Localized induction heating solder bonding for wafer level MEMS packaging, *Journal of Micromechanics and Microengineering*, 15, pp. 394–399.
21. Su, Y. C. and Lin, L. (2005). Localized bonding processes for assembly and packaging of polymeric MEMS, *IEEE Transaction on Advanced Packaging*, 28(4), pp. 635–642.
22. Tan, A. W. Y. and Tay, F. E. H. (2005). Localized laser assisted eutectic bonding of quartz and silicon by Nd: YAG pulsed-laser, *Sensors and Actuators A*, 120(2), pp. 550–561.
23. Tan, A. W. Y., Tay, F. E. H., and Zhang, J. (2006). Characterization of localized laser assisted eutectic bonds, *Sensors and Actuators A*, 125(2), pp. 573–585.
24. Bardin, F., Kloss, S., Wang, C., Moore, A. J., Jourdain, A., Wolf, I. D., and Hand, D. P. (2007). Laser bonding of glass to silicon using polymer for microsystems packaging, *IEEE/ASME Journal of Microelectromechanical Systems*, 16(3), pp. 571–580
25. Wu, Q., Lorenz, N., and Hand, D. P. (2009). Localised laser joining of glass to silicon with BCB intermediate layer, *Microsystem Technology*, 15, pp. 1051–1057.
26. Lorenz, N., Millar, S., Desmulliez, M., and Hand, D. P. (2011). Hermetic glass frit packaging in air and vacuum with localized laser joining,*Journal of. Micromechanics Microengineering*, 21, pp. 045039–045045.
27. Wu, Q., Semenova, Y., Wang, P., and Farrell, Numerical investigation on a laser based localised joining with a glass frit intermediate layer, *Microsystem Technology*, **18**, pp. 87–95 (2012).
28. Wissinger, A., Olowinsky, A., Gillner, A., and Poprawe, R. (2013). *Microsystem Technology*, 19, pp. 669–673 (2013).
29. Jiang, X., Wang, C., and Liu, W. (2015). A laser-assisted bonding method using a liquid crystal polymer film for MEMS and sensor packaging, *IEEE Transactions on Components, Packaging and Manufacturing Technology*, 5(5), pp. 583–591.
30. Qiu, X. and Wang, J. (2008). Bonding silicon wafers with reactive multilayer foils. *Sensors and Actuators A*, 141(2), pp. 476–481.
31. Qiu, X., Zhu, J. Oiler, J., and Yu, H. (2009). Reactive multilayer foils for MEMS wafer level packaging, *IEEE 59th Electronic Components and Technology Conference*, pp. 1311–1316.
32. Braeuer, J. Besser, J., Wiemer, M., and Gessner, T. (2012). A novel technique for MEMS packaging: Reactive bonding with integrated material systems *Sensors and Actuators A*, 188, pp. 212–219.

33. Braeuer, J. and Gessner, T. (2014). A hermetic and room-temperature wafer bonding technique based on integrated reactive multilayer systems, *Journal of Micromechanics and Microengineering*, 24, pp. 115002–115010.
34. Leung, A. M., Jones, J., Czyzewska, E., Chen, J., and Woods, B. (1998). Micromachined accelerometer based on convection heat transfer, *IEEE International Conference on MEMS*, pp. 627–630.
35. Hsu, C. H. and Schmidt, M. A. (1994). Micromachined structures fabricated using a wafer-bonded sealed cavity process, *Solid State Sensor and Actuator Workshop*, Hilton Head, pp. 151–155.
36. Cohn, M. B., Liang, Y., Howe, R., and Pisano, A. P. (1996). Wafer to wafer transfer of microstructures for vacuum package, *Solid State Sensor and Actuator Workshop*, Hilton Head, pp. 32–35.
37. Tilmans, H. A. C., Van de Peer, M. D. J., and Beyne, E. (2000). The indent reflow sealing (IRS) technique — A method for the fabrication of sealed cavities for MEMS devices, *ASME/IEEE Journal of Microelectromechanical Systems*, 9(2), pp. 206–216.
38. Lin, L., Pisano, A. P., and Carey, V. P. (1998). Thermal bubble formation on polysilicon micro resistors, *Journal of Heat Transfer*, 120(3), pp. 735–742.
39. Bower, R. W., Ismail, M. S., and Roberds, B. E. (1993). Low temperature Si_3N_4 direct bonding, *Applied Physics Letter*, 62(26), pp. 3485–3487.
40. Rude, T., Subramanian, J., Levin, J., Van Heerden D., Knio, O., and Powers, M. (2005). Hermetic sealing of microelectronics packages using a room temperature soldering process. *IMAPS Symposium*, Philadelphia.
41. Liang, Z. H., Cheng, Y. T., Hsu, W., and Lee, Y. W. (2006). A wafer-level hermetic encapsulation for MEMS manufacture application *IEEE Transactions On Advanced Packaging*, 29(3), pp. 513–519.
42. Decharat, A., Yu, J., Boers, M., Stemme, G., and Niklaus, F. (2009). Room-temperature sealing of microcavities by cold metal welding *ASME/IEEE Journal of Microelectromechanical Systems*, 18(6), pp. 1318–1325.
43. Tummala, R. (2001). *Fundamentals of Microsystems Packaging*, 1st ed. New York: McGraw-Hill, pp. 346–353.
44. Kim, J., Jeong, B., Chiao, M., and Lin, L. (2009). Ultrasonic bonding for MEMS sealing and packaging *IEEE Transactions on Advanced Packaging*, 32(2), pp. 461–467.

Yu-Ting Cheng received the M.S. degree in Materials Science and Engineering from Carnegie Mellon University in 1996 and the Ph.D. degree in Electrical Engineering from the University of Michigan, Ann Arbor, in 2000. After his graduation, he served IBM Thomas J. Watson Research Center, Yorktown Heights as a RSM. Currently, he is a Professor at the Department of Electronics Engineering at National Chiao Tung University. His research interests include Sensors and Actuators and Microsystem Integration. He has served as a TPC member in many international conferences including IEEE Transducers, Sensors, NEMS, ISMM, ISSNIP, and APCOT. He is a senior member of IEEE and the member of IOP, and Phi Tau Phi.

Yu-Chuan Su received the B.S. and M.S. degrees in Power Mechanical Engineering from the National Tsing Hua University, Taiwan, in 1993 and 1995, respectively. In 2003, he received the Ph.D. degree in Mechanical Engineering from the University of California, Berkeley, with an emphasis on MEMS technology. Since 2004, he joined the National Tsing Hua University, Taiwan, and is currently a professor at the department of Engineering and System Science. His research interests are in design and fabrication of polymer-based nano- and microfluidic systems for biomedical and energy related applications, including 3D printing, droplet-based microfluidics, energy harvesting, power MEMS and controlled drug delivery. He is a member of IEEE, ASME, and Phi Tau Phi.

Liwei Lin joined UC-Berkeley in 1999 and is now a Professor at the Mechanical Engineering Department and Co-Director at the Berkeley Sensor and Actuator Center. His research interests are in design, modeling and fabrication of micro and nano sensors and actuators. Dr. Lin is the recipient of the 1998 NSF CAREER Award and the 1999 ASME Journal of Heat Transfer best paper award. He is a subject editor for IEEE/ASME *Journal of Microelectromechanical Systems* and the North and South America Editor for *Sensors and Actuators — A Physical.* He is the founding chairman of the ASME MEMS division and an ASME Fellow.

Chapter 12

Microsprings for High-Density Flip-Chip Packaging

Eugene M. Chow and Christopher L. Chua

12.1 Introduction and Motivation

The basic goal of electronics packaging is to connect different parts, making the most fundamental component, the electrical interconnect. The properties of the interconnect determine the possible properties of the rest of the system. Solder has been the dominant interconnect for decades and the industry has spent, and continues to spend, significant resources to work around inherent solder problems, such as brittleness, high-temperature assembly, non-reworkability, toxicity, and poor size scalability. In addition, the low-margin business model dominant in the packaging backend supply chain significantly hinders introduction of innovative solutions, limiting the industry to slow, incremental improvements. While significant research and intellectual property continues to go into front-end device development, backend packaging limitations continue to constrain the overall system performance and size. During the last two decades, an additional $400 billion has been invested in the front end compared to the backend [1]. This work attempts to challenge these trends, to try and simplify packaging and enable new possibilities by introducing a new fundamental packaging element, a micro-scale spring interconnect.

While flip-chip is leading high-performance packaging, there are still many challenges for that technology. Current flip-chip interconnects, such as solder balls, gold bumps, and conductive adhesives, all have very limited compliance. Without adequate compliance, coefficient of thermal expansion (CTE) mismatches within the package between the silicon-integrated circuit (IC) and the package substrate leads to failures during thermocycle testing. This is particularly important for large-die flip-chip on organic board applications such as memory or advanced ASICs. More expensive low CTE boards or intermediate structures (ball grid arrays) are thus pursued. Processors are moving toward low-k dielectrics, which can be damaged by thermal expansion mismatches, increasing the demand for compliance. Interposers and added thermal expansion matched layers help to reduce the relative movement near the flip-chip interconnects, but they also increase package height, size, and cost. The demands of lead-free solder and the associated high reflow temperature exacerbates the problem.

Other challenges for flip-chip include pitch, testing, and known-good-die (KGD). Increases in the number of signals and continuing shrinkage in die size have increased the demand for higher pin counts and smaller pitch in markets such as display driver ICs for chip-on-glass and microprocessors for chip-in-package. Chip testing and thus the cost of KGD remains expensive, as the industry relies on expensive probe cards that cannot probe an entire wafer. For many high-performance dies, KGD, which passes parametric testing, does not exist due to testing parasitics. Finally, the recent popularity of portable electronics has driven more integration and more multichip system-in-package solutions. The economics of multichip integration would greatly benefit from a reworkable interconnect technology, which current approaches such as solder do not readily offer.

Compliant interconnects provide an alternative to solder. The ideal process for a compliant interconnect would keep costs low with batch fabrication steps, few masks, no expensive materials, and conventional processing tools. Solutions proposed for this application include compliant layers under the solder [2], metallization formed over organic elastomers [3, 4], springs formed from wirebonds [5], bent printed circuit board (PCB) metal [6], and machined bulk metal for land grid array (LGA). Compliant layers under solder is established for memory packaging, but does not improve integrated testing for MCM (no rework), low-profile, and high-density capability. The wire-bond approach, due to its serial manufacturing process, is inherently more expensive than a batch process. The metalized elastomer requires more lithography steps than the stress-engineered approach. The bent PCB metal is recently commercialized and the LGA well established for sockets, but both are millimeter scale and not scalable due to the use of thick films. In contrast, the stress-engineered spring approach presented here has been demonstrated down to high-density interconnect regimes, such as 6- and 20-μm pitch. The stress-engineered spring process inherently includes redistribution, as the spring metal layers are also traces, which reduces cost.

We have developed a spring-based flip-chip interconnect to address these challenges. It provides compliance, which enables basic pressure contact testing (gold–gold) and also permanent packaging. This integrated test and packaging capability addresses the KGD problem and reduces the cost of building multichip modules (MCM). The springs also address differential thermal expansion mismatches between substrates because of its mechanical compliance and have demonstrated excellent bare-die silicon interconnect on organic substrates, enabling larger chip to board assemblies. As a sliding pressure contact, low k dielectric vulnerability is reduced. Unlike solder, the flip-chip gap height can be tuned to be very small ($<1\,\mu$m as planarization allows) for compact applications or chip stacking, as well as to high gaps (hundreds of μm) to provide space for other components. The springs are fabricated with wafer-scale processing, are photolithographically defined, and they self-assemble by rising off the surface during fabrication. Their structure enables very fine pitches ($6\,\mu$m demonstrated). Less than $100\,\mathrm{m}\Omega$ per contact has been demonstrated for springs with over $30\,\mu$m of compliance; lower resistance designs are possible. The two-mask spring fabrication process also includes routing and thus can possibly replace the top IC metal routing, pads, UBM, and solder deposition. The springs can be soldered and underfill is not required.

This chapter reviews our spring work. In the next section, we summarize the fabrication and assembly. We then review the motivation and prototypes demonstrated for various applications, including high-end MCMs, low profiles, chip on board (COB), and high density. Finally, we will briefly review package test approaches and then describe more fundamental single-spring studies.

12.2 Spring Fabrication and Assembly

Standard wafer-scale thin-film processing is used to make the springs. The springs are metal traces that protrude from the surface. Fabrication (Fig. 12.1) consists of thin-film metal depositions of a sacrificial layer followed by a stress-engineered layer, which has a large balanced stress gradient ($1\,\mathrm{GPa}/\mu$m). The springs and traces are then patterned with photolithography and etching. A second resist mask defines release windows around springs and protects traces from being released (Fig. 12.1(b)). The inherent stress gradient in the spring metal relaxes when the underlying

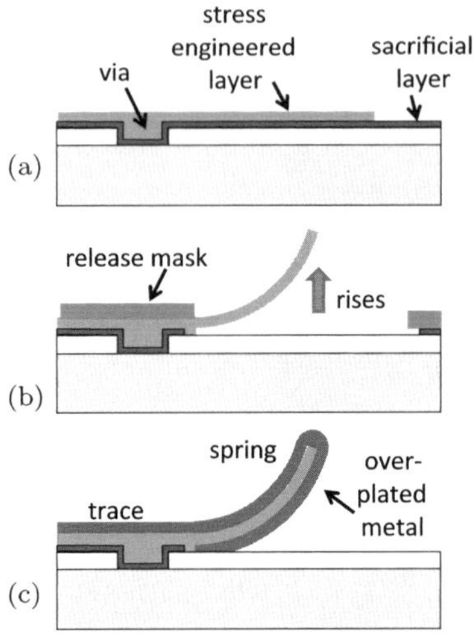

Fig. 12.1. Spring fabrication process.

sacrificial layer is etched, causing the spring beams to rise up from the wafer surface. If needed, subsequent overplating of metals is used to mechanically strengthen and increase conductivity (Fig. 12.1(c)). The portions of the metal traces that do not release serve as anchors and redistribution layers. In contrast, solder does not inherently include a redistribution layer.

Typical materials used are MoCr or Ni for the spring metal and Cu, Ni alloys, and gold for the overplate layers. Spring layer thickness is typically 1–2 μm and overplate layers are typically 1–5 μm thick, depending on the spring dimensions. The process is very uniform, as tip heights (vertical distance above the substrate) and spring thickness vary by less than 5% and yields of 99.99% have been demonstrated in PARC's general-purpose research lab environment [7]. The process takes place at low temperature, so if the substrate accommodates thin-film processing, then springs can be fabricated. Springs have been fabricated on flexible substrates, as well as on organic and ceramic package substrates, following polishing to planarize the surface [8]. The process has also been demonstrated on foundry BiCMOS wafers to make RF inductors and found to not damage the circuits [9].

The process enables a large design space and scales with electronics miniaturization trends. Very high densities can be readily achieved because the films are thin and photolithography is used. Springs as small

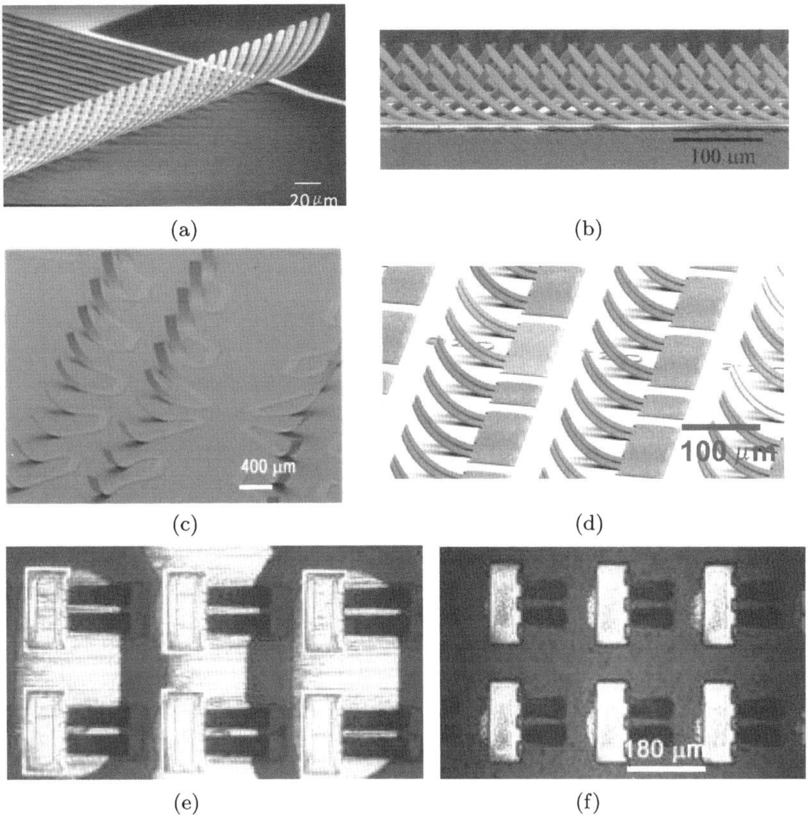

(a) (b) (c) (d) (e) (f)

Fig. 12.2. Fabricated springs. (a) 6-μm linear pitch (VCSEL array), (b) 20-μm pitch linear array (LCD driver chip), (c) >650-μm pitch (memory), and 180 μm × 180 μm pitch (ASICs) on (d) silicon, (e) organic, and (f) ceramic substrates.

as 4-μm wide on a 6-μm linear pitch have been demonstrated for high linear density optoelectronics applications (Fig. 12.2(a)) [10]. Larger springs, 100 μm wide, and 400 μm long have been demonstrated for memory chips (Fig. 12.2(c)). For 2D arrays, work has focused on 180 μm × 180 μm arrays (Figs. 12.2(d)–(f)). Smaller pitch 2D arrays are possible, depending on the vertical compliance requirement of the application. In general, there is a design trade-off between the required compliance, pitch, and resistance. More details on the electrical properties are in Section 12.5.

To assemble a chip with springs, flip-chip tooling is used to align springs to the pad chip (Fig. 12.3). Here the chip with springs can be an active IC, a silicon interposer, or even an organic package. Similarly, the pad chip

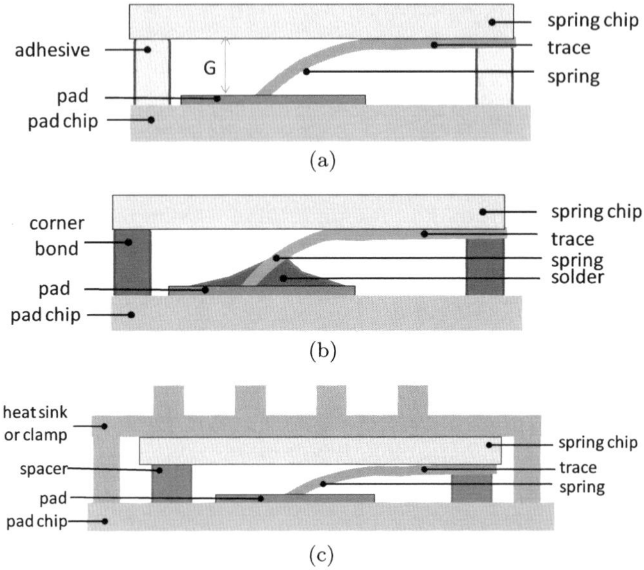

Fig. 12.3. Assembly options. (a) Pressure contact, (b) soldered spring, and (c) heat sink clamped.

can be an organic, ceramic, flexible, or silicon substrate. During assembly, the spring tips are aligned to the pad and compressed, causing a natural scrub as the tip slides laterally. The springs can be retracted and recompressed a couple times to clean the tips and pads, similar to probing techniques. While taking into account any substrate nonplanarity issues and anticipated vertical expansion of the package during environmental testing, the springs should be compressed a minimum distance to make sure they are in the resistance plateau region to ensure uniform resistance values (see Section 12.5). The minimum compression varies by spring design, but for the 180-μm pitch springs, with 50-μm initial height, the minimum compression is $\sim$20 μm, and the final gap G is 15–20 μm. If the springs are not designed to tolerate full compression, a gap stop, for example employing tape, polyimide or balls, can be used to limit the maximum compression.

For the pressure contact version (Fig. 12.3(a)), the springs need to be bonded to hold the chip down and maintain spring contact force. Small amounts of UV-curable adhesive on the edges of the chip are spot cured to secure the chip and insure the gap is stable. If needed, additional thermally curable adhesive is then applied. Packages with adhesive on the edges only or throughout the entire package (like an underfill) have both been shown to pass basic reliability tests [7]. Unlike solder, the springs do not need underfill for stress distribution. For packages that require heat sinks, we plan to use clamps to hold the spring chip down. A clamped package, without adhesive or a heat sink, has already been demonstrated (Fig. 12.3(c)) [11]. Spring tips can also be soldered to the substrate, such as by placing tips into solder already on pads (Fig. 12.3(b)). Corner bonds are needed for this solder package to pass shock testing, but underfill is not needed. The COB section 12.3.3 provides more details of this type of package. A unique option for silicon-to-silicon packages is to use a ball and pit alignment system (Fig. 12.7(a)). Due to self-alignment abilities, this approach provides 1-μm final accuracy in x, y, and z axes while using relatively crude assembly precision ($\sim$100 μm).

Overall, the spring-based flip-chip has the potential to simplify packaging compared to current approaches. For example, for the case of placing pressure contact springs on an IC wafer and then assembling them to a pad substrate, there are multiple steps that the traditional solder-based flip-chip would need, which the spring would not (Fig. 12.4). The spring metal distribution capability could obviate the need for the top global IC metal.

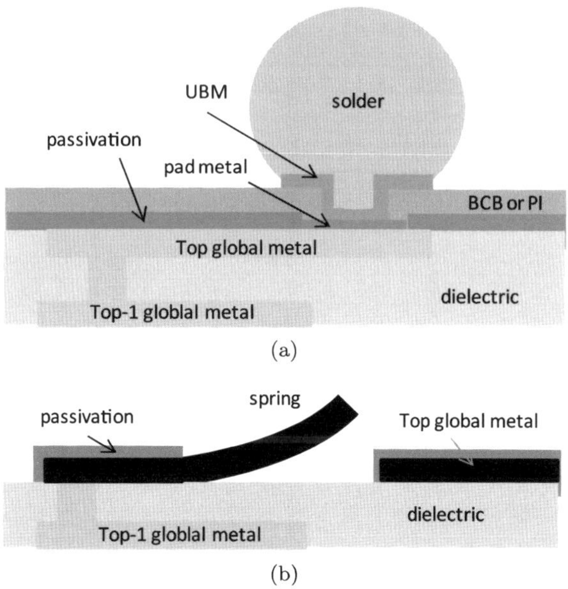

Fig. 12.4. Preparing an IC for flip-chip for (a) solder and (b) springs.

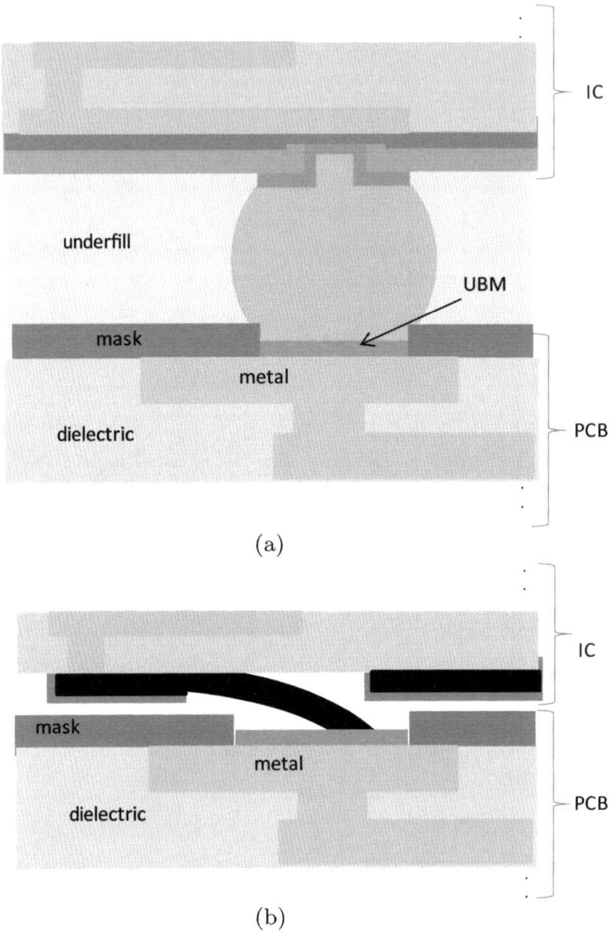

Fig. 12.5. Cross-section schematic of (a) solder flip-chip and (b) spring flip-chip.

Conversely, the spring trace distribution layer can be placed above the top global IC metal and used to further reduce impedance. Pad metal, under bump metallurgy (UBM), consisting of an adhesion, diffusion barrier, and wetting layer, and solder printing could all be avoided. In the package cross-section (Fig. 12.5), the substrate and underfill is shown. Because underfill is not needed in the spring case, more springs can be placed in parallel and larger PCB metal pads could be used, helping to reduce impedance and improve thermal design. It is possible that the PCB mask might not be needed. Table 12.1 summarizes the differences.

Table 12.1. Comparing fabrication and assembly steps starting just before the last IC metal layer. The spring process uses fewer steps.

Traditional Flip-Chip	**Spring Flip-Chip**
1. Top layer IC metal routing	1. Spring metal
2. UBM metal	2. Assemble
3. Solder/Cu Pillar	3. Bond
4. Assemble	
5. Underfill	

12.3 Applications

Here we review different electronics applications and demonstrated prototypes of spring-based flip-chip packaging.

12.3.1 *Multichip Modules*

Multichip modules (MCM), sometimes called multichip packages (MCP), use multiple bare dies in the same package. They enable the smallest form factor for system-in-package approaches. They can also enable the highest performance package because parasitics between chips are minimized. However, in general, this type of packaging is considered very expensive and thus not used widely. There are multiple reasons for this. Assembling and bonding the die with nonreworkable approaches like solder flip-chip makes KGD required, or else the module yields are low. If just one of the many die in an assembled MCM are bad, then the entire module, along with the other good die, are rejected. Unfortunately, KGD are generally much more expensive, and often do not exist, particularly for high-performing or analog chips. While functional yield can be determined with available probe card technology, parametric yield cannot be because of the associated parasitics. Optimal design of MCMs is hindered because the parasitics cannot be readily modeled without parametric KGD data.

These problems can be addressed with a readily reworkable interconnect, such as a solder-free, pressure contact-based springs, enabling new MCM assembly and test strategies (Fig. 12.6). The springs can be placed on the most expensive die, all the dies, or just the substrate. After assembly, but before permanent bonding, the entire module is tested. If any one of the chips or substrates is found to be faulty, only that component is rejected and replaced with a new one. Permanent bonding is only

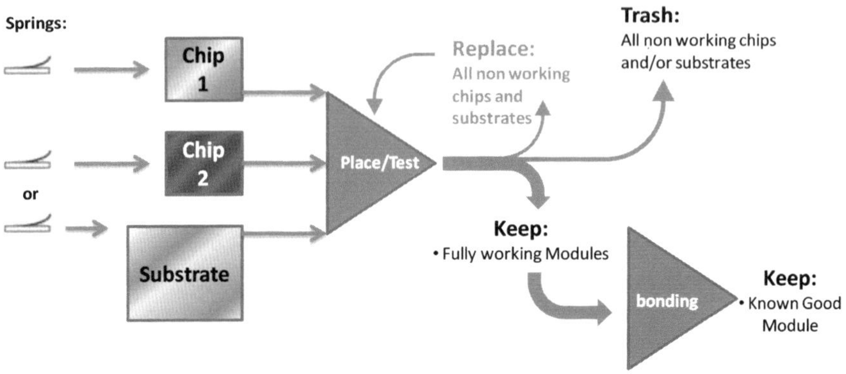

Fig. 12.6. Springs enable a rework step, where bad die or substrates are replaced by new ones.

applied after a successful module test. Cost models for this reworkable MCM process suggest significant savings (>10×) for all the three scenarios, whenever at least one die is low yield, there are many dies, or KGD is expensive. The cost of the module is also found to be less sensitive to the die yields.

Sun Labs, at Oracle, is pursuing an advanced MCM consisting of multiple high-end processors and wireless chip-to-chip communication. Capacitive coupling between input/output circuit pads of neighboring or face-to-face chips is used to enable high bandwidth chip-to-chip communication [12]. A reworkable spring interconnect is critical for making these arrays of processors feasible. For this application, a spring contact was designed to achieve both large compliance ($>30\,\mu$m) and low resistance ($<100\,$mΩ). The test array has ∼2500 contacts and is $180\,\mu$m × $180\,\mu$m pitch (Fig. 12.2(c) and Fig. 12.7(b)). A unique ball and pit alignment scheme was used to enable $1\,\mu$m alignment, important for the capacitive coupling (Fig. 12.7). Daisy chain test packages have been shown to pass over 1000 thermocycles (0–100°C) and 1000 humidity hours (85°C/85% RH). High current tests show stable resistances and <1 deg temperature rise for 250 mA per contact (Fig. 12.16) [13]. Many packages have also been assembled without the balls (Fig. 12.3(a)) and shown to be reliable with or without an underfill [10]. In addition to silicon substrates, the springs have been fabricated on organic and ceramic substrates (Figs. 12.2(e) and (f)) [8]. A custom IC designed to characterize the contacts was assembled, enabling resistance measurements of thousands of contacts. Characterization of the packages show resistances that match a solder based package. The

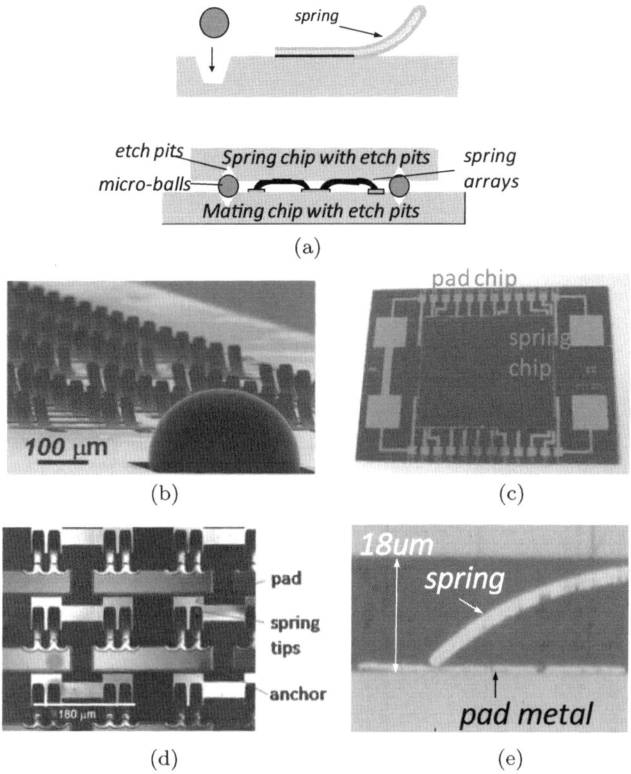

Fig. 12.7. (a) Schematic of a spring with a pit/ball and assembled into a package. The package was secured with either a clamp or adhesive. (b) Springs at 180 μm × 180 μm pitch. (c) An assembled package. (d) IR image into a package, and (e) cross-section of a contact [11].

package testing section below reviews testing and [7], [8], and [11] provide more detail.

12.3.2 *Low-Profile Chip Stacking*

Miniaturization trends in mobile electronics make minimizing the vertical gap between the chips critical. Low-profile solder joints fail more quickly than taller joints due to their lack of compliance. This couples the height to the pitch, so a 250-μm pitch solder tends to be at least 150-μm tall. Low profile solder joints are unreliable and difficult to fill with underfill. In contrast, a sliding spring contact can have a very low height profile, independent of the pitch requirement. The spring package in Fig. 12.7(e) has a gap of only 18 μm. Smaller gaps are possible, limited by the planarity of the substrate, as no underfill is required for the springs.

12.3.3 *Large Chip on Board*

Another industry trend is to assemble die directly to the PCB termed wafer level packaging (WLP) or chip on board COB. No added packaging structures such as other organic or ceramic substrates are used, saving space and cost, as the die is essentially flip-chip packaged directly to the PCB. Generally, this is only possible for small die (<7 mm) and large pitches (>400 μm), primarily because the solder cannot tolerate the large thermal expansion mismatch between silicon (2.7 ppm/°C) and PCB (12–20 ppm/°C). Larger dies have larger induced solder stresses and fail.

The large compliance available in a spring-based package can enable large die assembly to PCB. To demonstrate this, large dies were directly assembled to a board [14]. In the test vehicle, the springs were fabricated on glass substrates with the same thermal expansion coefficient as silicon. The springs are 100-μm wider, 400-μm long, and a couple of hundred microns tall prior to assembly (Fig. 12.2(c)). The dies were flip-chip assembled to a board, which had lead free-solder paste on the pads (Fig. 12.8). During a subsequent thermal reflow step, the solder wets the spring and bonds it to the pad (Fig. 12.3(b)). After application of corner adhesive, the package was complete (Fig. 12.9).

This package passed over 20,000 thermocycles, which is an order of magnitude more than typical industry tests. This suggests improved reliability or that even larger dies are possible. Humidity and shock tests were also passed. No underfill was used in these packages. In contrast, all solder-based approaches to WLP/COB require underfill to distribute the stress. The die size was 11.5 mm × 6.5 mm, and laid out in a daisy chain pattern similar to a DDR2 memory chip. For this market, high-throughput

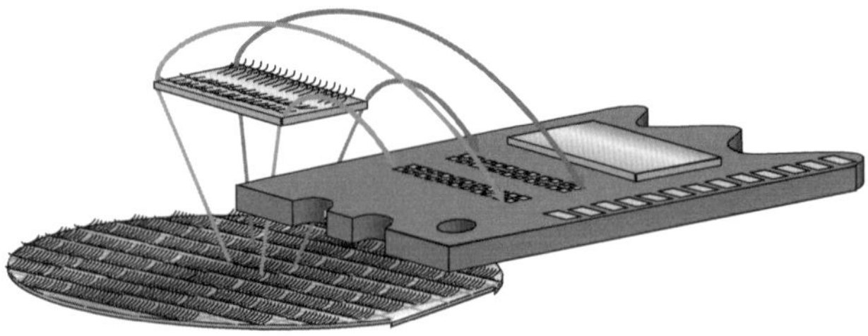

Fig. 12.8. Schematic of a die with springs being flip-chip packaged directly to an PCB.

Fig. 12.9. Photo of a demonstrated package of a large die directly to a PCB. Looking through the glass substrate, the spring wetted with solder is visible.

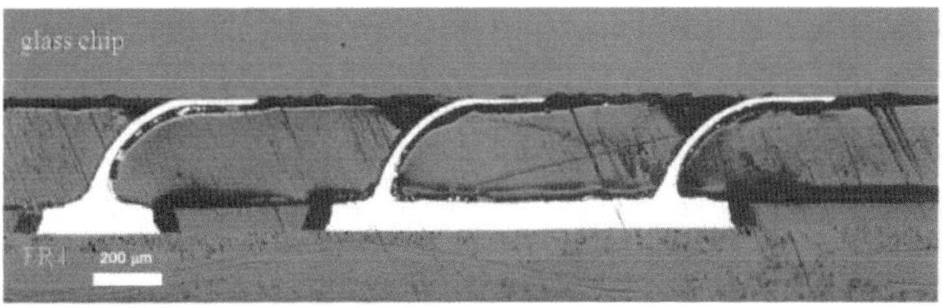

Fig. 12.10. Cross-section of soldered spring. Wax was applied around the springs only for the cross-section sample to aid in polishing. The tested packages were open-air, and had no underfill.

testing and burn-in is required. A wafer-level test solution is possible with the springs, without the need of an expensive probe card, as the springs on the wafer can contact pads on a flat contactor. The test strategy created less need for rework, so solder, and the associated unreworkability, were acceptable for this application.

12.3.4 *High Density*

Applications with significant amounts of data often use high numbers of signals. Examples include high-density imaging or detector arrays, such as advanced optoelectronics applications. For an industrial xerographic laser printer system, Xerox explored using high-density arrays of vertical cavity surface emitting lasers (VCSELS). Lasers on a 6-μm pitch were required, but there was no known way to connect the ASIC driver chip signals at this pitch. Two hundred springs at this pitch (Fig. 12.2(a)) were fabricated on both sides of a glass carrier, forming a bus connector with gold-to-gold pressure contacts at each end (Fig. 12.11). This was the first packaging application for the springs [10].

Large flat panel displays require many driver chips around their borders. High-density connections enable the driver chips to be smaller area and lower cost. For this application, 800 contacts at 20-μm pitch were demonstrated in a prototype package (Fig. 12.2(b)). For these springs, the stress-engineered layer was electroplated. Springs were embedded in adhesive to bond them to the substrate (Fig. 12.12). Daisy chain test

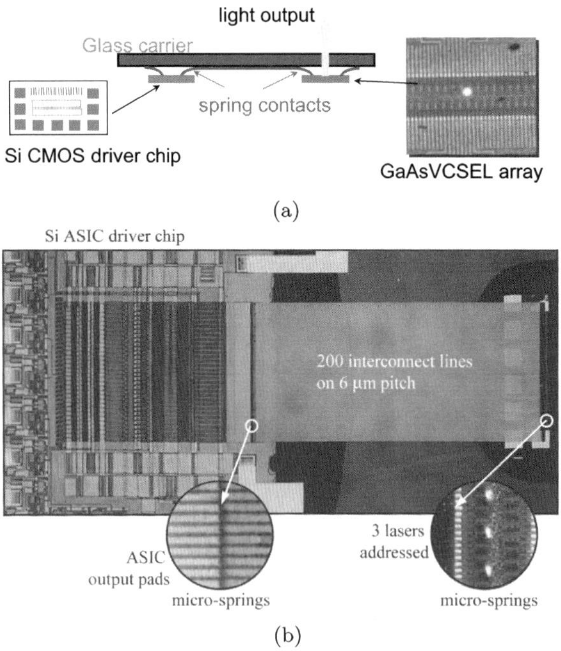

Fig. 12.11. Packaging of a 6-μm pitch laser array: (a) schematic and (b) photo.

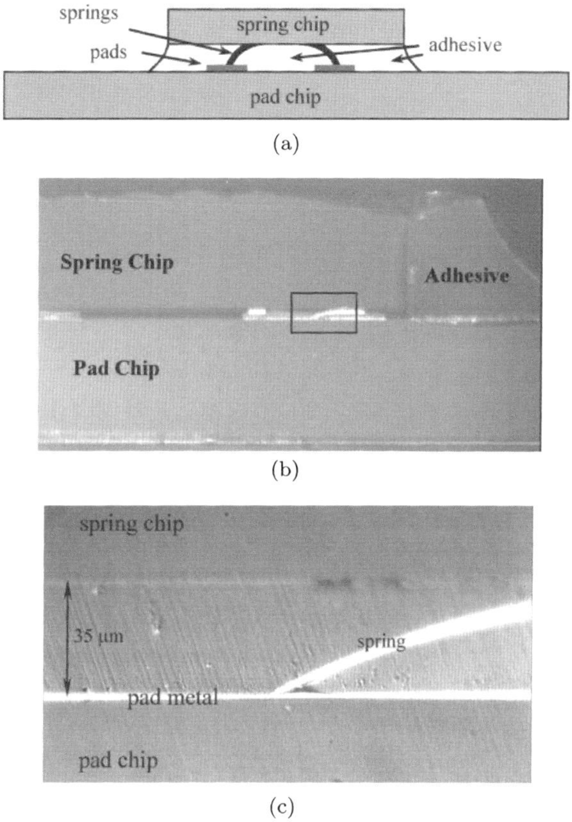

Fig. 12.12. Cross-section (a) schematic, (b) and (c) photos of a 20-μm pitch gold–gold pressure contact package.

packages passed basic thermocycle (0–100°C) and humidity (60°C, 90% RH) tests [15].

12.3.5 *Other Applications*

The large compliance of a spring package provides stress isolation to the die, which could be particularly advantageous for sensitive die like MEMS inertial sensors such as accelerometers and gyros. The measurement principle of these chips rely on accurate stress measurements. They can generally not be flip-chip packaged because the solder joints transfer too much stress to the chip, corrupting the signal. Similarly, as the IC industry moves to fragile low-k dielectrics to improve signal speed and power metrics, and stress transferred from the stiff solder into the IC is a key concern. Spring packaging approaches could help alleviate this problem.

Microspring technology can also be considered for markets that already use springs, such as LGA sockets and connectors. LGA are used in sockets for large, high-value ICs and are made from precise cutting and bending of bulk metal films, or recently from PCB technology [6]. These approaches have not shown the ability to scale below 0.5-mm pitches. Insertable connectors are generally even larger, as their manufacturing relies on bulk metal machining. A thin film-based microspring could enable finer pitch connectors and further miniaturization of cabling between boards.

12.4 Package Testing

Basic package reliability tests performed for the above prototypes are briefly summarized. More details can be found in the detailed papers of each prototype.

12.4.1 *Daisy Chains*

Thermocycle, humidity, high current, and shock tests were performed with daisy chain packages. The goal is to show stable resistance values under different conditions. Daisy chain structures test structures based on four-wire measurements provided resistance measurements of a series of spring contacts, pads, and associated traces (Fig. 12.13). The device under test (DUT) for each measurement could have two contacts in series or many more. For thermocycle and humidity tests, all the chains are measured before and periodically throughout the test (Fig. 12.15). In addition, the entire package daisy chain was connected to a high-speed event detcctor *in-situ* during thermocycle tests (Fig. 12.14). This measurement helps to show that there were no intermittent opens.

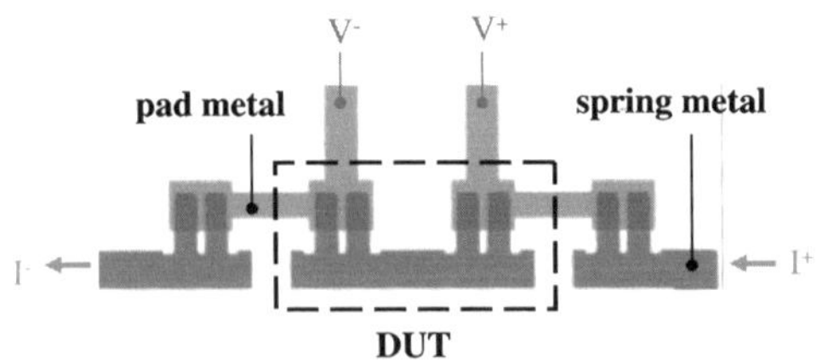

Fig. 12.13. Four-wire daisy chain package test structure schematic for a two-contact chain.

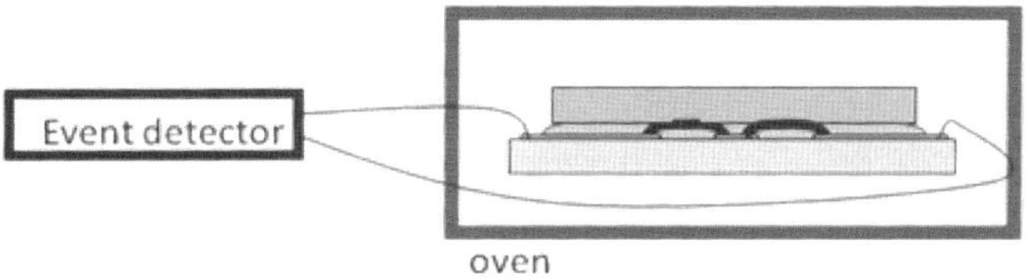

Fig. 12.14. Schematic of oven-thermocycle test setup with *in-situ* glitch monitoring.

number of contacts	4-wire resistance [ohms] initial	598 cycles	1020 cycles	% change
2	0.341	0.335	0.328	-3.6
2	0.334	0.330	0.326	-2.4
2	0.334	0.331	0.327	-2.2
2	0.343	0.349	0.338	-1.4
2	0.341	0.335	0.335	-1.8
2	0.347	0.335	0.333	-4.0
2	0.234	0.233	0.231	-1.3
2	0.233	0.229	0.224	-4.0
2	0.238	0.234	0.233	-2.4
2	0.236	0.241	0.241	2.1
134	34.567	34.069	33.790	-2.2
134	34.668	34.098	33.926	-2.1
246	70.444	69.654	68.917	-2.2
246	71.043	70.065	69.787	-1.8
384	100.936	99.963	98.602	-2.3
384	102.327	101.813	100.898	-1.4
530	137.244	135.688	133.295	-2.9
530	140.191	139.879	138.359	-1.3

Fig. 12.15. Sample thermocycle test data for a single package, showing stable or decreasing values for all chains. Each row corresponds to a different chain in the package.

For the 180-μm pitch springs, the target IC operating at maximum power (355 W) would have a 90 mA per contact, corresponding to an average current density of 7×10^4 A/cm^2. Over five times this current density has already been shown to be reliable for high-powered MEMS switches. This suggests that if the contact design avoids high local current densities, and corresponding hot spots, then it should be reliable. Experimentally, packages were heated to simulate an active IC and subjected to high DC current while the voltage was monitored. For the case of polished silicon chips, IR imaging can simultaneously be performed (Fig. 12.16) [13]. The lack of significant hot spots, as observed, is desirable for minimizing electromigration risks.

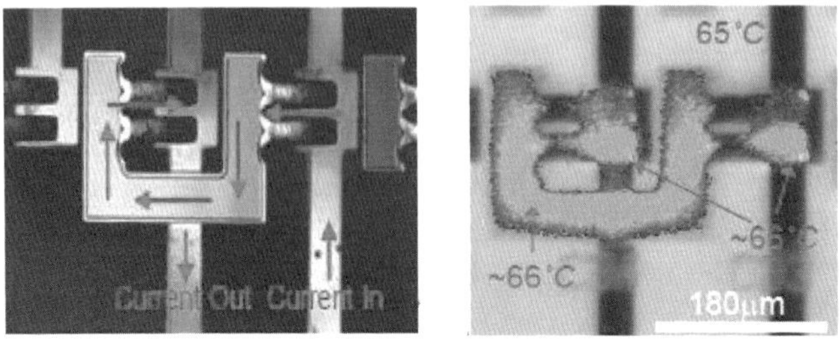

(a) (b)

Fig. 12.16. (a) IR image into a package marking current flow along the pad chip, up a spring contact, through an anchor trace, back down a spring contact, and then out a pad trace. (b) Thermal image of the same contact while conducting 250 mA, showing no hot spots more than 1° above the ambient 65°C.

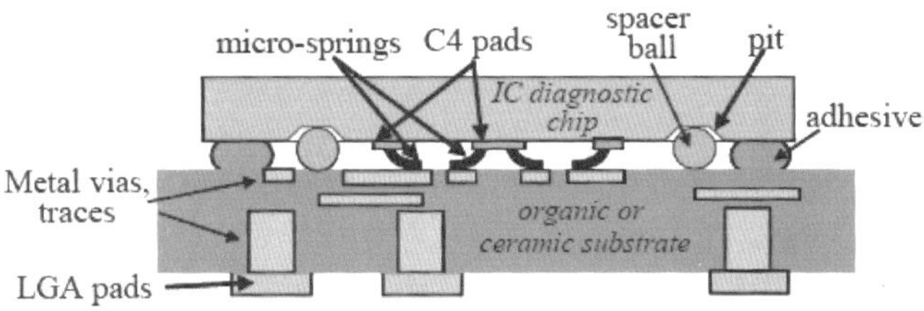

Fig. 12.17. Schematic of the IC spring package. Springs were fabricated on the organic/ceramic substrate.

12.4.2 *Active IC*

Recently testing has begun with a custom diagnostic IC chip (0.18 μm CMOS), designed to characterize package properties [16]. Springs were fabricated on both organic and ceramic packages (Fig. 12.2(e) and (f)). The IC, with gold pads, was assembled onto the substrates (Fig. 12.17). Test structure resistances of all 1701 spring contact test sites were measured showing chip functionality. This is the first time compliant spring interconnects have directly replaced traditional C4 solder bumps in an IC package [11].

12.5 Spring Contact Properties

Multiple single spring studies have been performed to study the properties of the microsprings, including force or resistance vs compression, high-frequency modeling, and fretting with *in situ* electrical event detection.

12.5.1 *Force*

To guide packaging design, such as the target gap, new spring designs are experimentally characterized by measuring the force as a function of compression [17]. Though the spring is curved, they are relatively thin, and force models generally follow straight beam mechanics and linear force curves (constant stiffness) are observed. As springs get thicker or near full compression, the springs enter a superlinear regime and the effective spring constants increase. Generally, springs are designed to be fully compliant, meaning they can compress all the way down without yielding. The lift height of the springs might lower a few microns after the first full compression, due to plating underneath the spring plastically deforming, but subsequent compressions do not change the lift height. Unlike probe cards, packaging springs do not need to show a high number of full compressions ($>10^5$), as package reliability data is more relevant; however, force curves have been measured to be stable for over 500 repetitions.

The goal of the spring is to provide enough force for stable electrical properties during the package lifetime. Macroscopic gold-to-gold pressure contacts are known to be very stable in the industry, but microscale pressure contacts are new to the packaging industry. The above prototypes all used spring contact operation at forces below 25 mg (~0.25 mN), corresponding to typical average pressures of 10–20 MPa or higher. This pressure is believed to be similar to macroscopic pressure contacts. MEMS switches have been shown to be reliable at ~0.1 mN [18]. Much higher forces and pressures are possible with the microspring if desired, as springs with many grams of force have been demonstrated. The trade-offs for higher force springs include more wear or more gold required, more bonding force required, and more compromises to either the pitch or compression range. Sample force curves are given in Fig. 12.18.

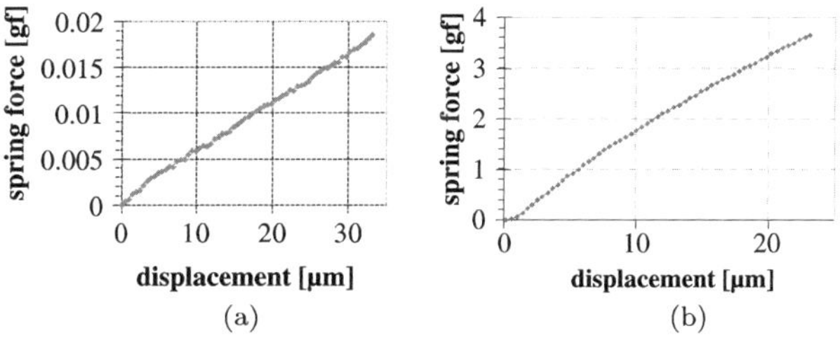

Fig. 12.18. Measured force (gram force) vs displacement curves for (a) a low-force 180-μm pitch 2D array spring and (b) a high-force 40-μm linear pitch spring.

12.5.2 *Resistance*

Single-contact resistance measurements are made with a four-wire test structure as a function of compression [7]. As the spring compresses, the resistance lowers and then plateaus. With subsequent scrubs, the resistance tends to decrease slightly. The target gap should be well within the plateau region to ensure uniform resistances. For the springs in Fig. 12.19, compressions should be at least 10–20 μm, plus any anticipated package nonplanarities. Note that the thermal expansion of the adhesive is relatively negligible [15], but clamped package designs should be studied individually.

To better understand the resistance, finite element modeling (Fig. 12.20) and experimental data were combined to give a detailed breakdown of the 180-μm pitch spring (Table 12.2) [7]. The lead trace resistance is due to the test structure and is significantly lower for the practical case of the spring on a via. The spring body resistance is determined by the spring geometry and can be lowered by shortening the spring

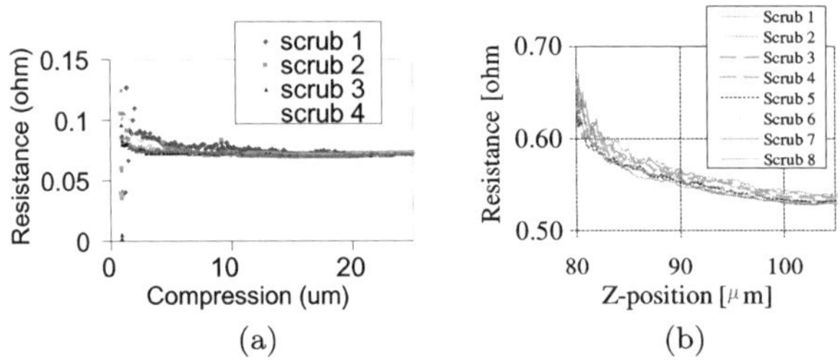

Fig. 12.19. Measured resistance vs compression for (a) 180-μm pitch 2D array spring [7] and (b) 20-μm pitch linear array spring [15].

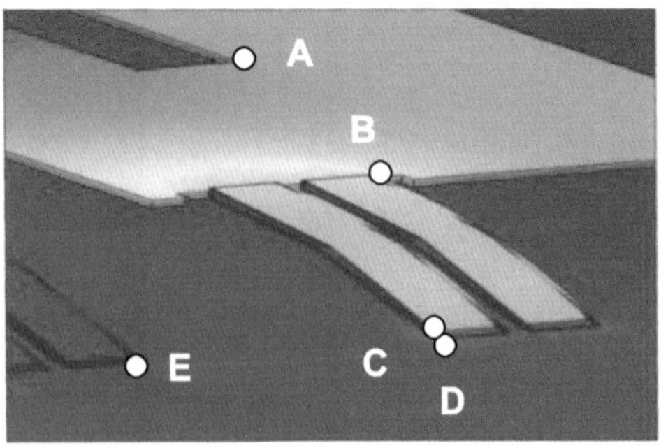

Fig. 12.20. Model of 180-μm pitch spring contact (two beams in parallel), corresponding to measurement in Fig. 12.19.

Table 12.2. Resistance values with reference to Fig. 12.20.

Resistance Component	Value (mohm)
Lead trace R_{A-B}	23
Spring body R_{B-C}	36
Tip-pad R_{C-D}	11–41
Pad spreading resistance R_{D-E}	~0
Total	70–100

or thickening the spring and adding a gap stop. A 10-mΩ resistance spring design for another application is currently under development.

The target resistance depends on the application and the layout details of the substrates. For exactly the same pad area, the springs will inherently have less conductivity than a solid bump, because providing compliance requires using less metal. However, there are important differences between the springs and bumps that minimize this trade-off. Unlike bumps, the springs are traces that can move signals laterally significant distances, so it is more appropriate to compare the resistance between points on the two substrates. This means comparing a trace plus a bump to a spring. In addition, bumps have very sharp current density changes due to the many layers required, while the spring anchor is a gentle curve with less smooth current density variations and the tip is gold to gold (no material change), helping to minimize hot spots. The spring process includes a redistribution layer that can provide common power and ground planes. Finally, smaller gaps are possible with the springs, reducing vertical travel requirements. General comparisons of spring resistance to bumps should be studied for a particular design.

12.5.3 *Bandwidth*

Simulations suggest the springs can support high-frequency applications. A 20-μm pitch coplanar waveguide design (Fig. 12.21) has less than 1 dB of insertion loss at 70 GHz for a tip-pad interface resistance of less than 10 Ω, well above the resistances we expect (see Section 12.5.2). Similarly, negligible dependence on gap is observed, as 3 to 50 μm gaps were all less than 1 dB of loss at 70 GHz. The inductance is balanced with the capacitance to achieve controlled impedance to the matched loads. Absolute inductance values for isolated springs will vary by the dimensions and materials, but generally the springs are thin film traces with a small bent tip contact and will have properties similar to other thin-film technologies.

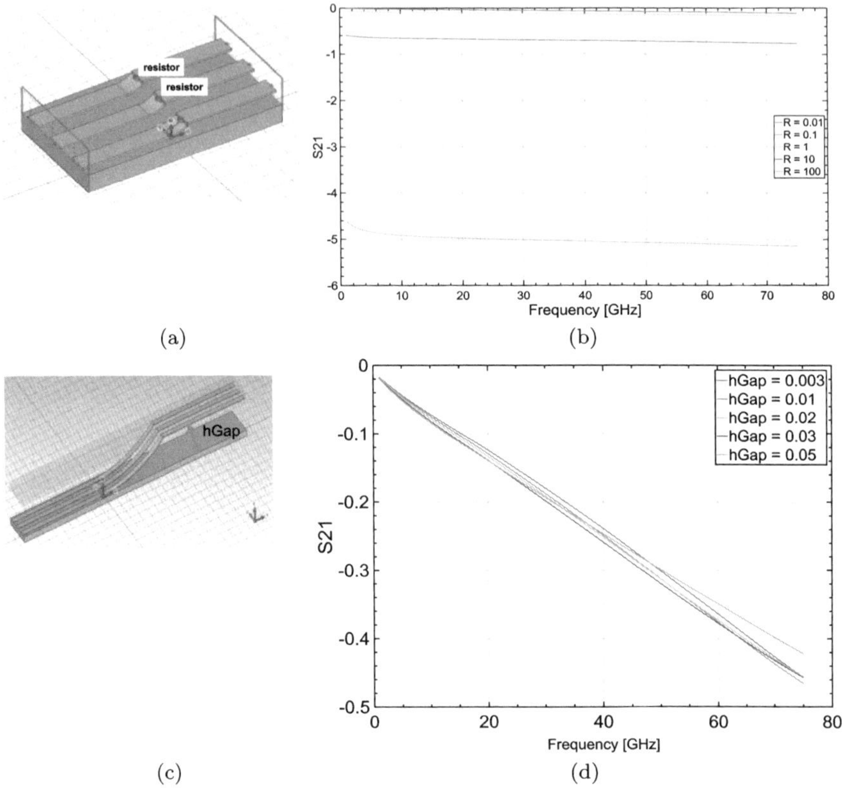

Fig. 12.21. Frequency response simulation of 20-μm pitch spring design. (a) Schematic of ground-signal-ground traces. (b) S21 parameters as a function of interface resistance (R_c) between the tip and the pad. (c) A function of assembly gap. (d) At 70 GHz, insertion loss is <1 dB for $R_c < 10\,\Omega$ and gap $<50\,\mu$m. Results used 3D electromagnetic simulations (CST Microwave Studio).

12.5.4 *Fretting*

An accelerated lifetime study used in the connector industry is fretting, where the contact is shaken repeatedly. A key goal of the test is to study possible intermittent behavior, where the resistance momentarily increases. Placing the spring operating point safely in the plateau of the resistance curve helps to ensure stable force and DC resistance, but the interface of the tip and pad under movement is more difficult to predict, so the high-speed electrical events are tracked during the test (Fig. 12.22) [17]. Fretting tests have shown no glitches until after 10,000 shake cycles (Fig. 12.23). Possible failure mechanisms such as intermittent stiction have not been observed. Figure 12.24 shows the tip wear after a fretting test, showing an increase in

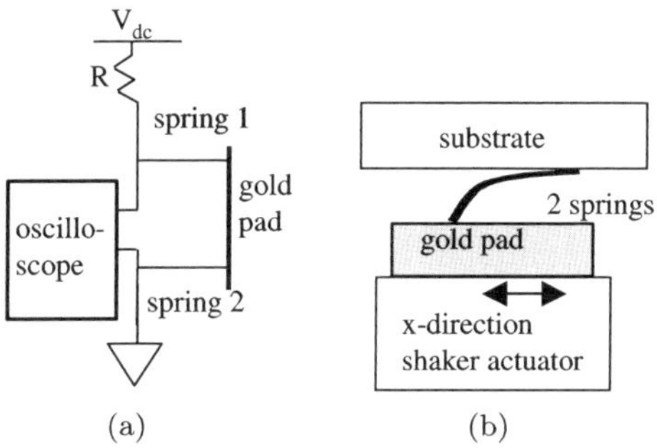

Fig. 12.22. (a) Electrical and (b) physical schematic of the fretting experimental setup. Two springs contacting a gold pad form two electrical contacts in series. An oscilloscope (4 ns sensitivity) monitors the junction when a dc current is applied and the contact pad is cyclically displaced in the x-direction.

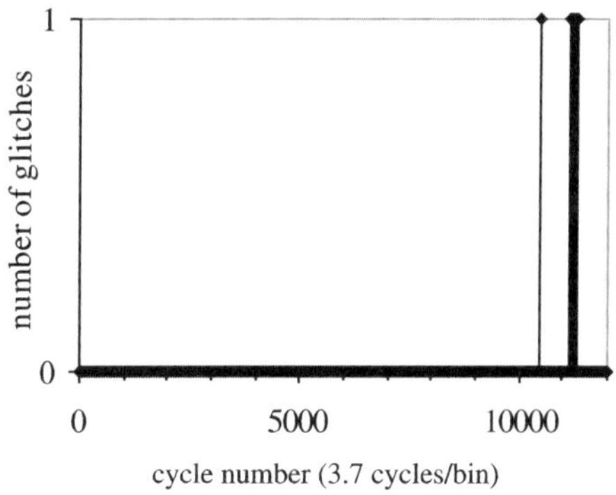

Fig. 12.23. Fretting histogram for 50 μm scrubs at 10 mA. The first glitch, corresponding to a 10 Ω intermittency, is observed at cycle 10458 (3 s scrub period).

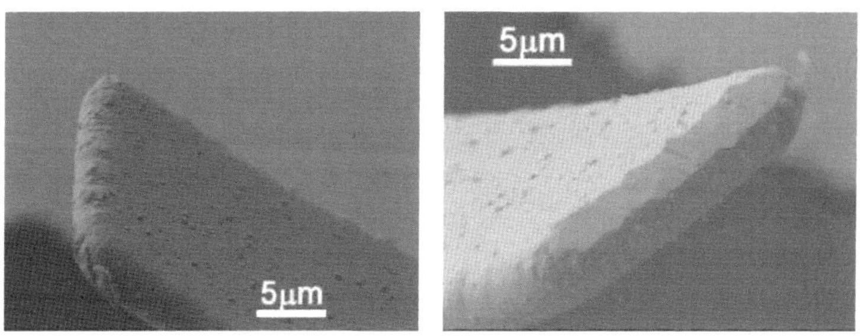

(a) (b)

Fig. 12.24. SEM image of spring tip (a) before and (b) after fretting testing.

contact area due to gold compression and wear, but no underlying metals are exposed and no significant material buildup is observed. Such properties are desirable for a spring-based interconnect.

12.6 Summary and Conclusions

A photolithographically defined microspring for electronics packaging has been developed. A variety of applications and prototype demonstrations were described. The technology has the potential to provide fundamental new capabilities for packaging, by improving testing options, stress management, and miniaturization. Key future challenges include matching the technology to the appropriate business plan to enable commercial development.

Acknowledgments

At PARC, many people have contributed to the springs work over the years, including Don Smith, Andy Alimonda, Dave Fork, Dirk DeBruyker, Thomas Hantschel, Koenraad van Schuylenbergh, Lai Wong, Vicki Geluz-Aguilar, Greg Anderson, Bowen Cheng, Kunal Sahasrabuddhe, Kevin Klein, Tom Birgersson, Ty Jagerson, John Knights, Jennifer Ernst, Randall Paquette, Armin Volkël, George Daniel, and Eric Peeters. Sun Labs, at Oracle, enabled the 180-μm pitch work, with help from Ivan Shubin, Alex Chow, Jack Cunningham, Matt Giere, Jing Shih, Ying Luo, Nyles Nettleton, John Simons, Ashok Krishnamoorthy, Jim Mitchel, and David Douglas. NIST and Xerox partially funded some early spring work. Significant portions of this chapter was originally published in the International Wafer Level Packaging Confcrence Proceedings, Santa Clara, California, October 11–14, 2010.

References

1. Worldwide Semiconductor Equipment Market Statistics (WWSEMS) Historical Report (1991–2009), SEMI, May 2010.
2. Guilian, G., Haba, B., Oganesian, V., Honer, K., Ovrutsky, D., Rosenstein, C., Axelrod, E., Hazanovich, F., and Aksenton, Y. (2007). Compliant wafer level package for enhanced reliability, *High Density Packaging and Microsystems Integration*, 26–27 June, pp. 1–5.
3. Bakir, M. S., Reed, H. A., Kohl, P. A., Martin, K. P., and Meindl, J. D. (2002). Sea of leads ultra-high-density compliant wafer-level packaging technology, in *Proc. 52nd ECTC*, pp. 1087–1094.

4. Hedler, H., Meyer, T., Leiberg, W., and Irsigler, R. (2003). Bump wafer level packaging, a new packaging platform (not only) for memory products, in *Proc. 2003 Int. Symp. Microelectron.*, Boston, MA, November 18–23.
5. Novitsky, J. and Miller, C. (2000). MicroSpring contacts on silicon: Delivering Moore's law-type scaling to semiconductor package, test and assembly, in *Proc. 2000 Int. Conf. High-Density Interconnect Syst. Packag.*, pp. 250–255.
6. Chen, P., Kalakkad, D., Matsubayashi, K., Maynard, W. and Williams, J. (2009). Assessing design tradeoffs with a customizable area-array connector, *Proceedings of the SMTA International Conference*, San Diego, CA, October 4–8.
7. Cheng, B., Chow, E. M., DeBruyker, D., Chua, C., Sahasrabuddhe, K., Shubin, I., Cunningham, J., Luo, Y., and Krishnamoorthy, A. V., Microspring characterization and flip chip assembly reliability, *42th International Microelectronics and Packaging Society* (IMAPS), November 1–5.
8. Shubin, I., Chow, E. M., Cunningham, J., Debruyker, D., Grier, M., Cheng, B., Nettleton, N., Anderson, G., Simons, J. and Douglas, D. (2010). A package demonstration with solder free compliant flexible interconnects, *60th Electronics Components and Technology Conference* (ECTC 2010), Las Vegas, 1–4 June, pp. 1429–1435.
9. Chua, C. L., Fork, D. K., Schuylenbergh, K. V., and Lu, J.-P. (2003). "Out-of-plane high-Q inductors on low-resistance silicon," *Journal of Microelectromechanical Systems*, 12, pp. 989–995.
10. Chua, C. L., Fork, D. K., and Hantschel, T. (2002). Densely packed optoelectronic interconnect using micromachined springs, *IEEE Photon. Technol. Lett.*, 14(6), pp. 846–848.
11. Shubin, I., Chow, E. M., Cunningham, J., De Bruyker, D., Chua, C., Cheng, B., Knights, J. C., Sahasrabuddhe, K., Luo, Y., Chow, A., Simons, J., Krishnamoorthy, A. V., Hopkins, R., Drost, R., Ho, R., Douglas, D., and Mitchell, J. (2009). Novel packaging with rematable spring interconnect chips for MCM, *IEEE 59th Electronic Components and Technology Conference* (ECTC 2009), May 26–29, pp. 1053–1058.
12. Drost, R., Hopkins, R., Ho, R., and Sutherland, I. (2004). Proximity communication, *IEEE Journal of Solid-State Circuits*, 39, pp. 1529–1535.
13. Cheng, B., Chow, E. M., DeBruyker, D., Shubin, I., Cunningham, J., Chow, A., Shi, J., Giere, M. and Böhringer, K. F. (2010). High current properties of a microspring contact for flip-chip packaging, *Solid State Sensor and Actuators Workshop*, Hilton Head, pp. 206–209.
14. Chow, E. M., Fork, D. K., Chua, C. L., Schuylenbergh, K V., and Hantschel, T. (2009). Wafer-level packaging with soldered stress-engineered microsprings, *IEEE Transactions on Advanced Packaging*, 32, pp. 372–378.
15. Chow, E. M., Chua, C., Hantschel, T., Van Schuylenbergh, K., and Fork, D. K. (2006). Pressure contact micro-springs in small pitch flip-chip packages, *IEEE Transactions on Components and Packaging Technologies*, 29, pp. 796–803.

16. Schauer, J., Pinckney, N., Nettleton, N. and Popovic, D. (2009). A test platform for the thermal, electrical, and mechanical characterization of packages, *Proceedings of the 42nd International Microelectronics and Packaging Society* (IMAPS).
17. Chow, E. M., Hantschel, T., Klein, K., Fork, D. K., Chua, C. L., Wong, L. and Van Schuylenbergh, K. (2003). Intermittency study of a stressed-metal micro-spring sliding electrical contact, presented at *IEEE 53rd Electronic Components and Technology Conference*, New Orleans, May 27–30, pp. 1714–1717.
18. Maciel, J., Majumder, S., Morrison, R. and Lampen, J. (2003). Lifetime characteristics of ohmic MEMS switches, *Proceedings of the SPIE — The International Society for Optical Engineering*, 5343, pp. 9–14.

Eugene M. Chow received the B.S. degree in Engineering Physics from the University of California, Berkeley, and the M.S. degree in Electrical Engineering, the M.S. degree in Management Science and Engineering, and the Ph.D. degree in Electrical Engineering from Stanford University, Stanford, CA. He researched microelectromechanical systems at Stanford University, focusing on through wafer electrical interconnects, atomic force piezoresistive cantilever arrays, and deep plasma etching. He is currently a Principal Scientist and manages the Microsystems Group in the Electronic Materials and Devices Laboratory at PARC, in Palo Alto, CA. He is involved in commercial and government device research involving microsprings, microassembly, and biodevices (drug delivery). He has also worked on novel printing concepts and integration of thin-film transistor and microelectromechanical systems. He has over 80 issued patents and over 80 technical publications.

Christopher L. Chua received B.S. degrees in Physics and Electrical Engineering from Massachusetts Institute of Technology and the Ph.D. degree in Electrical Engineering from Cornell University, Ithaca, NY. He is currently a Principal Scientist at PARC in the Electronic Materials and Devices Laboratory. There he has developed ultra-high density interconnect technology for interconnecting densely-packed surface-emitting lasers (VCSELs) with silicon driver electronics. He pioneered a VCSEL architecture employed by leading manufacturers. He has more than 100 US patents and has authored many technical papers in leading scientific publications.

Chapter 13

MEMS Reliability

Chien-Ming Huang, Arvind Sai SarathiVasan, Yunhan Huang,
Ravi Doraiswami, Michael Osterman and Michael Pecht

13.1 Introduction

Microelectromechanical systems (MEMS) devices (e.g. accelerometers, gyroscopes, pressure sensors, resonators, relays, switches, and micro-pumps and valves) have been widely adopted for use in mobile phones and gaming consoles owing to their high throughput, cost efficiency, small size, and ability to integrate well with electrical circuits. MEMS devices are also being adopted for use in applications such as RF/microwave or optical communication, energy scavenging, and in the biomedical field. Texas Instruments' Digital Micromirror Device (DMD) for digital projectors [1], Analog Device's accelerometer for airbag deployment [2], STMicroelectronics' gyroscopes in iPhones [3], and MEMS printer heads for inkjet printers [4] are only a few examples of commercial uses of MEMS technology. More than 100 billion devices are expected to be shipped by 2020 [5].

MEMS devices are expected to be incorporated in large quantities into mission-, safety-, and infrastructure-critical systems [6, 7]. However, these applications' reliability requirements (both service life and stress conditions) are strict compared to those for current mobile phones and gaming console applications. MEMS manufacturers can repeat their success with mobile and gaming applications in mission- and safety-critical applications only if long-term performance can be assured.

One of the interesting aspects of current MEMS testing practices is that reliability qualification of MEMS devices is still based on the standards established for the qualification of microelectronic packages. Unlike regular microelectronic packages, MEMS devices typically include both integrated circuits and movable mechanical parts. Depending on the application environment and purpose, the materials and the manufacturing steps used might be different between each device. Due to this diversity in design, materials, and manufacturing methods, the dominant failure mechanisms in MEMS are not the same across all MEMS devices and not the same as regular microelectronic packages. Thus, standards developed for qualifying microelectronic packages will not be sufficient to assure long-term performance of MEMS devices — this assurance can only be provided by understanding the dominant failure mechanisms and by performing appropriate accelerated stress tests. In this chapter, the known failure mechanisms of MEMS devices are reviewed with the intention to help understand the types of reliability issues encountered with MEMS devices.

The following list summarizes the failure mechanisms covered in this chapter. Each failure mechanism will be discussed individually in the subsequent sections. Table 13.1 summarizes the knowledge level of various MEMS failure mechanisms.

- Mechanical fracture
  - Overload, shock
  - Vibration
  - Corrosion
  - Fatigue
- Stiction
  - Van der Waals force
  - Capillary force
  - Chemical bonding

Table 13.1. Knowledge level of various failure mechanisms.

Failure mechanism	Involved MEMS devices	Knowledge level
Mechanical fracture	Accelerometer, gyroscope, resonator, vibration energy harvester	Very good
Stiction	RF-MEMS capacitive switch, RF-MEMS ohmic switch	Very good
Creep and fatigue	Resonator, RF-MEMS capacitive switch	Very good
Electrical short/open	RF-MEMS ohmic switch	Good
Charge accumulation	RF-MEMS capacitive switch	Good
Contamination	Resonator	Limited
Wear	RF-MEMS capacitive switch	Limited

  - Electrostatic charging
  - Residual stress

- Charge accumulation
  - Electrical stress
  - Radiation
  - Improper handling

- Wear
  - Adhesion
  - Abrasion
  - Corrosion
  - Surface fatigue

- Creep and fatigue
  - Intrinsic stress
  - Applied stress
  - Thermal stress

- Electrical shorts and opens
  - Dielectric material degradation
  - ESD, high electrical field
  - Electromigration
  - Oxidation

- Contamination
  - Intrinsic (e.g. crystal growth)
  - Manufacturing-induced
  - Usage environment-induced.

13.2 Mechanical Fracture

Mechanical fracture is the local separation of an object or material into two or more pieces under the action of stress. In MEMS devices that contain movable structures, including accelerometers, gyroscopes, and micromirrors, fracture can result from various causes: mechanical shock and overload [8], corrosion [9], stress corrosion cracking (SCC) [10–13], and material fatigue [14]. An example of a fractured MEMS cantilever is presented in Fig. 13.1.

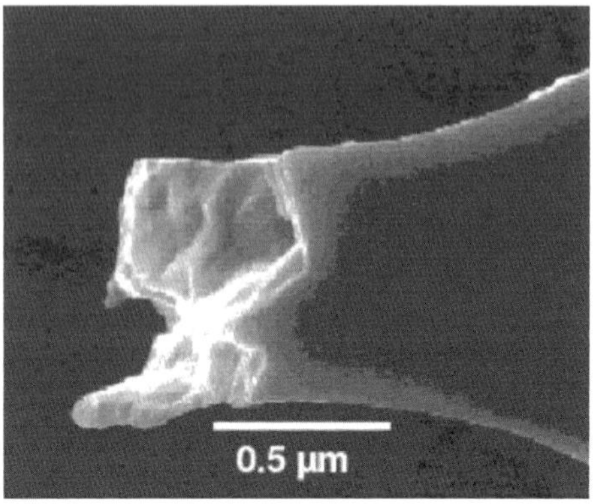

Fig. 13.1. Mechanical fracture of a silicon cantilever. A clear brittle fracture signature is observed [15] (Courtesy of Sandia National Laboratories).

13.2.1 *Shock*

Mechanical shocks can induce highly dynamic loads on structures, causing cracking, chipping, and fractures. Researchers have indicated through experiments the need for robust models to predict and explain the various reported failures in MEMS devices and structures [17]. There have also been attempts to study the shock reliability of particular MEMS devices at Sandia National Laboratories [18, 19] and at other laboratories [20]. Béliveau *et al.* [21] characterized experimentally the response times and linearity of output signals of several commercial accelerometers to shock loads up to 70,000 g (where "g" is the Earth's gravitational acceleration). They concluded that the survivability of the devices was related to their scale size features, which reduced their susceptibility to shock loading. Béliveau *et al.*'s study indicates that element design is key to determining the g survivability of devices.

Brown *et al.* [22, 23] studied MEMS sensors subjected to conditions over 10,000 g. They successfully demonstrated the survivability of the sensor and telemetry (TM) unit at about 50,000 g in a flight test. Brown *et al.*'s works demonstrate the feasibility of using a MEMS sensor in a high-g environment. Tanner *et al.* [24] tested MEMS microengines against shock pulses of various time durations and amplitudes and observed broken mechanical components (e.g. gear anchor, pin joint, and linkage arm in the comb-drive actuators). They found electrical shorts induced by debris from the die edges and alignment marks in actuators at levels greater than 10,000 g. They also found that cleaning up the edge of the die or encasing the moving parts could mitigate the debris' effect. Tanner *et al.*'s research raises more global concern for MEMS reliability at high shock levels within the package. The limiting factor will reside in the survival of the package itself in overall functionality. To reduce or prevent these failures from occurring, plastic or non-ceramic packages are utilized.

Younis *et al.* [8] presented a Galerkin-based reduced-order model that is capable of accurately capturing the dynamic behavior of micro-cantilevers and clamped–clamped microbeams under shock pulsing at various amplitudes (low-g and high-g). Avoiding stress concentration by filleting sharp points, lines, and corners is an effective way to prevent fracture. Younis *et al.* thought that the shock response of regular-shaped microstructures could be investigated by reduced-order models. Their research on analytical formulation of the shock mode is insightful for MEMS designers. For example, the static response of irregular-shaped microstructures can be obtained numerically by using finite element models, which can be done by commercial software.

MEMS gyroscopes are widely used in automotive, consumer electronics, and aerospace applications [25–27]. Owing to their small size, light weight, and low power consumption, they are especially present in portable electronic products. The first cell phone equipped with a 3-axis MEMS gyroscope was introduced to the market in 2010, and in the next year more phones and tablet computers were equipped with MEMS gyroscopes [28]. Shock impact reliability of electronic devices is a challenge for the automotive and aerospace applications owing to the harsh environments (up to 20,000 g shock loading). With the success of consumer-level MEMS gyroscopes, shock impact reliability remains a major concern because portable consumer electronics are prone to accidental drops. In a normal operating environment, a cell phone experiences acceleration of around a few to dozens of g's [29]. However, when a cell phone accidentally falls from a height of 1.2 m, it can experience accelerations ranging from 500 to 5,000 g depending on the impact orientation and the strike surface [30–33]. Internal collisions often occur inside the cover of modern portable electronic products because of the dense packaging. These sudden impacts can induce acceleration levels of tens of thousands of g's on the printed wiring board (PWB) [34, 35].

When designing robust MEMS gyroscopes, it is helpful to understand all the possible failure modes. Shock-loaded MEMS can fail in different modes including fracture [19, 35, 36], delamination [19, 37], and stiction [38, 39]. For comprehensive MEMS design and impact reliability assessment, both efficient reliability tests and advanced simulations are necessary. Several experiments and simulations have been performed by researchers in this field. For example, Wagner *et al.* [40, 41] studied the mechanical reliability of epi-polysilicon MEMS structures under shock loads and of micro-machined membranes under particle impact. They found that the reliability of silicon nitride membranes was related to the absolute distance to the edge under the particle impact. The lifetime of the membrane could be increased tenfold under particle bombardment by optimizing the edge region. Wagner *et al.* also found that slight deviations from optimum loading conditions in epi-polysilicon test structures can significantly increase the maximum experienced stress. Based on the deviation assessments, the dynamic failure stress was estimated to be 50% less than the quasi-static fracture strength. Wagner *et al.*'s research provides useful information for reliable MEMS structure design under external shock and particle impact.

Brown [42] carried out extensive experiments in harsh military environments (up to 20,000 g shock loading) on MEMS sensors and showed the robustness of some devices. Brown's works show that MEMS devices in military applications will be able to provide the essential requirements for a successful munition transition. Choa [43] focused on vacuum packaging of MEMS gyroscopes and provided their drop and vibration qualification test results. Choa proposed three methods to enhance the vacuum integrity and then improve the fatigue life of MEMS gyroscopes under cyclic mechanical shock. This research provides another point of view for improving the reliability of MEMS devices by enhancing the degree of vacuum. Al Saleem *et al.* [44] indicated that the PCB motion has a significant influence on the dynamic response of MEMS devices under shock impacts by performing a theoretical and experimental investigation. They found that the experimental data were in good agreement with the simulation results, which confirmed their assumptions. Saleem *et al.*'s research emphasizes the importance of taking PCB effects into account while evaluating the design of MEMS devices under shock load.

Hauck *et al.* [45] predicted the probability of crack initiation due to stress concentrations by conducting simulations of dropping MEMS accelerometers under a 20,000 g impact. They proposed a quantitative assessment of failure probabilities in the MEMS structure. Hauck's works are transferable and can be applied to a large range of MEMS designs. Mariani *et al.* [46] simulated crack evolution induced by impact in polysilicon MEMS accelerometers. They concluded that the failure mode of a uniaxial MEMS accelerometer was marginally modified by the package, but strongly influenced by polysilicon topology. Mariani *et al.*'s research shows that the formulation of materials is key to the reliability of MEMS devices. Yang *et al.* [47] presented simulation results for MEMS accelerometers inside quad-flat no-leads (QFN) packages. They predicted that stiction may occur in the QFN packaged two-axis MEMS accelerometer during the drop test, but fracture failure should not happen. Yang *et al.*'s works can be applied to other MEMS devices as well as different packaging schemes to analyze potential failure modes of packaged MEMS products.

The studies on MEMS gyroscopes, especially 3-axis gyroscopes, are limited because most of the investigations focused on MEMS accelerometers. As a result, new experiments and simulations should be carried out for better understanding of MEMS gyroscope behavior under shock impacts. To study shock reliability, shock impact tests at several levels should be considered. The acceleration tolerance of the package and the functional system should be studied separately. For the package tolerance, non-functional dummy devices are employed and failures are defined as the visible package failure. For the functional system, the output changes in offset, sensitivity, or noise in any axis of the device can be characterized as failures if they exceed the predefined values. Examples of acceleration tolerance tests for gyroscopes are presented in [48]. The characterization of the acceleration tolerance began with a shock impact with 1,000 g peak acceleration, and the acceleration was increased by 500 g every time if no failure was recorded. If the gyroscope failed, the next one would be tested at an initial acceleration level of 1,000 g below the previously recorded smallest acceleration tolerance. Sixteen gyroscopes were tested per impact orientation in each failure criterion. The devices were in a power-off state, and health monitoring was conducted between the shock impacts via a rotation device [49] during the shock impact tests. If failures happened more than two times within five consecutive evaluations in the health monitoring system, the failure was considered permanent. Otherwise, the failure was regarded as transient.

All the presented results provide deep insight into the reliability of the MEMS gyroscope while sustaining shock impacts. The comb structure in Li *et al.*'s study [48] was extremely sensitive to the X-orientation motion due to the nature of the gyroscope. A shock level of 1,500 g is considered enough to induce internal contacts (surface-to-surface contact) between the movable and stationary parts. These contacts might lead to short circuits, errors in the signal output, or even stiction. Similarly, the spring structure is sensitive to the Y-orientation motion due to the nature of the gyroscope. However, the contact failures that occurred in the X- or Y-orientation impact might also happen in the Z-orientation impact because this impact would also excite the in-plane movements of the comb structures. Li's study [48] explains why the average shock levels of the functional failures are close regardless of the impact orientations. It can help the MEMS design avoid or alleviate contacts during shock impacts by introducing new distance stoppers into the structure. Moreover, the internal contacts can be predicted during the shock impacts in different orientations.

13.2.2 *Vibration*

Vibration is a repetitive mechanical motion. While developing a vibration-sensing instrument, a number of attributes should be considered. First, an oscillating motion often contains both linear and rotational components. Second, most vibration-sensing instruments focus on the magnitude of the oscillation, not on tracking the absolute position. Thus, linear sensors such as MEMS accelerometers are sufficient for capturing motion information. When the motion is primarily linear, understanding the direction of vibration can be essential, especially by using single-axis sensors. On the other hand, three-axis sensors can offer more mounting flexibility because the orthogonal orientation enables the selection of one or more axes regardless of the vibration direction.

Brown *et al.* [50, 51] performed extensive experiments on MEMS accelerometers, such as shock, vibration, temperature cycling, and flight tests on artillery projectiles. They saw promising results on automobile-grade accelerometers. Lee *et al.* and Huang [53, 54] monitored a clamped micromirror with reflected optical patterns and determined it to be error free over a range of frequencies from 200 Hz to 10 kHz during vibration experiments. They claimed there was no vibration effect on the clamped mirror at this scale. On the other hand, vibration happens when an actuator is not clamped and free to move.

The working principle of MEMS accelerometers is based on the piezoresistive effect of semiconductor materials. The piezoresistors and beam structures are connected to form a Wheatstone bridge. The acceleration value can be obtained according to the change of output voltage on the Wheatstone bridge. Piezoresistive MEMS accelerometers may suffer structural damage, surface adhesion, or fracture in an external vibration environment. The main failure mechanism is beam fatigue of the accelerometer because long-time environmental vibration eventually leads to fracture of the cantilever beam. When MEMS accelerometers fail under long-time constant vibration stress, it can be concluded that long-time vibration can be destructive. Thus, it is necessary to examine the vibration reliability of MEMS accelerometers.

The fatigue failure of MEMS accelerometers under a vibration environment is a gradual and cumulative process that partially consumes the life of an accelerometer. Moreover, fatigue properties are based on the cumulative damage theory. When the cumulative damage reaches a certain degree, the structure of the MEMS accelerometer will be fractured. The relationship between the stress and the fatigue life can be described. The breaking strength decreases gradually while the cycling times are increased. In order to assess and predict the fatigue life of a MEMS accelerometer quantitatively, it is necessary to establish the relationship between external stress and device life, and to analyze the feasibility of accelerated testing under a vibration environment according to the testing results.

Some design parameters can be constrained for reliability analysis and life prediction of MEMS accelerometers; this kind of constraint facilitates the research and development of inertial sensors and provides higher reliability in the vibration environment. Studies on MEMS accelerometers are as follows: Zhou *et al.* [56] presented a fully symmetrical double-sided H-shaped beam structure for MEMS capacitive accelerometers. The sensitivity of the device is 0.24 V/g, and the nonlinearity is 0.29% over the range of 0–1 g. Zhou *et al.*'s H-shaped beam structure shows better resonance frequency, quality factor, sensitivity, and less nonlinearity than other structures in the literature [57–59]. Peng *et al.* [60] designed a new accelerometer with axially deformed tiny beam structure. They used mirrored triply periodic photonic crystal to replace the original pressure-sensitive resistor as the sensing element. The magnitude of the acceleration can be calculated while measuring the transmissivity of the photonic crystal. Peng *et al.*'s research provides acceleration estimation on an axially deformed beam structure for device reliability analysis.

Jiao *et al.* [61] adopted a wafer-level glass–silicon–glass structure and introduced a system-level aluminum shell filled with two-component epoxy resin as a new packaging method for MEMS accelerometers. The advantages of the packaging accelerometer are strong sealing, little residual stress, high-impact resistance, good stability, and suitability for small batch production. Li *et al.*'s works [62] analyze the relationship between the packaging materials and the performance of sensors. They verified that higher sensitivities and better anti-overload capabilities were found in accelerometers with a stainless steel package rather than a ceramic package. They also found that die adhesive thickness and potting materials can affect the sensitivity of the sensor.

Mohd-Yasin *et al.* [63] measured the combined noise and reliability of a 3-axis micro-accelerometer by developing a special measurement system. They aimed to provide empirical noise and reliability data on commercial 3-axis micro-accelerometers. They concluded that the devices were very reliable when subjected to individual excitation tests, including gravitational, thermal, and acoustical tests, but exhibited slight deviations during the combined tests. Mohd-Yasin *et al.*'s research provides the empirical data for real-world application of commercial devices. Ma and Li [64] quickly determined the weak links of micro-accelerometers in a vibration environment by examining the effects of step stress and sustained stress on micro-accelerometers. They summarized the stress limits, the relevant failure modes, and the failure mechanisms of the sensors in a vibration environment. Yuan *et al.* [65] evaluated the storage reliability and life of accelerometers using step-stress accelerated degradation testing. They calculated the storage life of each sample with the threshold of failure, and concluded that the storage life complied with the Weibull distribution pattern.

Referring to Yang's work [66] and statistical theory, the failure probability for the same sample under different acceleration stresses is equal if the failure mechanism does not change. Accordingly, the accelerated and the normal stress tests have the same testing ability statistically. This finding makes these two testing methods equivalent. Thus, accelerated stress testing can be applied to the MEMS accelerometer under a vibration environment [67, 68].

The reliability relationship of the MEMS accelerometer can be predicted and evaluated by statistical methods. According to Miner's principle, the cumulative damage of variable stress amplitude is expressed as:

$$D_r = \sum_i \frac{\sigma_{p_i}^k n_i}{C} \tag{13.1}$$

Assume the same damage for the same testing times under the equivalent constant stress amplitude and variable stress amplitude, as is shown:

$$D_r = \sum_i \frac{S_r^k n_i}{C} \tag{13.2}$$

Combining expressions (1) and (2) leads to the following equation:

$$S_r = \left(\frac{1}{D_r}\right)^{\frac{1}{k}} \left(\frac{\sum_i \frac{\sigma_{p_i}^k n_i}{C}}{\sum_i n_i}\right)^{\frac{1}{k}} \tag{13.3}$$

The accelerated model of random vibration follows the fatigue characteristics of the S-N curve and can be approximated as follows:

$$NS^k = C \tag{13.4}$$

where k and C are material constants, S is the root-mean-square value of the stress, and N is the life under S stress.

Assume the equation $\eta = N/N_f$ (N_f is a constant value and the average life per unit time), then the equation $\eta = CS^{-k}$ can be obtained from formulas (3) and (4). Take the natural logarithm on both sides of the equation to obtain the following equation:

$$\ln \eta = \alpha + \beta \ln S \tag{13.5}$$

where $\alpha = \ln C$, $\beta = -k$. The failure equation of a MEMS accelerometer under vibration stress can be described by the inverse power law model, and its coefficients can be determined as α and β according to the testing data.

13.2.3 *Corrosion*

Corrosion is disintegration of a material into its constituent atoms due to chemical reactions with its surroundings. Corrosion can be a contributor to mechanical fracture and is generally categorized into three classes according to its causes: pitting, intergranular, and crevice corrosion. In pitting corrosion, the localization begins at microscopic heterogeneities such as scratches and inclusion, whereas in crevice corrosion the localized aggressive environment is in a macroscopic crevice [69]. Zhang *et al.* [70] proved that the maximum sustainable load of a micro-sized nickel cantilever beam decreases as the time of exposure to a corrosive environment (3.0% NaCl solution at room temperature) increases. They also pointed out that a corrosion-induced fracture first occurs in the range of a notch of a higher tensile stress concentration and exhibits clear evidence of a trans-columnar fracture. Several other factors, such as moisture and temperature, have been found to increase the corrosion rate of some metals, such as aluminum, and the junctions of dissimilar metals.

Stress corrosion cracking (SCC) is a form of crevice corrosion that occurs on a material in an area of high stress concentration. It has been

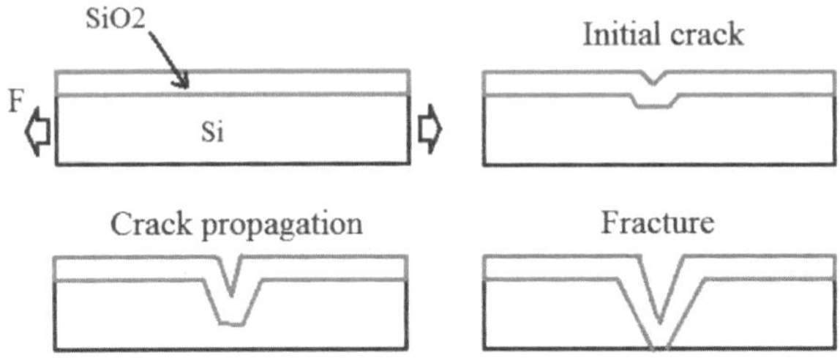

Fig. 13.2. SCC failure mechanism in silicon substrate when subjected to tensile strength [71].

found that metal structures with severe SCC can appear bright and shiny even while being filled with microscopic cracks, which makes it difficult to detect failure in advance [71]. SCC can also be a concern with silicon because silicon in air is always covered by a thin native oxide layer and results in crack growth when it is under tensile stress. As the crack propagates, the silicon comes closer to the SiO_2/air interface, thereby facilitating the oxidation of the silicon deeper into the structure. This process continues until the remaining part of the structure can no longer stand the concentrated stress and finally fractures. The process is illustrated in Fig. 13.2. Additionally, fatigue is one of the main causes of material fracture, especially for metals and alloys. Fatigue-induced failure mechanisms are discussed in Section 13.5.

Diagnosis of fracture failure can be conducted by visual inspection with an optical microscope or scanning electron microscope (SEM). An SEM is a type of electron microscope that images a sample by scanning it with a finely focused, high-energy electron beam in a raster pattern. In conventional SEM imaging, specimens must be electrically conductive at the surface and electrically grounded to prevent the accumulation of electrostatic charge at the surface. For nonconductive samples, an ultrathin coating of electrically conductive material (e.g. gold, tungsten, graphite) either by sputtering or by evaporation is needed. Characterization of the fracture surfaces using SEM will produce clues about crack initiation sites and fracture propagation, which are helpful for determining the root cause of a fracture.

13.3 Stiction

Stiction is one of the main factors limiting the broad applications of MEMS. Stiction is the unintentional adhesion or friction of compliant microstructure surfaces, making the restoring forces unable to overcome the interfacial forces. Once the contact is made, the effect of the two forces is sufficient to deform the structures and pin them to the substrate, which results in device failure. This kind of failure is one of the major yield losses in MEMS fabrications.

Adhesion may be triggered by either capillary condensation or van der Waals forces [72]. Capillary condensation is related to moisture and surface contamination, and van der Waals forces are affected by surface roughness. Because the devices are minute, gravity and other body forces do not play a significant role.

MEMS stiction problems can be classified into two categories: release-related stiction and in-use stiction [73]. Release-related stiction forms while removing the sacrificial layer and is caused primarily by capillary forces. In-use stiction usually occurs while exposing the released microstructures to a humid environment. Several techniques have been developed to prevent this undesirable stiction, including supercritical drying [74], freeze-drying [75], and separation dimples and sharp corners on contact surfaces [76]. However, in-use stiction becomes more serious when the microstructures are exposed to a humid environment [77]. Capillary force is the strongest force to cause surface stiction and is caused by water droplets between two surfaces in a humid environment. Water vapor can condense in small cracks or pores to form a thin film between the contact points of the surfaces. Water vapor also can condense as water droplets in the narrow gaps between the surfaces. If the surfaces of the microstructure are hydrophilic, the microstructure would be pulled toward the substrate by the attractive force exerting from water droplets, thereby causing the microstructure to eventually collapse and be adhered permanently. In a humid environment, the capillary force exerting between MEMS surfaces can be determined by surface hydrophobicity and temperature.

Van der Waals forces cause materials to be attracted in short range at the molecular level. Usually, the van der Waals force is weaker than the capillary force. However, when the surface roughness is very low ($<1\,\mathrm{nm}$) [77], a large proportion of the surfaces can be attracted to contact with each other and thus play a significant role in the stiction behavior. On the other hand, when the surfaces are very rough ($>100\,\mathrm{nm}$) [73], only a small proportion of the surfaces would be attracted because the van der Waals force is weak. Depending on the MEMS fabrication conditions, electrostatic force, Casimir force [78], and hydrogen-bridge also can cause stiction problems [79].

The effect of stiction can be significantly reduced by coating surface passivation and the use of moisture-free packaging [80]. To greatly reduce moisture, is the preferred means of encapsulation hermetic packaging. Surface passivation layers are also commonly used to lower the surface interaction energy and prevent stiction. These layers provide hydrophobic surfaces that prevent water condensation and capillary stiction. However, the reliability and reproducibility of these layers is affected by the high temperatures that occur in MEMS packaging steps. The best way to avoid stiction failures is to eliminate the presence of contacting surfaces by using adequate design or enhancing the restoring force.

Due to the high ratio of surface area to volume in some MEMS devices, surface forces can play a dominant role relative to volume forces (e.g. gravity) and sometimes cause microscopic structures to stick together when they come into contact. Therefore, stiction is a serious problem for MEMS devices that contain moving parts. The stiction force can be determined by calculating the adhesion energy, which is defined as the energy required for separating surfaces once they are in contact [81]. The adhesion energy, $E(z)$, of two perfectly flat surfaces at a separation distance z per unit area due to surface forces can be approximated by [82]

$$E(z) = \sum_i \frac{c_i}{z^{n_i}} \tag{13.6}$$

where c_i is a constant for force i depending on the properties of the surfaces and their environment, and n is the power of the interaction ranging from 0 to 2 for different forces. Apart from the highest parts (asperities) on the two surfaces touching each other, the other points also contribute the overall adhesion energy. Hence, a distance distribution function $h(z)$ is used in the model to describe the distance between two points located on two surfaces [83]. Given the distance distribution function, the adhesion energy equation can be rewritten by taking into account all contributions of different forces as follows:

$$E = \int_0^\infty \frac{c_i}{z^{n_i}} h(z)dz \tag{13.7}$$

A number of experimental techniques have been developed to detect the interactions involved in microstructures. Maboudian [84] reviewed different experimental methods used for examining adhesion and friction at the micrometer scale. Many published articles have focused on the adhesion problem. However, the friction coefficient and wear characteristics of microstructures would be affected by various surface treatments, which were designed to reduce adhesion, but they remained undiscovered. Maboudian's work indicates that the effect of various surface treatments on mechanical properties, such as fracture strength, is related to surface interaction energy. Spengen *et al.* [85] examined the influence of sticking surfaces on the reliability of a MEMS accelerometer. They concluded that calculating the reliability of devices was possible when the restoring force and/or the surface interaction energy were changing over time. Spengen's work means the reliability of micro-accelerometers can be estimated by using accelerated reliability tests with enhanced stressing conditions.

The sources of stiction can be classified into five categories: van der Waals adhesion, capillary adhesion, chemical bonding, electrostatic charge, and residual stress [86]. Figure 13.3 shows images of different types of stiction that occur in MEMS devices.

13.3.1 *Van der Waals Force*

Van der Waals force is caused by the interaction of atoms or molecules arising from permanent or induced dipoles at surfaces in close contact [88]. For two adjacent closely spaced plates, the van der Waals force can be described by the following equation:

$$F_{vdW} = \frac{A}{6\pi D^3} \tag{13.8}$$

where D is the distance between the two plates and A is the Hamaker constant (e.g. 1.6 eV for silicon), which is dependent on the permittivity, refractive index, and main absorption frequency of the surface materials [89].

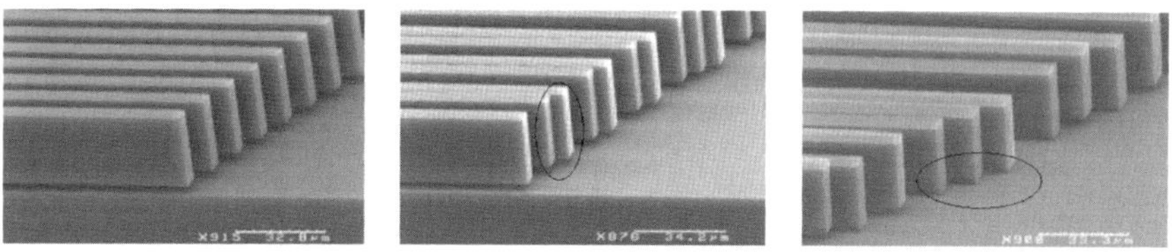

Fig. 13.3. Examples of mixed adhesion (the areas in the circles): (Left) no adhesion (good release); (Middle) only lateral adhesion; and (Right) lateral and vertical adhesion [87]. (Courtesy of IOP Publishing Ltd.)

A positive Hamaker constant corresponds to an attractive force. The van der Waals force decreases very quickly at distances larger than about 20 nm. On the other hand, when two surfaces get so close that their electron shells are deformed, the repelling forces between atoms will reduce the strength of the van der Waals force. Therefore, to simplify the calculation, only the inter-atomic distance range, which is between 0.165 nm and 20 nm, is considered, Eq. (1). Thus, Eq. (7) can be rewritten as [82]:

$$E_{vdW} = \int_{0.165\,\mathrm{nm}}^{20\,\mathrm{nm}} \frac{A}{24\pi z^2} h(z)dz \tag{13.9}$$

Minimizing the real contact area and optimizing the surface roughness distribution can minimize the van der Waals force and, thus, the associated stiction. How the above mentioned factors minimize the stiction will be discussed in detail at the end of this section.

13.3.2 *Capillary Force*

The capillary force in MEMS devices can be derived from two sources: in-use processes and manufacturing processes. For in-use process, almost all surfaces are covered with a monolayer of water. Although the thin water film formed by the monolayer on surfaces can reduce surface wear by functioning as a lubricant, monolayers are usually unfavorable for MEMS structures, especially those with moving parts, because of the induced capillary force. To make matters worse, most dielectric layers, such as SiO_2, are hydrophilic, which further contributes to capillary stiction [90]. The second source of capillary adhesion is the manufacturing process. When a device is removed from an aqueous solution after the wet etching of an underlying sacrificial layer during a release process, the liquid meniscus formed on hydrophilic surfaces pulls the microstructure toward the substrate and stiction can occur. This release-stiction problem can be alleviated by dry HF etching or supercritical CO_2 drying.

One interesting finding by Bowden and Tabor [91] is that the adhesion energy due to capillary condensation depends only on the amount of surfaces wetted, and not on how close together the surfaces are in the wetted area. Thus, the equation for the adhesion energy due to capillary forces can be expressed as

$$E = \int_0^{d_{cap}} 2\gamma_1 \cos(\theta) h(z)dz \tag{13.10}$$

in which d_{cap} is the characteristic distance closer than which water will fill all spaces, γ_1 is the surface tension of water, and θ is the contact angle [92, 93].

For MEMS micromirrors, excessive adhesive force between the landing tip and lending site may lead to stiction failure of the device. While applying the electronic reset sequence, adhesive force is sufficiently high and may obstruct the movement of the mirror. Capillary water condensation induces the landing tip of the mirror to stick to an adequate landing site. A partial vacuum is created at the interface due to the surface tension, and increasing forces are required to pull the tip apart from the landing site. For this reason, the common method to eliminate stiction in a MEMS micromirror is to implement springs on the landing tips of the mirror [80]. When the mirror's landing tip contacts the landing site, the spring bends and stores energy to assist the mirror in taking off from the surface when the reset pulse is being applied and bias voltage is being removed.

13.3.3 *Chemical Bonding*

Another common contributor to stiction is chemical bonding between contact surfaces [94]. One example is hydrogen bonding, which is associated with water on surfaces. The first monolayer of water on a surface is chemisorbed to the surface in the form of hydroxyl. Subsequent layers are physisorbed as water. Both molecules have strong dipoles due to the positive charge located in the hydrogen and the negative charge in the oxygen. When surfaces come into contact, the dipoles on either surface create an attractive force. Hydrogen bonding can only occur between molecules with a permanent dipole containing strong electronegative atoms such as oxygen, fluorine, and nitrogen. The hydrogen bonding strength is generally smaller than covalent bonding but larger than van der Waals bonding. The total adhesive force between the surfaces depends on the number of these bonds, and thus, on the relative humidity as well as the hydrophobicity of the surfaces. Large surface areas and high surface energies of the released microstructures would increase the probability of in-use stiction by high adhesion forces associated with capillary, van der Waals, electrostatic, and hydrogen bonding effects [95]. A simple analysis of an ideally smooth silicon surface with intervening medium of water and 1 V applied voltage is presented in Fig. 13.4. The results show that capillary F_{cp}, electrostatic F_{el}, and van der Waals F_{vdW} attractive forces exceed the typical range of micromachine restoring forces when the surface gap is less than 10 nm [95]. The

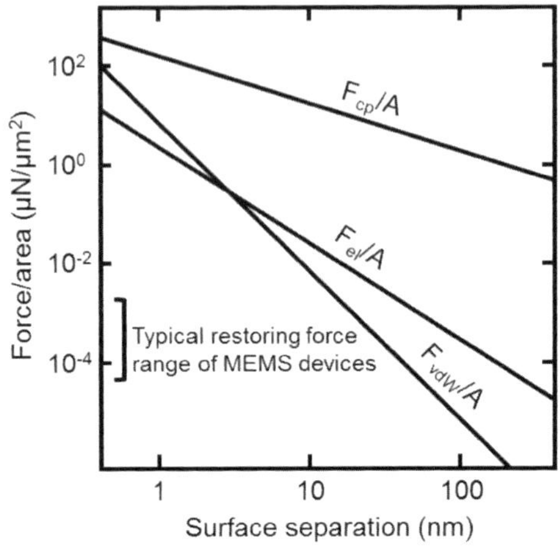

Fig. 13.4. Variation of interfacial attractive forces with surface separation distance for smooth silicon surfaces [95].

reason is that capillary, electrostatic, and van der Waals forces are inversely proportional to the first, second, and third/fourth (short/long range) power of the surface separation distance, respectively. The reduction of hydrogen bonding can be accomplished by changing the hydrophobic/hydrophilic nature of the surfaces [96, 97].

13.3.4 *Electrostatic Charging*

Stiction can be caused by electrostatic force on two neighboring surfaces with different contact potentials, tribocharging of rubbing surfaces, and ion and/or electron injection and trapping in dielectric layers. Stiction is especially common in electrostatic MEMS devices requiring high operating voltages, such as RF-MEMS, whose typical equivalent structure is shown in Fig. 13.5. F_{e} is the electrostatic force generated by normal operating voltage V. F_{m} is the mechanical restoring force due to the deformation of the membrane or cantilever in RF-MEMS, assuming that the spring constant is k. A bias force F_e' is generated by the accumulated charge in the dielectric layer during the operation. Its direction can be either downward or upward.

The operating voltage (typically 20–50 V) over a thin dielectric layer can subject the switch to an ultrahigh electrical field, where electrons and ions are potentially prone to be injected and trapped inside the dielectric. A bias force is generated by these accumulated charges [98]. When the bias

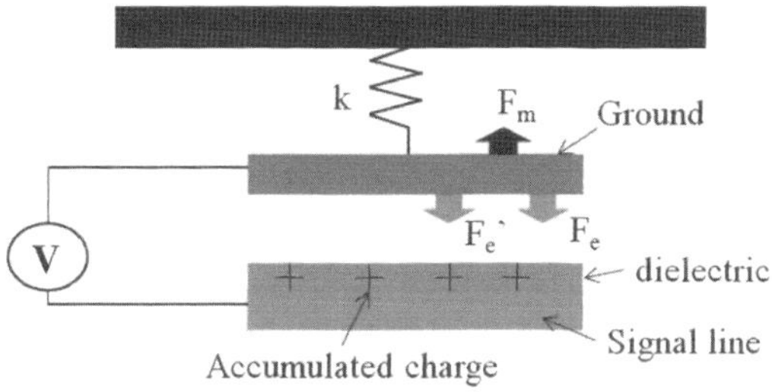

Fig. 13.5. Parallel-plate model of a typical RF-MEMS.

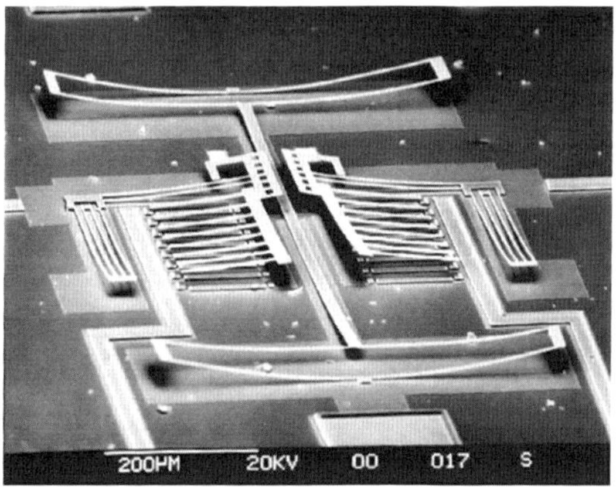

Fig. 13.6. SEM image showing MEMS structures bending up after release. Vertical residual stress was generated due to the absence of the annealing process [103]. (Courtesy of Prof. Kristofer S. J. Pister.)

force becomes large enough to overcome the mechanical restoring force of the membrane, it causes the membrane to remain in the down position even after the removal of the operating voltage.

13.3.5 *Residual Stress*

Many MEMS devices need to be released from substrates during the microfabrication process. Residual stress can be introduced in the thin-film deposition step through evaporation, sputtering, and chemical vapor deposition (CVD). Residual stress comes into play during the release process (removal of the sacrificial layer). After being released, the microstructures are free to move, deform, or bend, which can cause adjacent surfaces to come into contact, and, with the help of the surface forces described above, stiction occurs. Figure 13.6 depicts an example of deformation due to residual stress. DeBoer *et al.* [99–101] first distinguished two kinds of adhered beams due to residual stress. They indicated the S-shaped cantilever beam configuration should be studied much more for adhesion forces in surface micromachining because this configuration provided great latitude in measuring auto-adhesion. The equilibrium is deep and easily attainable from either side of the beam structure. Yee *et al.* [102] derived a stiction model on freestanding thin-film beams that takes into account the net internal bending moment caused by the residual stress gradient in structural polysilicon. They analyzed the effect of residual stress gradient quantitatively by comparing various suspended microstructures with different surface energies, which induced stiction. Yee's work provides a formation technique for substrate polysilicon to observe the effects of residual stress gradient on stiction under low adhesion force condition.

Adhesion and stiction can be reduced by modifying the structural stiffness, interfacial topography, and surface chemistry characteristics [104]. Van Spengen *et al.* [82] reviewed approaches for preventing stiction in MEMS, including the use of rough surfaces, the use of hydrophobic surfaces, prevention of water from coming in contact with MEMS structures, deposition of an anti-stiction coating, and design of a moving structure that is stiff enough to overcome stiction force. They found stiction can be diminished by lowering the resistivity of a dielectric that is prone to charging due to electrostatic causes, and the restoring forces degraded faster due to creep at high temperatures. According to Van Spengen's work, it is possible to calculate the expected failure rate of the devices during their intended life in a shock profile. Man *et al.* [105] demonstrated the elimination of the adhesion of polysilicon microstructures to their substrates by depositing a relatively conformal, 10–20-nm-thick hydrophobic fluorocarbon (FC) coating formed by plasma polymerization of C_4F_{10} precursor on a field-freezone.The coating proved to be very resistant to chemicals, high temperatures, and wear. However, this approach has a drawback, namely, it is challenging to find coating materials that are compatible with the microfabrication processes and packaging techniques. Taii *et al.* [106] presented a surface treatment technique that effectively avoided both process stiction and in-use stiction of MEMS devices by using vapor HF (hydrofluoric) acid for sacrificial release, UV ozone for oxidation, and vapor HMDS (hexamethyldisilazane) for SAM (self-assembled monolayer) coating of small surface free energy. HMDS-SAM-coated silicon surfaces have been found to have an adhesive strength that is 20% of that between the oxidized silicon surfaces.

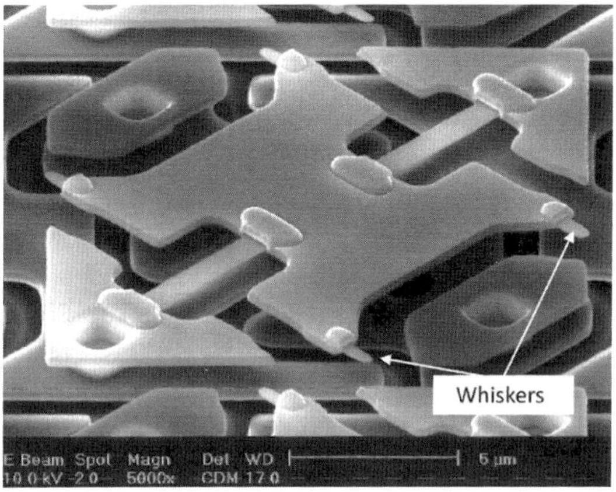

Fig. 13.7. Spring whiskers on a micromirror [107]. (Courtesy of Texas Instruments.)

Another way to combat the stiction problem is to add miniature springs to the edges of movable parts. Van Kessel *et al.* [107], presented a way of depositing alumina whiskers on the edges of the mirror landing tips in order to prevent microstructures from contact, as shown in Fig. 13.7.

The use of rough surfaces by altering surface topography [86, 92] is another option for preventing stiction because it reduces the surface energy by reducing the amount of surface area in intimate contact. Adding dimples using reactive ion etching (RIE) on the back side of the to-be-released structures is also a method widely used in the design stage to prevent stiction [108]. An example of the use of dimples in the design stage is presented in Fig. 13.8. The idea is to separate the freestanding structures and substrates so that the surfaces of these structures will not reach a distance where the surface adhesion is larger than the restoring force.

Lastly, hermetic packaging is a good solution for preventing stiction by keeping the water vapor levels low and preventing capillary condensation. MIL-STD-883F, method 1014 [109] specifies how to determine the hermeticity of MEMS packaging.

13.4 Wear

Wear is a phenomenon associated with rubbing or impacting surfaces that can be a concern in MEMS devices with sliding elements. Wear is defined as the removal of material from a solid surface as the result of mechanical action [111]. Figure 13.9 shows severe wear that occurs in one location

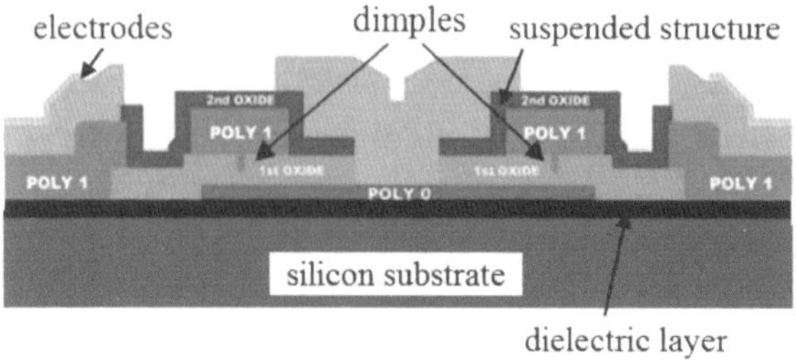

Fig. 13.8. Design of a rotary MEMS with multi-user MEMS processes (MUMPs). POLY 1 is designed to be suspended over POLY 0 substrate to perform the rotating function. However, malfunctions may occur when POLY 1 sticks to a POLY 0 either during the release process or in usage. Therefore, dimples are designed to prevent the POLY 1 part from touching the POLY 0 substrate. (POLY 1 is the rotor of an electrostatic MEMS motor; the 1st and 2nd OXIDE will be removed to release the structure at the end of the fabrication process) [110].

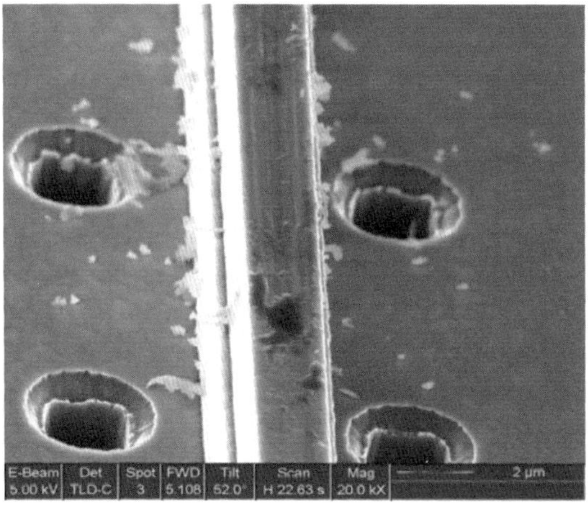

Fig. 13.9. Severe wear on a friction pad ultimately caused device failure [114]. (Courtesy of Sandia National Laboratories.)

on a friction pad. Wear can be attributed to four main causes: adhesion, abrasion, corrosion, and surface fatigue. Adhesive wear is caused by one surface pulling fragments off of another surface while sliding due to the surface forces between them. DiBenedetto [111] showed that the volume of a material fractured by adhesive wear is determined by the relationship

$$V_{\mathrm{AW}} = \frac{k_{\mathrm{AW}} F x}{3\sigma_y} \tag{13.11}$$

where σ_y is the yield strength of the material, k_{AW} is the material-dependent wear constant, x is the sliding distance, and F is the load on the material. Experiments show that the adhered length, rest time, temperature, relative humidity, and sliding velocity of a micro-cantilever can affect the adhesion forces between two contacting surfaces, and in turn affect the adhesive wear [112, 113].

Abrasive wear occurs when a hard, rough surface slides on top of a softer surface and strips away material from the softer surface. Corrosive wear occurs only when two surfaces chemically interact with one another and the sliding process strips away one of the reaction products. In air, the most dominant corrosive medium is oxygen. This type of wear could cause failure in chemically active MEMS devices, especially microfluidic and biological MEMS. Corrosive wear depends upon the chemical reactions involved, but it can be modeled as [111]

$$h_{\mathrm{CW}} = \frac{k_{\mathrm{CW}} x}{3} \tag{13.12}$$

where h_{CW} is the depth of wear, k_{CW} is the corrosive wear constant (on the order of 10^{-4} to 10^{-5}), and x is the sliding distance.

Surface fatigue wear mostly occurs in rolling applications where smooth surfaces are subjected to cyclic loading instead of sliding. Cracks due to material fatigue initiate and propagate parallel to the surface, resulting in up to hundreds of nanometers particles, which is larger than any other wear type [111]. In the failure analysis process, wear debris is sometimes impossible to observe using a conventional optical microscope or SEM because the rubbing surfaces lie below other structures. Therefore, focused ion beam (FIB) has emerged as a valuable tool that produces clean cross sections of an area of interest in MEMS with submicron accuracy. While the SEM uses a focused beam of electrons to image a sample, an FIB instead uses a focused beam of ions (usually gallium). When the high-energy ions strike a sample, they will sputter atoms from the sample surface, and thus, FIB is used as a micro-machining tool to cut or modify materials on a micrometer or nanometer scale.

Polysilicon, a commonly used structural material in MEMS, exhibits high friction and poor wear properties at polysilicon–polysilicon interfaces [115–119]. It has been demonstrated that humidity is a strong factor in the wear of rubbing surfaces in polysilicon micro-fabrication. The operation of MEMS in a very low humidity environment can lead to much greater amounts of wear debris than operation at higher humidity [120, 121]. As wear debris agglomerates, third-body wear can be initiated, which is often evidenced by scratches on the surface. Finally, severe wear would ultimately cause the device failure.

A proven way to reduce wear failure is by applying protective interfaces to MEMS structures. Such interfaces include solid films [122–136] (some of which are also used as anti-stiction coatings), gas-phase lubrication [137], and combining a bound monolayer and a mobile phase on the surface [138]. Mayer *et al.* [131] demonstrated less wear particle generation with a 10-nm-thick Al_2O_3 coating on polysilicon structure via atomic-layer deposition (ALD) than a native oxide layer in a MEMS micro-engine. ALD is a good method to deposit friction- and wear-reducing films on MEMS structures. Smallwood *et al.* [139] proposed applying a diamond-like carbon (DLC) coating to protect MEMS structures against wear. Tests have shown that DLC-coated devices that run in air have a 16-fold increase in reliability performance over their uncoated counterparts. Applying a DLC coating to MEMS structures can prevent them from failing quickly and catastrophically, especially in a harsh environment. Bandorf *et al.* [140] developed a disc setup to investigate the friction and wear of various DLC coatings deposited on substrates of varying hardness. He found that friction and wear were greatly reduced when a substrate of high elastic modulus was used. Bandorf's work indicates that the substrate material and the topography are key factors for friction and wear optimization. Guo *et al.* [141] proposed another category of thin films capable of being fabricated on MEMS surfaces: organic film, such as PFTS (perfluorodecyltrichlorosilane, $C_6F_{13}CH_2SiCl_3$). To conclude from Bandorf and Guo's works, an effective way to reduce the friction is to coat a solid thin film on the surface of MEMS devices. The film can be either organic or inorganic.

13.5 Fatigue and Creep

Fatigue is progressive, and localized structural damage occurs when a material is subjected to cyclic loadings that are less than its yield or ultimate tensile stress limit. Therefore, fatigue is mostly limited to active components such as resonating membranes, cantilevers, and comb drives, while passive structures, by design, normally are not exposed to enough mechanical stress cycles to induce fatigue. Fatigue starts with a crack that initiates at an area of high stress concentration and slowly propagates through the material until failure occurs. Muhlstein and Brown [142], and Shrotriya *et al.* [143] observed significant changes in surface topography in polysilicon during cyclic actuation and attributed them to stress-assisted interactions between water molecules and the SiO_2 layer on the surfaces of the notched specimens in the regions where the cyclic stresses were highest. Muhlstein and Brown's studies and other micro- and nano-scale test structures represent a unique platform for materials characterization and fundamental studies in both MEMS and traditional structural materials. Tsuchiya *et al.* [144] reported that no fatigue was observed in silicon when cyclic loading was less than 50% of the silicon fracture strength. Tsuchiya's work points out that the magnitude of cyclic loading is a key factor to induce the fatigue failure of microstructures.

The lifetime of MEMS devices can be affected by fatigue. Repeated motions can induce stress, significantly lower the material strength, and lead to crack growth and eventually failure. Both silicon and polysilicon are unresistive to fatigue. Stress engineering during the design phase and materials selection can reduce the problem, but humidity control is the key factor for fatigue elimination. Micromirrors are often affected by fatigue. Each micromirror is hinged so that it can rotate. Because each mirror is switched thousands of times per second, hinge fatigue should be taken into consideration. To avoid fatigue, micromirror hinges are usually fabricated using thin-film technology. The fatigue properties of thin-film layers are different from those of bulk materials. Metal thin films exhibit much less fatigue than their macroscopic counterparts because they are only a few grains thick and do not have internal crystal structure [145]. Thin films are also less prone to breaking because they are less stiff than their macroscopic counterparts. Fatigue induces dislocation movement to the material surface to form the fatigue crack after sufficient damages have accumulated. At that point, not enough damage will be accumulated on the thin-film surface to form fatigue cracks. However, the fatigue properties of thin films are usually undetermined, so the fatigue predictions are prone to error. Hinge components should have higher material strength that far exceeds the expected maximum stress.

Creep is the inelastic time-dependent deformation of a solid material under the influence of an applied mechanical stress. Creep can occur below the yield strength of a given material. Creep occurs when strain varies with time under the constant stress. Dislocation movements and atom diffusion induce the deformation, which depends on the material grain size, temperature, and initial stress [146]. Surface flatness becomes affected by creep as well as by the parameters of mechanical parts over time. Metals

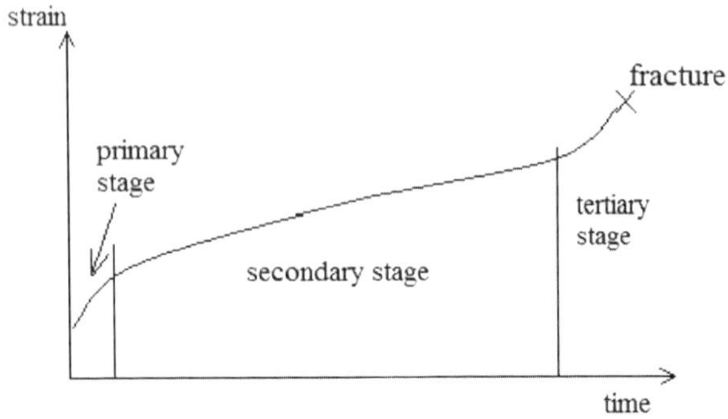

Fig. 13.10. Illustration of material creep. The creep curve consists of three stages. The time scale is linear. The value of strain varies from material to material.

are known to creep under stress, whereas silicon and polysilicon are more robust against creep as brittle materials. Figure 13.10 shows the typical three-stage strain–time relationship of creep. In the primary stage, the strain rate decreases due to work hardening, which refers to the strengthening of a metal by plastic deformation resulting from dislocation movements in the crystal structure of the material. In the secondary stage, the strain–time relationship reaches a minimum and becomes roughly constant as a result of the balance between work hardening and annealing (thermal softening). The characterized creep strain rate typically refers to the rate in this secondary stage. In the tertiary stage, the strain rate exponentially increases with strain due to a reduction in cross-sectional area caused by phenomena such as necking or internal void formation. Tertiary creep is typically followed by separation or fracture of the material by tensile stress, which is referred to as "creep rupture".

In optical MEMS devices, silicon is often coated with a thin metal film. Reflective metal coatings on micromirrors are required for the desired optical performance [146]. However, micromirrors can become deformed by annealing. After cooling, micromirrors can have significant curvature due to the coefficient of thermal expansion (CTE) mismatch between silicon and metal. In single-sided metallization, the curvature will slowly decrease as the metal creeps but not the underlying silicon. For symmetrical mirrors, both sides of which are metalized by using two metal films deposited under different conditions, an uncontrolled increase in mirror curvature can be expected. However, by increasing silicon thickness, flatter micromirrors can be obtained, and that would affect the micromirrors' resonant frequency, response time, and susceptibility to mechanical shock.

For micromirror hinges, creep-induced hinge memory, unlike hinge fatigue, can significantly threaten MEMS micromirror device reliability. A life-limiting failure mode can occur when a micromirror operates in the same direction for a long period of time. When the bias voltage is removed, the mirrors should return to a flat state. Returning to a non-flat state is known as a "hinge memory effect" [80]. The angle between the flat and non-flat state is called a "residual torque" angle. When this angle increases, at one point, the mirror will not be able to land on the other side. The main factors for hinge memory failure are duty cycle and operating temperature, but creep is the main cause of this type of failure [145]. Structural mirror beam materials with high-melting-point compounds are being used, such as Al_3Ti, AlTi, and AlN, because high-melting-point metal often has low creep. Since it is obvious that temperature affects the lifetime of micromirror devices, thermal management is essential. Heat sinks are used in order to keep the temperature in the device within an acceptable range. Adequate thermal management significantly influences the lifetime of the device and allows mirrors to be efficiently controlled over a longer period of time.

Creep can be a reliability concern for some types of MEMS, such as RF-MEMS and DMDs [148, 149], where long hold times at stressed states occur. For example, Sontheimer's work [150, 151] shows hinge memory effect of DMDs is reversible and not a permanent degradation, which differs from the irreversible failure mechanisms in competing display technologies. Tregilgas [152] reported on the creep effect in aluminum thin films, which often serve as functional structures, such as hinges and joints in MEMS devices. Since the creep effect in aluminum thin films is relatively large for applications where billions of operating cycles are required, alternative materials must be found to replace them. Modlinski [153–155] studied the creep resistance of a number of aluminum compounds — $Al_{98.3}Cu_{1.7}$, $Al_{99.7}V_{0.2}Pd_{0.1}$, $Al_{93.5}Cu_{4.4}Mg_{1.5}Mn_{0.6}$, and $Al_{99.6}Cu_{0.4}$ — and found that the creep resistivity of these alloys can be changed by a hardening process in which the grain interior is strengthened by producing extremely small and uniformly dispersed precipitate. They demonstrated that $Al_{93.5}Cu_{4.4}Mg_{1.5}Mn_{0.6}$ is the most resistant to long-term deformation and is a very promising bridge material for RF-MEMS switches. Another important functioning material is polysilicon. Its plastic deformation under high temperature was studied by Tuck *et al.* [156], who developed a micro-test and measured the high-temperature plastic deformation within polysilicon. They found that no measurable creep occurred when the temperature was below 1072 K and that the creep rate increases as temperature and stress increase at higher temperatures. In many cases, MEMS structures are subjected to stress, either thermally or mechanically induced, before final fracture.

13.6 Electrical Shorts and Opens

After decades of improving the reliability of semiconductor devices, electrical short failure rates have been reduced to a reasonable range. However, the influence of dielectric properties and behavior on MEMS devices is different from that on semiconductor devices and is still not fully understood. This difference in influence is primarily due to three factors. First, there are various contributors to electrical shorts in MEMS, some of which are dielectric degradation, charge injection and accumulation, ohmic contact, and electromigration. Second, the characteristic length scale and electrical stress of dielectrics in MEMS are often an order of magnitude or more than in semiconductor devices. The typical lateral dimensions of dielectrics in MEMS are tens to hundreds of micrometers, which is much larger than those in semiconductor devices. Thus, MEMS devices inevitably have a higher probability of introducing defect sites in dielectrics than in semiconductor devices. Third, higher applied voltages in MEMS can also increase the probability of dielectric breakdown, which is the major cause of electrical shorts in MEMS devices. The inspection tools for electrical failure include, but are not limited to, probes with electrical test equipment such as an ohm meter, photo emission microscopy (PEM), SEM, and X-ray diffraction. PEM provides photoelectron distribution by collecting the electrons originate from the sample when the sample is excited by appropriate radiation, ranging from UV light to X-ray. By comparing a malfunctioning device to a properly operational unit, one can pinpoint the dielectrics that have been damaged by electrostatic discharge.

13.6.1 *Dielectric Breakdown Due to Operational Voltage*

The electrical field generated by the normal driving voltages of MEMS devices is one cause of electrical shorts. Although MEMS devices do not necessarily operate under high voltages, their thin dielectric layer makes them experience a high electrical field. For example, in a typical capacitive RF-MEMS, a dielectric of 500 nm with an actuation voltage of 50 V across it will experience an electrical field of 1×10^8 V/m. This kind of ultrahigh electrical field is a concern for the dielectric layer because conduction is no longer governed by Ohm's law, but by nonlinear conduction, such as Schottky-type and Poole–Frenkel-type conduction, which is due to charged particle injection and via traps [157]. With such an ultrahigh field strength, the dielectric layer can be broken down, resulting in an electrical short. Zafar *et al.* [158] presented a model for calculating the trapped charge density. Zafar concluded the threshold voltage shifted, but the subthreshold slope remained constant with stressing time for both Al_2O_3 and HFO_2.

13.6.2 *Dielectric Breakdown Due to Electrostatic Discharge*

Electrostatic discharge (ESD) involves the transfer of electrical charge between two objects at different potentials, either through direct contact or through an induced electrical field. ESD can result in a very high current passing through a dielectric material within a very short period of time and can cause material breakdown and sometimes burn out, which is the partial or complete melting of a large device area. Despite a great deal of effort during the past decade, ESD still affects production yields, manufacturing costs, product quality, and product reliability. In semiconductor devices, ESD accounts for more than 25% of failures, including gate oxide breakdown, junction spiking, and latch-up [159]. In MEMS devices, where dielectrics play more roles, including sacrificial, structural, optical, masking, passivation, etch stop, insulator, and encapsulation layers, ESD problems in the dielectric are especially destructive [160–164] and an important area for improvement of reliability. Common sources of ESD are the human body [165], charged devices, charged environment, and cosmic rays. A human body routinely develops an electrical potential in excess of 1,000 V, which can have a catastrophic effect on a MEMS device [166].

A standard procedure has been established for testing and classifying micro-circuits as well as some MEMS devices according to their susceptibility to damage or degradation by exposure to a defined ESD [167]. Olszewski *et al.* [168] created an experiment to test the effects of ESD with and without moisture on an RF-MEMS switch by measuring the changes in the capacitance–voltage curve. If the MEMS devices are designed to operate in an ionizing (nuclear) radiation environment, such as outer space or a nuclear power plant, the possibility of charge accumulation due to radiation in the dielectrics should be taken into account. Knudson *et al.* [169] and Lee *et al.* [170] tested the performance of MEMS accelerometers under nuclear radiation and found that the change in output voltage was caused by charge buildup in dielectric layers beneath the moving mass. The changes in the electrical field distribution, which are caused by radiation-induced charge trapping in the insulating layers, produced changes in the output voltage because the mass was coupled electrically to the substrate. Charge trapping in dielectric layers can affect the performance of the sensor. Thus, it is necessary to consider mechanical sensor construction while designing the mechanical sensor of the IC component. McClure *et al.* [171] proposed a mechanism for creating a charge distribution in the dielectric of the MEMS device exposed to nuclear radiation. They concluded the total dose effect depended on actuator geometry and materials. McClure's studies suggest these kinds of devices should be thoroughly characterized for radiation effects prior to their usage in a space and/or nuclear radiation environment. It is better to have test environments that are similar to the application environment. Newman *et al.* [172] presented the contact surface of a wafer-packaged RF-MEMS switch damaged by ESD. It exhibited the damaged surface after 914 billion switch cycles at 20 kHz with an incident RF frequency of 10 GHz without degradation. However, the expected lifetime of a switch depends on its application and may range from only 100 million switch cycles for some short-lived applications up to several hundred billion switch cycles in tunable filter or space-based applications.

Researchers have developed many ways to solve the ESD problem in dielectrics. Low field strengths, trap-free dielectric, and leaky dielectric layers have been proposed as solutions [173]. Wairaven *et al.* [174] proposed two approaches to solve two classes of problems: manufacturing-related and in-use-related approaches. Some of the ESD precaution and proper handling procedures include: leave ESD-sensitive components in their anti-static packaging until they are ready to be used; dissipate static electricity before handling any ESD-sensitive components by touching a grounded metal object, such as an unpainted metal chassis; and use wrist straps and floor mats or other antistatic devices to prevent possible ESD from damaging the components.

13.6.3 *Electrical Failure Due to Ohmic Contact*

Electrical failures are prone to occur at ohmic contacts that are the interfaces of two different conductive materials with different Fermi levels. As two contacts come together, only the highest peaks touch and are subject to a full current load, which causes a sudden rise in temperature and a "melt-down" of those peaks. Contact resistance at an interface can be calculated by the following equation [175]:

$$R_c = \frac{\rho_e}{na} + \frac{\rho_e}{D} \tag{13.13}$$

where R_c is the contact resistance, ρ_e is the electrical resistivity, a is the average contact asperity radius, n is the number of asperities per unit area, and D is the diameter of the area over which the contacts are distributed. Once the temperature at an interface caused by Joule heating reaches a high enough level, the contact material will melt, usually causing severe problems in MEMS, such as superheating of the surrounding air and its ionization, arcing, material transfer, or welding [176].

Other causes of ohmic contact failure include surface contamination, erosion, and surface changes due to absorption or oxidation. The quality of ohmic contact can be represented by contact resistance. Kwon *et al.* [177] studied the electrical behavior and lifetime of RF-MEMS contact switches made of Au, Pt, Ir, and AuPt alloys at different contact forces and current conditions and found that Au-to-Pt, Pt-to-Pt, and Au-to-Ir are alternative materials that can enhance the reliability of high-power RF-MEMS switches. Kwon's work indicates an effective method to enhance reliability for high-power RF-MEMS contact switches by contacting the dissimilar materials, such as Au-to-Pt or Au-to-Ir. Dickrell and Dugger [178] investigated the contact resistance degradation of a hot-switched low-force Au-Pt contact using a modified nano-indentation apparatus and found that the contact resistance increased by two orders of magnitude over the initial resistance in only five to ten cycles, indicating that the electrical contact surface was being contaminated *in situ*. Dickrell's work provides a possible solution to circumvent the contamination-forming process by using parallel discharge paths for stored electrical energy in the contact circuit. Recently, low-loss ohmic switches have been created by fabricating dimples on one of the electrodes during deposition [179]. It is believed that a more even distribution of the electrostatic force across the plate can result in uniform pressure over the contact area, yielding lower contact resistance. Gaddi *et al.* [180] proposed an approach for achieving low-loss contacts with a standard surface micro-machining technology. The insertion loss and isolation of the devices are compatible with multi-standard wireless RF-transceiver application specifications. Gaddi's studies point out that an even distribution of the electrostatic force across the plate can result in uniform pressure for all contacts, contributing to lower contact resistance. Ducarouge *et al.* [181] demonstrated a power RF-MEMS device capable of handling 3.8 W at 20 GHz by carefully designing the contact line. Sputter cleaning of the surface prior to deposition along with deposition of nickel as a first layer can provide for much improved ohmic contact stability and homogeneity [182].

13.7 Contamination

Many MEMS devices rely on the movement of mechanical components to perform their designated functions. However, the introduction of unintended materials, which is called "contamination", can render the device inoperable. Contamination can be introduced from various sources, including manufacturing processes such as surface cleaning, metal deposition, patterning, releasing, and annealing; dusty in-use operation environments where foreign particles penetrate through MEMS packages; and intrinsic problems such as crystal growth (e.g. tin whiskers). Figure 13.11 shows a MEMS actuator after being subjected to high mechanical shocks. A particle contaminant was found close to a comb-finger actuator, which resulted in an electrical short. In most cases, contamination has been attributed to the energetic decomposition of existing contaminant species into highly resistive compounds [183, 184].

Contamination can affect the mechanical aspect of the MEMS device by blocking moving parts. Contamination can also result in electrical degradation and failure, including limiting of the frequency response of high-frequency devices due to the increasing RC time constant, power

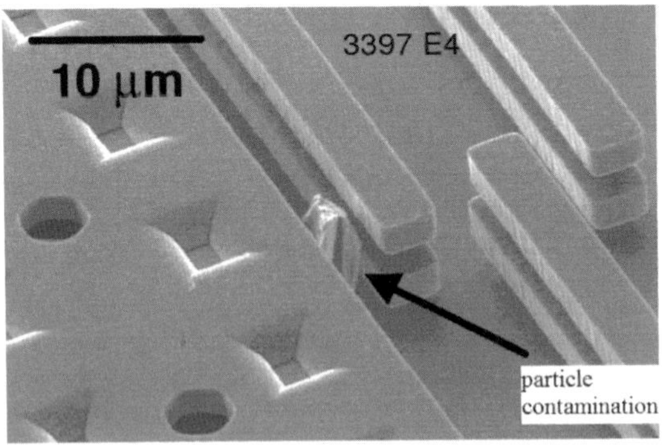

Fig. 13.11. Particle contamination found within the actuator region of a microengine subjected to a high shock top impact. Notice that the contact between the particle, stationary comb fingers, and the shuttle resulted in electrical short of the drive signal to the ground plane [18]. (Courtesy of Sandia National Laboratories.)

dissipation via Joule heating, electromigration, and delamination at contact areas [185]. All these effects lead to limitations on the lifetimes and to failures of MEMS devices.

Contaminants either become wedged between MEMS components or rest on top of surfaces. Hence, contamination-induced failure is not difficult to identify, as particles are relatively easy to diagnose using optical microscopes, SEM or FIB once the device is decapsulated and the MEMS component is exposed. Yet, precautions must be taken during the decapsulation process. Tabata *et al.* [186] have shown that contaminants can be easily detected using an interferometer, and dust particles that are greater than $0.3\,\mu$m in diameter can be prevented from entering a MEMS pressure sensor by using a dust collector at the edge of the package. In many cases, contaminants are induced via poor packages with voids or cracks. A scanning acoustic microscope (SAM) is useful for detecting lid seal integrity in hermetically sealed packages. It is a nondestructive technique that directs focused sound from a transducer at a small point on an object of interest, measures "time of flight" of the sound, and determines the object's geometry and microstructure of its surface. For identifying contaminants, an energy dispersive X-ray spectrometer (EDS) is a useful instrument that is commonly used in conjunction with the SEM. It allows users to measure compositional gradients, element concentrations, or distributions at a microstructural level, and thus, to identify contaminants and failure sites.

13.8 Reliability of Optical MEMS

Due to a growing market and stricter requirements, reliability of optical MEMS for telecommunications is identified as the next challenge facing manufacturers. Because of the diversity of device designs, materials, and functions for optical MEMS, it is necessary to understand both technology-related and external variables such as environmental and operational conditions. In MEMS reliability analysis, the failure mechanisms can be electrical and/or mechanical. However, the failure mechanisms of optical MEMS are more complicated than those in microelectronics for several reasons. First, optical MEMS devices are expected to interact with the environment under various conditions (e.g. temperatures). Second, some failures are impossible to predict for optical MEMS devices. Third, unlike ICs and microelectronics, reliability testing for MEMS devices is not standardized. Fourth, new testing procedures need to be developed for every new device.

Vibration in optical MEMS devices is a reliability concern [187, 188]. Due to the sensitivity and fragile nature of many MEMS devices, external vibrations can have a disastrous impact. They might cause failure by inducing surface adhesion or by destroying the device support structures. Long-term vibration can also contribute to fatigue. Moreover, shock can be another reliability issue. Shock is a single mechanical impact, not a rhythmic event, which creates a direct transfer of mechanical energy across the device. Shocks can lead to both adhesion and fracture.

The small size of optical MEMS devices is one of their greatest assets despite the fact that it makes them appear fragile. Small size makes optical MEMS devices robust and enables them to sustain low-frequency vibrations and mechanical shock without damage. Size may be related to another type of failure mechanism. The dimensions of MEMS devices are so small that foreign particles might cause non-functionality of one or more devices during fabrication. For this reason, manufacturers should detect and eliminate the source of each contaminating particle, especially during packaging, because particles sealed in the package may affect operation of the device during its lifetime. Hermetic packaging can solve this problem and can also eliminate capillary stiction effects. Packaging-introduced contamination is the most common failure mode of optical MEMS devices. This kind of contamination is still relatively unproven as well as not yet fully qualified for optical MEMS devices in telecommunications applications.

Several major reliability issues are related to stiction, friction, fatigue, and creep. However, development of reliable optical MEMS components is nontrivial and requires sophisticated design considerations and better control of micro-fabrication processes that are used in the realization of MEMS devices. One of the challenges to achieving this goal is to provide the process environment with constant temperature, particle-free, and stable equipment. Submicron alignment tolerances are usually required, and high-port-count optical MEMS require handling and packaging of large numbers of optical fibers, micromirrors, lenses, and electrical control lids. Light collimation and focusing; wavelength separation; precisely controlled, large, flat, and highly reflective microstructures; and significant control electronics are just some of the topics that need to be studied.

Reliable packaging does not prevent the mechanical action of the moving parts of the structure, but prevents the invasion of heat, moisture, and outgassing [189, 190]. Credible testing techniques during fabrication, assembly, and packaging are required as well as during the operational lifetime of the device. Because the number of ports in optical MEMS devices is

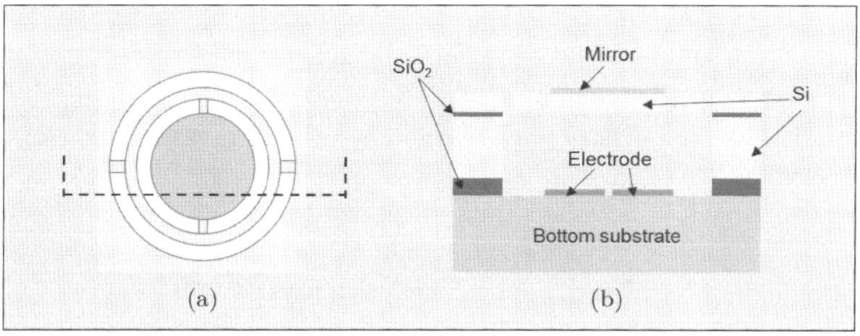

Fig. 13.12. (a) Top view of an SOI-based electrostatic MEMS mirror; (b) after structural release of the gimbaled mirror.

growing, expensive laser sources and flexible test architectures are required, which increases the challenges for testing requirements.

Optical MEMS can have a successful future in the telecommunications market if there are improvements in device structure, materials, and processing. Figure 13.12 is an illustration of a silicon on insulator (SOI) based electrostatic MEMS mirror. Flatter micromirrors and higher-quality lenses can lower yield losses. More ports are required to handle the expansion of the traffic. Reliable and cost-effective packaging is needed for housing thousands of tiny and fragile MEMS structures. Industrial standardization of MEMS technology has been established for several years, and the standards for optical MEMS devices should be customized according to customer requirements.

13.9 Reliability of RF-MEMS

MEMS are the integration of mechanical and electronic components on a common substrate. A common feature in MEMS is the application of suspended membranes in different geometries, such as beams, cantilevers, and bridges. This feature allows MEMS to obtain a unique and complex functionality [191]. Taking advantage of the high wide-band linearity, low insertion loss, small volume, and low batch fabrication cost of MEMS devices [192], an interesting application of MEMS is to switch the radio frequency (RF) signals in order to develop complex circuits, including phase shifters, varactors, tunable oscillators, filters, and antennas. Because the electrostatic actuation switch offers several benefits, such as compatibility with the integrated circuits fabrication process and low power consumption [191, 192], it is the most common among the various actuation methods of RF-MEMS switches.

In electrostatic RF-MEMS switches, a voltage is applied to the actuator pad, which is under the suspended membrane. The movement of this type of switch is driven by the contribution of the mechanical force, which tends to restore the membrane to its initial position, and the electrostatic force, which pulls down the membrane on the actuator. When the voltage is lower than a specific value (pull-down voltage), the electrostatic force is smaller than the mechanical force and the membrane remains in the up state (no actuation). On the other hand, when the voltage is greater than the pull-down voltage, the electrostatic force exceeds the mechanical restoring force and the membrane collapses to the down state (actuation). The electrostatic RF-MEMS switches can be differentiated by circuit configuration (series or shunt) or contact interface (ohmic or capacitive). A shunt capacitive RF-MEMS switch is showed in Fig. 13.13. The shunt capacitive RF-MEMS switches are characterized by an enhanced reliability and exhibit a higher isolation at the resonance frequency, which can be tuned by adjusting the geometric parameters of the moveable bridge [191, 192]. Much attention is focused on the fabrication of RF-MEMS switches because the mechanical and electrical properties of the materials are closely correlated to performance and reliability.

One of the metals that can be used for the fabrication of the membranes in MEMS switches is gold, which is widespread in both GaAs- and

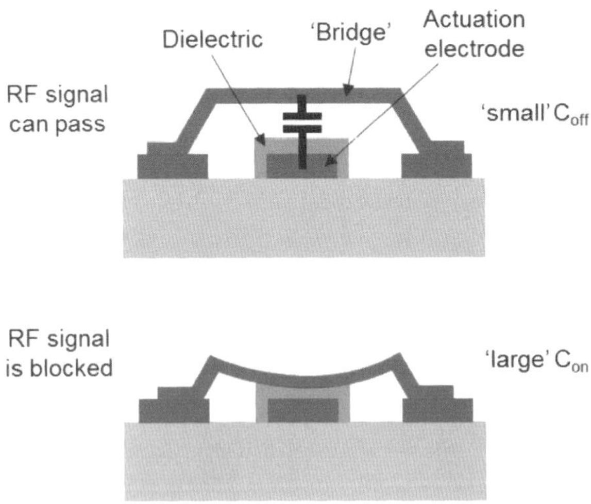

Fig. 13.13. Cross section of a typical capacitive shunt switch, with, as an example, a two-sided clamped beam as the moving 'bridge'. The position of the dielectric that is prone to charging is indicated.

silicon-based technologies. Gold is highly conductive and chemically inert, so it can reduce losses and prevent contamination and corrosion problems. Unfortunately, gold is not an excellent material from a mechanical point of view, and its yield, creep, and thermal relaxation properties are often induced high stresses and non-homogeneities in the stress distribution [193]. Prediction of the actual stress in gold membranes can improve the performance of MEMS switches, but it is not a trivial task due to the strong correlation to the fabrication process.

Dielectric layers play another role in device operation, especially in capacitive RF-MEMS switches, which deposit a dielectric layer on the actuator. In GaAs technology, Si_3N_4 is generally the material used for dielectric layers. On the other hand, SiO_2 is the most common dielectric material used in silicon-based technology. The reliability of capacitive RF-MEMS switches is limited by leakage currents and charging phenomena in the dielectric layer [194–199]. However, most research focuses on silicon-based switches [196–198], and few reports address the Si_3N_4 dielectric layers of GaAs-based switches [194, 199].

The prospect of a compact MEMS switch with reliable wide bandwidth for RF test instrumentation has been attractive for many years [200]. Attenuator switching, filter bank switching, and general RF path switching are typical applications of switches for RF instrumentation. Some operational characteristics, such as bandwidth, actuation lifetimes, and number of channels, can affect system performance. The demand for excellent long-term reliability performance encourages industries to develop more realistic switching solutions in RF applications [201].

MEMS technology for RF applications has verified the enormous fabrication potential for high-performance and widely reconfigurable RF passive components for more than a decade. Tunable filters [198], reconfigurable impedance matching networks [203], phase shifters [204], and high-order switching matrices [205] are applicable for wireless telecommunication in modern commercial RF systems (e.g. mobile phones). However, the reliability of such devices is a major concern. For example, in RF-MEMS switches, the sticking of the movable structural elements, which may cause malfunctions or failures, is a concern. The two primary effects that cause stiction in electrostatically actuated devices are micro-welding and dielectric charging. Micro-welding is due to high local current densities within the RF signal [206]. Dielectric charging, such as accumulated charges, is entrapped within the dielectric layers [207] to hinder the release of the switch.

Although RF-MEMS switches in GHz-range applications show performance advantages, the reliability concerns have prevented their widespread use. RF-MEMS switches can be found in cell phones, base stations, anti-collision car radar, and automatic test equipment (ATE) [208–210]. However, these devices have not been widely used except ATE. The reliability problems have existed for over 10 years, and significant research funding is invested in their further development because market studies clearly show RF-MEMS switches are promising. The market predictions indicate the business growth will be around USD 1 billion in annual sales in the upcoming 5 years [211]. The current major participants are Advantest, Omron, Radant MEMS, XCOM, Analog Devices, Wispry, and TDK-EPC [212].

Ohmic switches from manufacturers such as Omron and Radant MEMS can now be routinely ordered, partly even via standard distributors. They offer respectable lifetimes, typically up to 100 billion switch cycles cold switching at a power level of 30 dBm, and 1 billion cycles at 33 dBm. Hot switching characteristics are worse, with over 1 billion cycles of hot switching at a power level of –10 dBm, and a low 1,000 cycles at 20 dBm [213]. As a reliability comparison based on the number of switching cycles, small field-effect transistor (FET) switches in a microprocessor running at 3 GHz can accumulate 10^{17} cycles per year, while high-voltage (>100 V), high-power (>50 W) FETs in switched power supplies at 100 kHz still achieve more than 10^{12} cycles per year.

The maximum power levels of ohmic switches are limited because the devices use proprietary hard gold alloys as the contact material. The typical contact materials for high-power switching in macroscopic relays (such as gold–nickel, silver–nickel, silver–cadmium oxide, and palladium–tungsten bimetal [214]) cannot be used to boost the maximum power in MEMS switches. These materials would form isolating surface films, so MEMS switches cannot provide enough mechanical forces to punch through these films. In macroscopic relays, a scraping motion on the contacts is often employed, which is uncommon in MEMS. Therefore, it is not easy to build reliable high-power MEMS switch-based gold-alloy contacts. The inevitable wear of the relative soft metal-to-metal contact is the reason that MEMS switches cannot reach the ultra-high number of switching cycles common for semiconductor switches, such as transistors.

Capacitive RF-MEMS switches exchange the direct metal-to-metal contact for a variable capacitance, which acts as a switch at GHz frequencies. In principle, these switches would not suffer from this contact degradation mechanism. Also, because the variable capacitance does not contain a resistive element, such as the contact resistance of an ohmic switch, their insertion loss specification can be even better than that of metal-to-metal contact switches. This results in better filter specifications within specific bandwidths [215]. Therefore, the development of capacitive switches is pursued more than that of ohmic ones. However, the introduction of reliable capacitive RF-MEMS switches has been hampered by their own set of reliability threats.

RF-MEMS metal-contact switches are an emerging category of relays that are receiving increased attention for defense and industrial applications, such as wideband switching matrices, electronic test equipment, medical devices, and satellite systems [216–222]. These devices possess many advantages over conventional electromagnetic relays, such as their size, power consumption, and reliability.

The performance of a metal-contact RF-MEMS switch is primarily determined by the tribology of the contact area. The size of the contact spot and integrity of the contact material determine the switch resistance, insertion loss, power handling, and reliability [217, 223]. Increasing the static contact force to mN levels results in a larger contact spot size and improves the performance of the switch in all areas. However, increasing the transient impact force of the switch can reduce the integrity of the contact material and adversely affect the switch performance [224]. Moreover, the increment in adhesion between the metal contacts in mN-level contact force requires a corresponding mN-level restoring force to break the contact. The resulting devices have large lateral dimensions compared to the electrode gap [225, 226] and are susceptible to out-of-plane deflections due to stress effects and temperature. A well-designed device must balance all of these requirements satisfactorily. There are several examples of low and high contact force designs in the literature [227–232], and two of them are commercially available [233, 234].

13.10 Standards for MEMS Reliability Evaluation

During the early 1990s when companies first began to qualify MEMS devices, there were no qualification test standards for MEMS. MEMS accelerometers that were used in automotive airbag deployment modules represented the largest market [235]. Since there were no set reliability requirements, the original equipment manufacturers (OEMs) using these MEMS accelerometers expected these devices to pass the same stress tests that were used to qualify mechanical accelerometers. Hence, a subset of standardized microelectronic qualification tests were negotiated and agreed upon by the MEMS suppliers and the automotive OEMs [235]. As more companies entered the MEMS business, the qualification approach for MEMS accelerometers was adopted for other MEMS devices.

Table 13.2 lists some of the most common microelectronics qualification test standards that are currently used by MEMS suppliers for the reliability evaluation of MEMS devices. From Table 13.2, it can be seen that the most commonly used microelectronic qualification test standards are the military and the JEDEC standards. ISO/IEC's 60068-2 series is also widely used by MEMS manufacturers for accelerated stress testing, including high-temperature operating lifetime (60068-2-2), low-temperature operating lifetime (60068-2-1), thermal cycling (60068-2-14), vibrations (60068-2-6), and moisture cycling (60068-2-30). These standards all describe classical environmental tests used for accelerated aging. These standards, including MIL and JEDEC standards, were developed at least a decade ago and fail to keep up with ever-changing use conditions and the design complexity of MEMS products. As new MEMS products are introduced to the market, companies that qualify their new products based on

Table 13.2. Microelectronics qualification test standards used for testing MEMS devices.

Description	Standards
Thermal Shock	MIL-STD-883 Method 1011
High-Temperature Operating Life	MIL-STD-883 Method 1005 JESD22-A108C
High-Temperature Storage	MIL-STD-883 Method 1008 JESD22-A103C
Electrostatic Discharge (ESD) — Human Body Model (HBM)	MIL-STD-883 Method 015 ANSI/ESD STM5.1-2007 JESD22-A114F
Electrostatic Discharge (ESD) — Charged Device Model	ANSI/ESD STM5.3.1-2009 JESD22-C101E
Latch-up Test	JESD17 JESD78B
Temperature Cycling	JESD22-A104D
Early Life Failure Rate	JESD22-A108C
Low-Temperature Operating Life	JESD22-A108C
Low-Temperature Storage	JESD22-A119
Temperature-Humidity-Bias	JESD22-A101C
Autoclave	JESD22-A102
Highly Accelerated Stress Test	JESD22-A110C

standards that were meant for their products' predecessors or the closest technology may find themselves having field failures.

Few standards have been developed specifically for MEMS devices. The first standard dealing with RF-MEMS switches was created in 2011 [236]. It describes the terminology, definitions, symbols, and test methods used to evaluate and determine the essential ratings and characteristic parameters of RF-MEMS switches. The test switches were stressed under different environments until failure. The environmental conditions, switching rate, and applied power should be adjusted according to the device applications. Other standards dedicated to MEMS devices were issued recently by SEMI (SEMI MS 1 to MS 10). SEMI [237–239] focuses on the terminology and characterization of some key parameters of MEMS devices, such as Young's modulus and bonding strength. None of the existing MEMS standards address reliability evaluation or design qualification. Much of the MEMS industry still strongly supports microelectronics standards-based testing due to the lack of a better qualification testing approach. However, the reliability requirements for emerging mission- and safety-critical applications warrants a transition from a microelectronics standards-based qualification approach to a MEMS standards-based qualification approach with a more comprehensive qualification process that utilizes a variety of tools and techniques.

Although silicon-based MEMS devices are made using the same fabrication process as that of integrated circuits, the distinguishing features of MEMS devices are their mechanical elements [240] and the fact that many MEMS devices are not sealed; they must be open to the environment (e.g. microphones, bio-MEMS, moisture sensors). Both of these factors create challenges during the design, fabrication, packaging, and testing of MEMS devices. For example, MEMS accelerometers are generally housed in an inert environment with atmospheric pressure to achieve sufficient damping. On the other hand, biological MEMS sensors require access to the environment to detect the presence of a targeted biological substance.

MEMS sensors are also far more diverse than microelectronics. The dominant failure mechanisms in MEMS devices are not the same for all MEMS devices. For example, the top two failure modes for MEMS actuators are stiction and wear owing to the presence of moving parts that could come in contact with each other. However, MEMS sensors are more prone to electrical or package-induced failures [241]. Furthermore, for some failure mechanisms in MEMS devices such as clogging, surface modification, and micro-weld, there are no microelectronics qualification test methods to accelerate these failure modes. Thus, testing MEMS devices using the same methodologies as the standard procedures used for microelectronics packages might not always be possible, especially when quality and reliability need to be assessed.

Most original device manufacturers (ODMs) conduct a suite of tests to qualify their products based on the test conditions provided in standards. For instance, Analog Devices Inc. (ADI) qualifies their ADMP421 omnidirectional MEMS microphones based on the JESD 22 and MIL-STD-883 standards [235]. The qualification tests conducted by ADI require their microphones to survive for a specific amount of time in order to pass the qualification process. For example, MEMS microphones are required to survive for 1000 hours at +85°C and 85% RH in order to pass the temperature-humidity-bias (THB) test, as per JESD22-A101C [242]. Similarly, these microphones are required to survive for 1000 hours at +150°C and −40°C as per JESD22-A103C [243] and JESD22-A119 [244] in order to pass the high- and low-temperature storage tests, respectively. In these tests, the selection of the test temperature or time to survive is not based on the actual use conditions of the device. Hence, a pass/fail result from a standards-based test does not imply that the MEMS device will meet its reliability requirements, unless a relation has been established between the field use and the qualification test conditions [245]. In fact, a product that has failed the required standards-based qualification test may survive satisfactorily in the actual use conditions [246]. This leads to the rejection of reliable products that could have otherwise been shipped, thereby resulting in the loss of resources and money. On the other hand, a product that has passed the standards-based test may fail prematurely in the field, thus leading to the false acceptance of a defective product. This premature failure may be due to the fact that the qualification test conditions are not stringent enough to precipitate failures during the qualification process.

Another problem with microelectronics standards-based tests is that they fail to capture intermittent failures, which typically result in no-fault-found (NFF) issues [246]. Intermittent behaviors result in a loss of functionality for a specified duration of time. The functionality may be recovered when the stress condition that caused the intermittent behavior is removed. Intermittent failures can be expensive for manufacturers. According to a research report by Accenture in 2011, consumer electronic product manufacturers labeled 68% of their returned products as NFF. The NFF processing cost was US$2.5 billion, which accounted for 15% of the total cost of returns attributed to U.S.-based consumers [247]. Similar to microelectronic products, MEMS devices are also prone to intermittent failures because mechanical structures used in them are made of thin films suspended a few microns off the substrate, and these structures are susceptible to surface forces [248]. If the intermittent behavior is not captured during qualification testing, then this behavior may manifest itself as intermittent failures, or NFFs, in the field.

Proliferation of MEMS in consumer electronic products such as mobile phones and gaming consoles has led to an increase in consideration of MEMS devices for automotive, military, and space applications that have high service life expectations. For example, the required lifetime of an RF MEMS switch may range from 100 million cycles in short-lived applications to several hundred billion cycles in military and space applications [249].

According to a survey performed by the MEMS Industry Group (MIG) in 2011 [240], 70% of MEMS suppliers claim to perform life testing, and 80% claim that the lifetime of their MEMS devices is around 15 years. This raises the question: If 30% of the MEMS industry is not performing qualification testing, then how are they demonstrating their devices' reliability in terms of years of lifetime? This fuzziness in the way of reporting MEMS device reliability has led to a change in customer behavior towards MEMS suppliers, which ultimately has led to further testing challenges. MEMS manufacturers are often required by their customers to perform extended life tests to demonstrate their device lifetime, which in turn increases the cost of testing and time-to-market.

In order to qualify MEMS devices for such high reliability expectations, it is general practice to use accelerated life tests (ALTs). An ALT is intended to shorten the normal time of failure of a device due to a specific failure mechanism, and the test formulation has to be based on failure mechanism analysis [249]. However, accelerated tests are often conducted without knowledge of the failure mechanisms and without ensuring that the test accelerated the same mechanisms as those observed under normal operating conditions. For example, in the MIL-STD-883, MIL-STD-810, or JEDEC JESD22 standards, the failure mechanisms are not explicitly identified. Hence, results from microelectronic standards-based accelerated tests do not necessarily provide meaningful information, as the accelerated test conditions might not relate to the actual field conditions. Thus, failures caused by the ALTs recommended in these standards could be induced by failure mechanisms other than those that would occur under normal operating conditions [250]. A shift in failure mechanism can result in the wrong lifetime predictions.

For example, in microsystems (i.e. microelectronics or MEMS), thermal cycling or mechanical vibration induces cyclic stress or strain, which can lead to fatigue [251]. For high-cycle fatigue, material behavior can be represented by the following relationship between applied stress range and fatigue life [252]:

$$N = C\sigma_r^m \tag{13.14}$$

where N is the mean cycles to failure, σ_r is the cyclic stress range, and C and m are material constants to be determined through ALT. For low-cycle fatigue, the following equation has been employed to predict the thermal fatigue life [253]:

$$N = \frac{1}{2}\left(\frac{\varepsilon_{\mathrm{r}}}{2\varepsilon_{\mathrm{f}}}\right)^b \tag{13.15}$$

where ε_{r} is the plastic strain range, and ε_{f} and b are the fatigue ductility coefficient and exponent, respectively, which are determined through ALT. Figure 13.14 shows the curve generated by low- and high-cycle fatigue when the applied strain is plotted against the mean cycles to failure.

In accelerated fatigue tests, accelerating the mechanical loading or the thermal cycles elevates the stress or strain range. Elevating the stress or strain range through ALT will decrease the fatigue life. However, acceleration of the stress or strain range in testing may shift the failure mechanism from high- to low-cycle fatigue, in which case the prediction of fatigue life based on linear extrapolation of accelerated test data may result in overconservative predictions if the stress range is accelerated beyond a certain level.

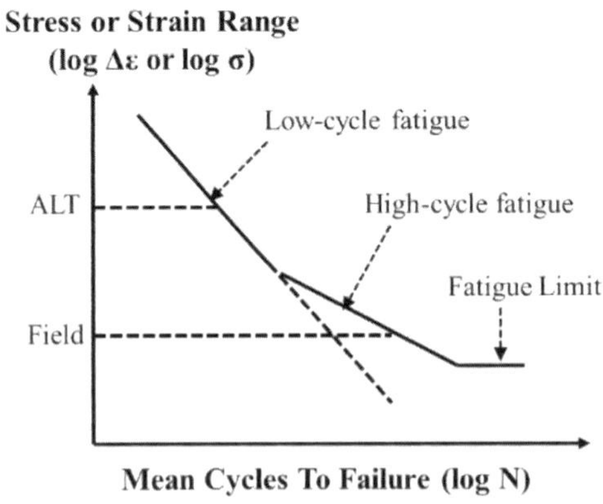

Fig. 13.14. Illustration of a typical S-N (strain vs. cycles to failure) curve.

In automotive and military applications, during some scenarios the margin to perform an accelerated test is small since the use conditions are closer to the limits of the device failure mechanism. In such cases, the only way to demonstrate device reliability is by extending the test time, which in turn increases the cost of testing and time-to-market. For example, extended lifetime tests on RF-MEMS switches take about 2 months, and 8 months are required to demonstrate 100 billion and 500 billion lifetime switching cycles, respectively [254]. Clearly, this has a significant impact on test cost and also causes delays in getting the product to market. Thus, standardized methods are needed that could aid in keeping test time short and still allow MEMS manufacturers to demonstrate their device reliability.

13.11 Summary

The ability to quantify the reliability of MEMS devices for anticipated use conditions is vital for their successful commercialization. Quantification can be achieved only if engineers have a clear understanding of the failure mechanisms and underlying physics of MEMS products. This chapter summarizes failure mechanisms that have been identified in MEMS devices, their physical root causes, their effects on device behavior and functionality, as well as test methods and inspection techniques. Additionally, various approaches to improve MEMS reliability are covered, including creep prevention through the use of creep-resistant materials for moving structural parts, wear prevention through the deposition of a lubricant layer, stiction prevention through the fabrication of dimples and whiskers, fracture prevention through filleting sharp lines and corners, and contamination-induced failure prevention via better hermetic sealing. MEMS engineers have borrowed many tools and methods from IC fabrication to test MEMS devices. However, due to the nature of MEMS interactions with the environment and their electromechanical complications, identification of potential failure sites and mechanisms is still a labor-intensive process. Fortunately, most MEMS devices, regardless of function and application, share basic components (e.g. membrane, cantilever) that exhibit common failure mechanisms (e.g. fatigue, stiction). Therefore, understanding those mechanisms provides a foundation for the lifetime models and field requirement-based test guidance that help MEMS engineers to design against possible failures and test MEMS devices more efficiently.

Acknowledgments

The authors would like to thank the Center for Advanced Life Cycle Engineering (CALCE), University of Maryland, for supporting this research. CALCE has provided a knowledge and resource base to support this study of MEMS reliability and testing evaluation. More than 100 national and international organizations and companies support the center. Also, the authors would like to thank Sandia National Laboratories, Texas Instruments, Dr. L. Hornbeck, Prof. K. S. J. Pister from UC Berkeley, IOP Publishing Ltd., M. Zimmerman, D. Mahadevo, B. Kwong, L. Krishnan, and G. Kumaraswamy for their valuable edits and comments.

References

1. Boysel, R. M., McDonald, T. G., Magel, G. A., Smith, G. C., and Leonard, J. L. (1992). Integration of deformable mirror devices with optical fibers and waveguides, in *Proceedings of the. SPIE*, vol. 1793, pp. 34–39.
2. Kuehnel, W. and Sherman, S. (1994). A surface micromachined silicon accelerometer with on-chip detection circuitry, *Sensors and Actuators A, Physical*, 45(1), pp. 7–16.
3. STMicroelectronics Inc. http://www.st.com/
4. FUJIFILM. https://www.fujifilmusa.com/products/industrial_inkjet_printheads/technology/silicon-mems/
5. Bouchaud, J. (2012). his iSupp–i — MEMS market trackehisiHS report. https://technology.ihs.com/api/binary/418487?attachment=true
6. O'Mahony, C., Olszewski, O., Hill, R., Houlihan, R., Ryan, C., Rodgers, K., Kelleher, C., Duane, R., and Hill, M. (2014). Reliability assessment of MEMS switches for space applications: Laboratory and launch testing, *Journal of Micromechanics and Microengineering*, 24, pp. 125009–125018.
7. Lessing, J., Glavan, A. C., Walker, S. B., Keplinger, C., Lewis, J. A., and Whitesides, G. M. (2014). Inkjet printing of conductive inks with high lateral resolution on omniphobic "RF Paper" for paper-based electronics and MEMS, *Advanced Materials*, 26, pp. 4677–4682.
8. Younis, M., Jordy, D., and Pitarresi, J. M. (2007). Computationally efficient approaches to characterize the dynamic response of microstructures under mechanical shock, *Journal of Microelectromechanical Systems*, 16(3), pp. 628–638.
9. Korb, L. J. (1987). *ASM Metals Handbook*, 9th ed., "Corrosion," Novelty, OH: ASM Int., vol. 13.
10. Van Arsdell, W. and Brown, S. B. (1999). Subcritical crack growth in silicon MEMS, *Journal of Microelectromechanical Systems*, 8(3), pp. 319–327.
11. Brown, S. B., van Arsdell, W., and Muhlstein, C. L. (1997). Materials reliability in MEMS devices, in *Proceedings of the. International. Conference. On Solid-State Sensors. Actuators Digest. Transducers*, vol. 1, pp. 591–593.

12. Brown, S. B. and Jansen, E. (1996). Reliability and long term stability of MEMS, in *Proceedings of the. Summer Topical Meetings Digest, Optical MEMS Applications*, pp. 9–10.
13. Brown, S. B., Povirk, G., and Connally, J. (1993). Measurement of slow crack growth in silicon and nickel micromechanical devices, in *Proceedings of the. IEEE Microelectromechanical Systems. Investigating. MicroStructure. Sensors. Actuators, Machine Systems*, pp. 99–104.
14. Connally, J. A. and Brown, S. B. (1991). Micromechanical fatigue testing, in *Proceedings of the. International. Conference. on Solid-State Sensors. Actuators Digest, Transducers*, pp. 953–956.
15. de Boer, M. P., Jensen, B. D., and Bitsie, F. (1999). A small area in-situ MEMS test structure to measure fracture strength by electrostatic probing, in *SPIE Proc. Mater. Dev. Character. Micro-Machining*, vol. 3875, pp. 97–103.
16. Srikar, V. T. and Senturia, S. D. (2002). The reliability of microelectromechanical systems (MEMS) in shock environments, *Journal of Microelectromechanical Systems*, 9, pp. 206–214.
17. Li, G. X. and Shemansky, J. R. (2000). Drop test and analysis on micro-machined structures, *Sensors and Actuators A, Physical*, 85(1–3), pp. 280–286.
18. Tanner, D. M., Walraven, J. A., Helgesen, K. S., Irwin, L. W., Brown, F., Smith, N. F., and Masters, N. (2000). MEMS reliability in shock environments, *Proc. IEEE Int. Rel. Phys. Symp.*, pp. 129–138.
19. Tanner, D. M., Walraven, J. A., Helgesen, K. S., Irwin, L. W., Gregory, D. L., Stake, J. R., and Smith, N. F. (2000). MEMS reliability in vibration environment, *Proc. IEEE Int. Rel. Phys. Symp.*, pp. 139–145.
20. Wagner, U., Muller-Fiedler, R., Bagdahn, J., Michel, B., and Paul, O. (2003). Mechanical reliability of epipoly MEMS structures under shock load, In *Proceedings of the. International. Conference. on Solid-State Sensors, Actuators Microsystems*, pp. 175–178.
21. Béliveau, A., Spencer, G. T., Thomas, K. A., and Roberson, S. L. (1999). Evaluation of MEMS capacitive accelerometers, *IEEE Design and Testing of Computers*, 16(4), pp. 48–56.
22. Brown, T. G. and Davis, B. S. (1998). Dynamic high-g loading of MEMS sensors: Ground and flight testing, in *Proc. SPIE-Int. Soc. Opt. Eng.*, vol. 3512, pp. 228–235.
23. Brown, T. G., Davis, B., Hepner, D., Faust, J., Myers, C., Muller, C., Harkins, T., Hollis, M., Miller, C., and Placzankis, B. (2001). Strap-down micro-electromechanical (MEMS) sensors for high-g munition applications, *IEEE Transactions on Magnetics*, 37(1), pp. 336–342.
24. Tanner, D. M., Walraven, J. A., Helgesen, K., Irwin, L. W., Smith, N. F., and Masters, N. (2000). MEMS reliability in shock environments, in *Proc. IEEE Int. Reliab. Phys. Symp.*, pp. 129–138.
25. Davies, B. R., Bateman, V. I., Brown, F. A., Montague, S., Murray, J. R., Rey, D., and Smith, J. H. (1998). Micromachined accelerometer design, modeling and validation, in *Proc. Conf. Model. Simul. Microsyst.*, pp. 552–556.

26. Li, G. X. and Shemansky, F. A. (2000). Drop test and analysis on micromachined structure, *Sensors and Actuators A, Physical*, 85(1–3), pp. 280–286.
27. Maluf, N. and Williams, K. (2000). *An Introduction to Microelectromechanical Systems Engineering*, Artech House.
28. Micro-Electro-Mechanical Systems (MEMS), *International Technology Roadmap for Semiconductors*, 2011 ed., ITRS, 2011.
29. Kwapisz, J. R., Weiss, G. M., and Moore, S. A. (2010). Activity recognition using cell phone accelerometers, *5 ACM SIGKDD Explorations Newsletter*, 12(2), pp. 74–82.
30. Tan, L. B., Ang, C. W., Lim, C. T., Tan, V. B. C., and Zhang, X. (2005). Modal and impact analysis of modern portable electronic products, in *Proc. 55th ECTC*, pp. 645–653.
31. Karppinen, J. S., Li, J., Pakarinen, J., Mattila, T. T., and Paulasto- Kröckel, M. (2012). Shock impact reliability characterization of a handheld product in accelerated tests and use environment, *Microelectronics Reliability*, 52(1), pp. 190–198.
32. Board Level Drop Test Method of Components for Handheld Electronic Products, JEDEC Standard JESD22-B111, 2003.
33. Subassembly Mechanical Shock, JEDEC Standard JESD22-B110A, 2004.
34. Meng, J., Mattila, T. T., Dasgupta, A., Sillanpää, M., Jaakkola, R., Anderson, K., and Hussa, E. (2012). Testing and multi-scale modeling of drop and impact loading of complex MEMS microphone assemblies, in *Proc. 13th EuroSimE*, pp. 1–8.
35. Meng, J., Mattila, T. T., Dasgupta, A., Sillanpää, M., Jaakkola, R., Luo, G., and Anderson, K. (2012). Drop qualification of MEMS components in handheld electronics at extremely high accelerations, in *Proc. 13th ITherm*, pp. 1020–1027.
36. McIntyre, D. G., Cunningham, S. J., Carper, J. S., Jaramillo, P. D., Tatic-Lucic, S., and Starr, L. E. (1997). Characterization of the influence of fabrication methods on microstructure failure, *Sensors Actuators A*, vol. 60, pp. 181–185.
37. Beliveau, A., Spencer, G. T., Thomas, K. A., and Roberson, S. L. (1999). Evaluation of MEMS capacitive accelerometers, *IEEE Design & Testing Computers.*, pp. 48–56.
38. Hartzell, A. and Woodilla, D. (1999). Reliability methodology for prediction of micromachined accelerometer stiction, in *Proc. IEEE Int. Reliability Physics Symp.*, pp. 202–205.
39. Brown, T. G., Davis, B., Hepner, D., Faust, J., Myers, C., Muller, P., Harkins, T., Hollis, M., Miller, C., and Placzankis, B. (2001). Strap-down microelectromechanical (MEMS) sensors for high-g munition applications, *IEEE Trans. Magnetics*, 37, pp. 336–342.
40. Wagner, U., Bernhard, W., Müller-Fiedler, R., Michel, B., and Paul, O. (2003). Reliability of micromachined membranes under particle impact, in *Proc. 12th Int. Conf. Solid-State Sens., Actuat. Microsyst.*, pp. 436–439.
41. Wagner, U., Müller-Fiedler, R., Bagdahn, J., Michel, B., and Paul, O. (2003). Mechanical reliability of epipoly MEMS structures under shock

load, in *Proc. 12th Int. Conf. Solid-State Sens., Actuat. Microsyst.*, pp. 175–178.
42. Brown, T. G. (2003). Harsh military environments and microelectromechanical (MEMS) devices, in *Proc. IEEE Sensors*, pp. 753–760, 2003.
43. Choa, S. H. (2005). Reliability of vacuum packaged MEMS gyroscopes, *Microelectron. Rel.*, 45(2), pp. 361–369.
44. Al Saleem, F. M., Younis, M. I., and Ibrahim, M. (2008). The effect of the PCB motion on the dynamic response of MEMS devices under mechanical shock loads, in *Proc. SEM 11th Int. Congr. Exposit. Experim. Appl. Mech.*, pp. 1–8.
45. Hauck, T., Li, G., McNeill, A., Knoll, H., Ebert, M., and Bagdahn, J. (2006). Drop simulation and stress analysis of MEMS devices, in *Proc. 7th EuroSimE*, pp. 1–5.
46. Mariani, S., Ghisi, A., Fachin, F., Cacchione, F., Corigliano, A., and Zerbini, S. (2008). A three-scale FE approach to reliability analysis of MEMS sensors subject to impacts, *Meccanica*, 43(5), pp. 469–483.
47. Yang, C., Zhang, B., Chen, D., and Lin, L. (2010). Drop-shock dynamic analysis of MEMS/package system, in *Proc. IEEE 23rd MEMS*, pp. 520–523.
48. Li, J., Broas, M., Makkonen, J., Mattila, T. T., Hokka, J., and Paulasto-Kröckel, M. (2014). Shock impact reliability and failure analysis of a three-axis MEMS gyroscope, *J. Microelectromech. Syst.*, 23(2), pp. 347–355.
49. Hokka, J., Broas, M., Makkonen, J., Raami, J., Li, J., Mattila, T. T., and Paulasto-Kröckel, M. (2012). Methods for reliability characterization of MEMS devices — Case studies of a MEMS microphone and a 3-axis MEMS gyroscope, in *Proc. 62nd ECTC*, pp. 62–69.
50. Brown, T. G. and Davis, B. S. (1998). Dynamic high-g loading of MEMS sensors: Ground and flight testing, in *Proc. SPIE-Int. Soc. Opt. Eng.*, 3512, pp. 228–235.
51. Brown, T. G., Davis, B., Hepner, D., Faust, J., Myers, C., Muller, C., Harkins, T., Hollis, M., Miller, C., and Placzankis, B. (2001). Strap-down micro-electromechanical (MEMS) sensors for high-g munition applications, *IEEE Trans. Magn.*, 37(1), pp. 336–342.
52. Patton, S. T., Cowan, W. D., and Zabinski, J. S. (1999). Performance and reliability of a new MEMS electrostatic lateral output motor, in *Proc. of IRPS*, pp. 179–188.
53. Lee, S. S., Motamedi, E., and Wu, M. C. (1997). G-Performance characterization of surface-micromachined FDDT optical bypass switches, in *Proc. SPIE*, 3226, pp. 94–101.
54. Huang, L. S., Lee, S. S., Motamedi, E., Wu, M. C., and Kim, C. J. (1998). MEMS packaging for micro mirror switches, in *Proc. 4th Electronic Compon. Technol. Conf.*, pp. 592–597.
55. Tanner, D. M., Walraven, J. A., Helgesen, K. S., Irwin, L. W., Gregory, D. L., Stake, J. R., and Smith, N. F. (2000). MEMS reliability in a vibration environment, *IEEE 00CH37059, 38th Ann. Int. Reliab. Phys. Symp.*, pp. 139–145.

56. Zhou, X. F., Che, L. F., Liang, S. L., Lin, Y. L., Li, X. L., and Wang, Y. L. (2015). Design and fabrication of a MEMS capacitive accelerometer with fully symmetrical double-sided H-shaped beam structure, *Microelectron. Eng.*, 131, pp. 51–57.
57. Xiao, F., Che, L., Xiong, B., Wang, Y., Zhou, X., Li, Y., and Lin, Y. (2008). A novel capacitive accelerometer with an eight-beam-mass structure by self-stop anisotropic etching of (1 0 0) silicon, *J. Micromech. Microeng.*, 18, pp. 1–7.
58. Hu, Q., Gao, C., Hao, Y., Zhang, Y., and Yang, G. (2011). Low cross-axis sensitivity micro-gravity microelectromechanical system sandwich capacitance accelerometer, *Micro Nano Lett.*, 6(7), pp. 510–514.
59. Yazdi, N. and Najafi, K. (2000). An all-silicon single-wafer micro-g accelerometer with a combined surface and bulk micromachining process, *Journal of Microelectromechanical. Systems*, 9(4), pp. 544–550.
60. Peng, P., Wen, T. D., and Xu, L. P. (2014). A new structure accelerometer based on meso-piezoptic, *Infrared Laser Eng.*, 43, pp. 3363–3367.
61. Jiao, X. Q., Chen, J. B., Yin, J. Y., and Meng, D. (2013). Novel packaging technology for high-G MEMS accelerometer, *J. Chin. Inertial Technol.*, 21, pp. 536–539.
62. Li, P., Gao, S. Q., Jin, L., and Shi, Y. B. (2012). Effects of package materials on performances of a piezoresistive MEMS accelerometer, *Explosion Shock Waves*, 32, pp. 623–628.
63. Mohd-Yasin, F., Zaiyadi, N., Nagel, D. J., Ong, D. S., Korman, C. E., and Faidz, A. R. (2009). Noise and reliability measurement of a three-axis microaccelerometer, *Microelectron. Eng.*, 86, pp. 991–995.
64. Ma, X. H. and Li, C. L. (2012). Test of reliability of micro-accelerometer in vibration environment, *Adv. Mater. Res.*, 588–589, pp. 1881–1884.
65. Yuan, H. J., Li, L. D., Duan, G., and Wu, H. (2012). Storage life and reliability evaluation of accelerometer by step stress accelerated degradation testing, *J. Chin. Inertial Technol.*, 20, pp. 113–116.
66. Yang, Y. C. (2013). Investigation of accelerated life testing of random vibration, *Harbin Inst. Technol.*, pp. 24–29.
67. Jinsuk, L. and Rong, P. (2010). Analyzing step-stress accelerated life testing data using generalized linear models, *IIE Trans.*, 42, pp. 589–598.
68. Li, C., Wang, J. N., and Yang, H. T. (2004). Accelerated mission test technology and reliability evaluation for highly reliable components, *Mech. Sci. Technol. Aerospace Eng.*, 23, pp. 876–882.
69. Davis, J. R. (1999). Corrosion of Aluminum and Aluminum Alloys, Novelty, OH: ASM Int.
70. Zhang, Q., Guo, X., Liu, Y., Dai, N., and Lu, P. (2007). Corrosion and fatigue in micro-sized Ni cantilever beams, *Trans. Nonferrous Met. Soc. China*, 17, pp. 213–217.
71. American Society for Metals, Metals Handbook (desk ed.), Chapter 32 (Failure Analysis), Novelty, OH: ASM Int., pp. 3224–3226, 1997.
72. Tadigadapa, S. and Najafi, N. (2001). Reliability of microelectromechanical systems (MEMS), in *Proc. Reliab., Test., Charac. MEMS/MOEMS Conf.*, pp. 197–205.

73. Zhao, Y. P., Wang, L. S., and Yu, T. X. (2003). Mechanics of adhesion in MEMS: A review, *J. Adhesion. Sci. Technol.*, 17(4), pp. 519–546.
74. Jafri, I. H., Busta, H., and Walsh, S. T. (1999). Critical point drying and cleaning for MEMS technology, in *Proc. SPIE*, vol. 3880 (MEMS Reliability for Critical and Space Applications), pp. 51–58.
75. Guckel, H., Sniegowski, J. J., Christenson, T. R., Mohney, S., and Kelly, T. F. (1989). Fabrication of micromechanical devices from polysilicon films with smooth surfaces, *Sens. Actuators*, 20(1–2), pp. 117–122.
76. Man, P. F., Gogoi, B. P., and Mastrangelo, C. H. (1997). Elimination of post-release adhesion in microstructures using conformal fluorocarbon coatings, *J. Microelectromech. S.*, 6(1), pp. 25–34.
77. Van Spengen, W. M., Puers, R., and de Wolf, I. (2003). On the physics of stiction and its impact on the reliability of microstructures, *J. Adhesion Sci. Technol.*, 17(4), pp. 563–582.
78. Chan, H. B., Aksyuk, V. A., Kleiman, R. N., Bishop, D. J., and Capasso, F. (2001). Quantum mechanical actuation of microelectromechanical systems by the Casimir force, *Science*, 291(5510), pp. 1941–1944, 2001.
79. Man, P. F., Gogoi, B. P., and Mastrangelo, G. H. (1997). Elimination of post-release adhesion in microstructures using conformal fluorocarbon coatings, *J. Microelectromech. S.*, 6(1), pp. 25–34.
80. Douglass, M. R. (2003). DMD reliability: A MEMS success story, in *Proc. SPIE 4980, Reliab., Test., Character. MEMS/MOEMS II*, vol. 4980, SPIE.
81. Mastrangelo, C. H. and Hsu, C. H. (1992). A simple experimental technique for the measurement of the work of adhesion of microstructures, in *Proc. IEEE Solid-State Sens. Actuator Workshop*, pp. 208–212.
82. van Spengen, W. M., Puers, R., and de Wolf, I. (2003). On the physics of stiction and its impact on the reliability of microstructures, *J. Adhesion Sci. Technol.*, 17(4), pp. 563–582.
83. van Spengen, W. M., de Wolf, I., and Puers, B. (2000). Auto-adhesion model for MEMS surfaces taking into account the effect of surface roughness, in *Proc. SPIE*, vol. 4175, pp. 104–112.
84. Maboudian, R. (1998). Adhesion and friction issues associated with reliable operation of MEMS, *MRS Bull.*, 23(6), pp. 47–51.
85. van Spengen, W. M., Puers, R., and de Wolf, I. (2003). The prediction of stiction failures in MEMS, *IEEE Trans. Device Mater. Rel.*, 3(4), pp. 167–172.
86. Komvopoulos, K. (1996). Surface engineering and microtribology for microelectromechanical systems, *Wear*, 200(1–2), pp. 305–327.
87. Raccurt, O., Tardif, F., Arnaud d'Avitaya, F., and Vareine, T. (2004). Influence of liquid surface tension on stiction of SOI MEMS, *Journal of Micromechanics and. Microengineering*, 14(7), pp. 1083–1090.
88. Delrio, F. W., de Boer, M. P., Knapp, J. A., Reedy, E. D., Jr., Clews, P. J., and Dunn, M. L. (2005). The role of van der Waals forces in adhesion of micromachined surfaces, *Nat. Mater.*, 4(8), pp. 629–634.
89. Bhushan, B. (1999). *Handbook of Micro/Nanotribology*, Boca Raton, FL: CRC Press.

90. DelRio, F. W., Dunn, M. L., and de Boer, M. P. (2008). Capillary adhesion model for contacting micromachined surfaces, *Script. Mater.*, 59(9), pp. 916–920.
91. Bowden, F. P. and Tabor, D. (1950). *Lubrication of Solids*, Oxford, U.K.: Clarendon.
92. Maboudian, R. and Howe, R. T. (1997). Critical review: Adhesion in surface micromechanical structures, *J. Vac. Sci. Technol. B*, 15(1), pp. 1–20.
93. Adamson, A. W. (1990). *Physical Chemistry of Surfaces*, 5th ed., New York: John Wiley.
94. van Spengen, W. M., Puers, R., and de Wolf, I. (2002). A physical model to predict stiction in MEMS, *Journal of Micromechanics and. Microengineering*, 12(5), pp. 702–713.
95. Komvopoulos, K. (1996). Surface engineering and microtribology for microelectromechanical systems, *Wear*, 200, pp. 305–327.
96. Herrmann, C. F., DelRio, F. W., Bright, V. M., and George, S. M. (2004). Hydrophobic coatings using atomic layer deposition and non-chlorinated precursors, in *Proceedings of the. IEEE International Conference on Microelectromechanical Systems*, pp. 653–656.
97. Houston, M. R. and Maboudian, R. (1995). Stability of ammonium fluoride-treated Si (100), *Journal of Applied. Physics*, 78(6), pp. 3801–3808.
98. Greaso, W. D. (2009). Analysis of the effect of ESD on the operation of MEMS, *IEEE Transactions on. Industrial Applications*, 45(6), pp. 2185–2191.
99. de Boer, M. P., Knapp, J. A., Mayer, T. M., and Michalske, T. A. (1999). The role of interfacial properties on MEMS performance and reliability, in *Proc. SPIE*, 3825, pp. 2–15.
100. de Boer, M. P. and Michalske, T. A. (1999). Accurate method for determining adhesion of cantilever beams, *Journal of Applied Physics*, 86(2), pp. 817– 827.
101. de Boer, M. P., Tabbara, M. R., Dugger, M. T., Clews, P. J., and Michalske, T. A. (1997). Measuring and modeling electrostatic adhesion in micromachines, in *Proceedings of the IEEE International Conference on. Solid State Sensors and Actuators*, 1, pp. 229–232.
102. Yee, Y., Park, M., and Chun, K. (1998). A sticking model of suspended polysilicon microstructure including residual stress gradient and postrelease temperature, *Journal of Microelectromechanical. Systems*, 7(3), pp. 339–344.
103. Pister, K. S. J. [Online Research Resources]. Available: www-bsac.EECS. Berkeley.edu/~pister/245
104. de Groot, W. A., Webster, J. R., Felnhofer, D., and Gusev, E. P. (2009). Review of device and reliability physics of dielectrics in electrostatically driven MEMS devices, *IEEE Trans. Device Mater. Rel.*, 9(2), pp. 190–202.
105. Man, P. F., Gogoi, B. P., and Mastrangelo, C. H. (1997). Elimination of post-release adhesion in microstructures using conformal fluorocarbon coatings, *Journal of Microelectromechanical Systems*, 6(1), pp. 25–34.

106. Taii, Y., Kakushima, K., Yokokawa, R., Ono, S., Takahashi, T., Isamoto, K., Chong, C., Fujita, H., and Toshiyoshi, H. (2007). Suppression of stiction force by all-vapor processes using HF, ozone, and HMDS for MEMS devices, *IEEJ Trans. Sens. Micromach.*, 127(4), pp. 221–227.
107. Van Kessel, P. F., Hornbeck, L. J., Meier, R. E., and Douglass, M. R. (1998). A MEMS-based projection display, in *Proc. IEEE*, 86(8), pp. 1687–1704.
108. Koester, D. A., Mahadevan, R., Shishkoff, A., and Markus, K. W. (1999). *MUMPs Design Handbook*, Durham, NC: Cronos Integrated Microsystems.
109. Method 1014, a United States Defense Standard, MIL-STD-883F, 2004.
110. EUROPRACTICE IC Service, http://www.europractice-ic.com/MEMS_memscap
111. Di Benedetto, A. T. (1967). *The Structure and Properties of Materials*, New York: McGraw-Hill.
112. Ali, S. M. and Phinney, L. M. (2004). Investigation of adhesion during operation of MEMS cantilevers, in *Proc. SPIE*, 5343, pp. 215–226.
113. Tambe, N. S. and Bhushan, B. (2004). Scale dependence of micro/nano-friction and adhesion of MEMS/NEMS materials, coatings and lubricants, *Nanotechnology*, 15(11), pp. 1561–1570.
114. Subhash, G., Corwin, A. D., and de Boer, M. P. (2005). Operational wear and friction in MEMS devices, in *Proc. 11th ICF.*
115. Lim, M. G., Chang, J. C., Shultz, D. P., Howe, R. T., and White, R. M. (1990). Polysilicon microstructures to characterize static friction, in *Proc. Micro Electro Mech. Syst. Annu. Int. Workshop MEMS Conf. Ser.*, pp. 82–88.
116. Tai, Y. C. and Muller, R. S. (1990). Frictional study of IC-processed micromotors, *Sens. Actuators A, Phys.*, 21(1–3), pp. 180–183.
117. Gabriel, K. J., Behi, F., and Mahadevan, R. (1990). In situ friction and wear measurements in integrated polysilicon mechanisms, *Sens. Actuators A, Phys.*, 21(1–3), pp. 184–188.
118. Wang, W., Wang, Y., Bao, H., Xiong, B., and Bao, M. (2002). Friction and wear properties in MEMS,*Sens. Actuators A, Phys.*, 97–98, pp. 486–491.
119. Beerschwinger, U., Mathieson, D., Reuben, R. L., and Yang, S. J. (1994). A study of wear on MEMS contact morphologies, *Journal of Micromechanical. Microengineering*, 4(3), pp. 95–105.
120. Tanner, D. M., Walraven, J. A., Irwin, L. W., Dugger, M. T., Smith, N. F., Eaton, W. P., Miller, W. M., and Miller, S. L. (1999). The effect of humidity on the reliability of a surface micro-machined microengine, in *Proc. IRPS*, San Diego, CA, pp. 189–197.
121. Patton, S. T., Cowan, W. D., and Zabinski, J. S. (1999). Performance and reliability of a new MEMS electrostatic lateral output motor, in *Proc. IRPS*, San Diego, CA, pp. 179–188.
122. Luo, J. K., Fu, Y. Q., Le, H. R., Williams, J. A., Spearing, S. M., and Milne, W. I. (2007). Diamond and diamond-like carbon MEMS, in *Journal of Micromechanical Microengineering*, 17(17), pp. 147–163.
123. Auciello, O., Birrell, J., Carlisle, J. A., Gerbi, J. E., Xiao, X., Peng, B., and Espinosa, H. D. (2004). Materials science and fabrication processes for

a new MEMS technology based on ultrananocrystalline diamond thin films, *J. Phys.: Condens. Matter*, 16(16), pp. 539–552.
124. Ashurst, R. W., Yau, C., Carraro, C., Maboudian, R., and Dugger, M. T. (2001). Dichlorodimethylsilane as an anti-stiction monolayer for MEMS: A comparison to the octadecyltrichlosilane self-assembled monolayer, *Journal of Microelectromechanical Systems*, 10(1), pp. 41–49.
125. Srinivasan, U., Houston, M. R., Howe, R. T., and Maboudian, R. (1998). Alkyltrichlorosilane-based self-assembled monolayer films for stiction reduction in silicon micromachines, *Journal of. Microelectromechanical. Systems*, 7(2), 252–260.
126. Maboudian, R. and Howe, R. T. (1997). Stiction reduction processes for surface micromachines, *Tribol. Lett.*, 3(3), pp. 215–221.
127. Lee, S. H., Kwon, M.-J., Park, J.-G., Kim, Y.-K., and Shin, H.-J. (1998). The surface modification with fluorocarbon thin films for the prevention of stiction in MEMS, in *Mater. Res. Soc. Symp. Proc.*, vol. 518, *Material Research Society*, pp. 143–148.
128. Kim, J.-M., Baek, C.-W., Park, J.-H., Shin, D.-S., Lee, Y.-S., and Kim, Y.-K. (2002). Continuous anti-stiction coatings using self-assembled monolayers for gold microstructures, *Journal of Micromechanical. Microengineering*, 12(5), pp. 688–695.
129. Senft, D. C. and Dugger, M. T. (1997). Friction and wear in surface micromachined tribological test devices, in *Proc. SPIE*, vol. 3224, pp. 31–38.
130. Radhakrishnan, S. G., Robertson, R. E., Adams, P. M., and Cole, R. C. (2002). Integrated TiC coatings for moving MEMS, *Thin Solid Films*, 420–421, pp. 553–564.
131. Mayer, T. M., Elam, J. W., George, S. M., Kotula, P. G., and Goeke, R. S. (2003). Atomic-layer deposition of wear-resistant coatings for microelectromechanical devices, *Applied Physics. Letters*, 82(17), pp. 2883–2885.
132. Patton, S. T., Cowan, W. D., Eapen, K. C., and Zabinski, J. S. (2001). Effect of surface chemistry on the tribological performance of a MEMS electrostatic lateral output motor, *Tribol. Lett.*, 9(3–4), pp. 199–209.
133. Maboudian, R., Ashurst, W. R., and Carraro, C. (2000). Self-assembled monolayers as anti-stiction coatings for MEMS: Characteristics and recent developments, *Sensors. Actuators A, Physical*, 82(1–3), pp. 219–223.
134. Sundararagan, S. and Bhushan, B. (2001). Static friction and surface roughness studies of surface micromachined electrostatic micromotors using an atomic force/friction force microscope, *J. Vac. Sci. Technol. A*, *Vac. Surf. Films*, 19(4), pp. 1777–1785.
135. Srinivasan, U., Houston, M. R., Howe, R. T., and Maboudian, R. (1997). Self-assembled fluorocarbon films for enhanced stiction reduction, in *Proceedings of the. International. Conference on. Solid-State Sensors and Actuators, Transducers*, Chicago, IL, pp. 1399–1402.
136. Zhu, X.-Y. and Houston, J. E. (1999). Molecular lubricants for silicon-based microelectromechanical systems (MEMS): A novel assembly strategy, *Tribol. Lett.*, 7(2–3), pp. 87–90.

137. Henck, S. A. (1997). Lubrication of digital micromirror devices, *Tribol. Lett.*, 3, pp. 239–247.
138. Eapen, K. C., Patton, S. T., and Zabinski, J. S. (2002). Lubrication of micro- electromechanical systems (MEMS) using bound and mobile phases of fomblin Z-Dol®, *Tribol. Lett.*, 12(1), pp. 35–41.
139. Smallwood, S. A., Eapenb, K. C., Pattonb, S. T., and Zabinskic, J. S. (2006). Performance results of MEMS coated with a conformal DLC, *Wear*, 260(11–12), pp. 1179–1189.
140. Bandorf, R., Lüthje, H., and Staedler, T. (2004). Influencing factors on microtribology of DLC films for MEMS and microactuators, *Diamond Relat. Mater.*, 13(4–8), pp. 1491–1493.
141. Guo, Z., Zhou, F., Fan, S., Zheng, D., and Zhuang, H. (2009). Research development of measuring methods on the tribology characters for movable MEMS devices: A review, *Microsyst. Technol.*, 15(3), pp. 343–354.
142. Muhlstein, C., Brown, S., and Failure Analysis Associates, Inc. (1997). Reliability and fatigue testing of MEMS, NSF/AFOSR/ASME Workshop Tribology Issues and Opportunities in MEM, Kluwer Academic, pp. 519–528.
143. Shrotriya, P., S. M. Allameh, and W. O. Soboyejo, On the evolution of surface morphology of polysilicon MEMS structures during fatigue, *Mech. Mater.*, 36(1–2), pp. 35–44.
144. Tsuchiya, T., Inoue, A., Sakata, J., Hashimoto, M., Yokoyama, A., and Sugimoto, M. (1998). Fatigue test of single crystal silicon resonator, in *Proc. Tech. Dig. 16th Sens. Symp.*, Kawasaki, Japan, pp. 277–280.
145. Tadigadapa, S. and Najafi, N. (2001). Reliability of Microelectromechanical Systems (MEMS), in *Proc. Reliability, Testing, and Characterisation of MEMS/MOEMS Conf.*, pp. 197–205.
146. Stanimirović, I. and Stanimirović, Z. (2013). Optical MEMS for Telecommunications: Some Reliability Issues, *Advances in Micro/Nano Electromechanical Systems and Fabrication Technologies*, Chapter 4, INTECH, 2013.
147. Van Spengen, W. M. (2003). MEMS reliability from a failure mechanisms perspective, *Microelect. Reliab*, 43(7), pp. 1049–1060.
148. Douglass, M. R. (2003). DMD reliability: A MEMS success story, in *SPIE Proc.*, vol. 4980, pp. 1–11.
149. Douglass, M. R. (1998). Lifetime estimates and unique failure mechanisms of the digital micromirror device (DMD), in *Proc. 36th Annu. Int. Reliab. Phys. Symp.*, Reno, NV, pp. 9–16.
150. Sontheimer, A. B. (2002). Digital micromirror device (DMD) hinge memory lifetime reliability modeling, in *Proc. 40th Annu. Int. Reliab. Phys. Symp.*, Dallas, TX, pp. 118–121.
151. Sontheimer, A. B. (2003). Effects of operating conditions on DMD hinge memory lifetime, in *Proc. 41st Annu. Int. Reliab. Phys. Symp.*, Dallas, TX, pp. 473–477.
152. Tregilgas, J. H. (1996). Micromechanical device having an improved beam, U.S. Patent No. 5 552 924, Sep. 3.

153. Modlinski, R., Ratchev, P., Witvrouw, A., Puers, R., and de Wolf, I. (2005). Creep-resistant aluminum alloys for use in MEMS, *Journal of Micromechanical. Microengineering*, 15(7), pp. S165–S170.
154. Modlinski, R., Witvrouw, A., Ratchev, P., Puers, R., den Toonder, J. M. J., and De Wolf, I. (2004). Creep characterization of Al alloy thin films for use in MEMS applications, *Microelectron. Eng.*, 76(1–4), pp. 272–278.
155. Modlinski, R., Witvrouw, A., Ratchev, P., Jourdain, A., Simons, V., Tilmans, H. A. C., den Toonder, J. M. J., Puers, R., and de Wolf, I. (2004). Creep as a reliability problem in MEMS, *Microelectron. Reliab.*, 44(9–11), pp. 1733–1738.
156. Tuck, K., Jungen, A., Geisberger, A., Ellis, M., and Skidmore, G. (2005). A study of creep in polysilicon MEMS devices, *J. Eng. Mater. Technol.*, 127(1), pp. 90–96.
157. Nataraj, D., Senthil, K., Narayandass, S., and Mangalaraj, D. (1999). Conduction studies on bismuth selenide thin films, *Cryst. Res. Technol.*, 34(7), pp. 867–872.
158. Zafar, S., Callegari, A., Gusev, E., and Fischetti, M. V. (2002). Charge trapping in high k gate dielectric stacks, in *IEDM Tech. Dig.*, pp. 517–520.
159. Duvvury, C. and Amerasekera, A. (1993). ESD: A pervasive reliability concern for IC technologies, in *Proc. IEEE*, 81(5), pp. 690–702.
160. Newman, H. S. (2002). RF MEMS switches and applications, in *Proc. IEEE IRPS*, pp. 111–115.
161. Rebeitz, G. M. and Muldavin, J. B. (2001). RF MEMS switches and switch circuits, *IEEE Microw. Mag.*, 2(4), pp. 59–71.
162. DeNatale, J., Mihailovich, R., and Waldrop, J. (2002). Techniques for reliability analysis of RF MEMS switch, in *Proc. IEEE IRPS*, pp. 116–117.
163. van Spengen, W. M., Mertens, R., and De Wolf, I. (2002). Experimental characterization of stiction due to charging in RF MEMS, in *IEDM Tech. Dig.*, pp. 901–904.
164. Becher, D., Chan, R., Hattendorf, M., and Feng, M. (2002). Reliability study of low-voltage RFNEMS switches, in *Proc. GaAs MANTECH Conf.*, San Diego, pp. 54–57.
165. Lee, J. C., Hoque, A., Croft, G. D., Liou, J. J., Young, R., and Bernier, C. (2000). An electrostatic discharge failure mechanism in semiconductor devices, with applications to electrostatic discharge measurements using transmission line pulsing technique, *Solid State Electronics*, 44(10), pp. 1771–1781.
166. B. Stark (1999). *MEMS Reliability Assurance Guidelines for Space Applications.* Pasadena, CA: JPL Publ.
167. Electronic Industries Association, Electrostatic Discharge (ESD) Sensitivity Testing Machine Model (MM), EIA/JESD22-A115A, 1997.
168. Olszewski, Z., Duane, R., and Mahony, C. O'(2005). Effect of Environment Humidy on the C–V Characteristics of RF-MEMS Capacitive Switch, *Tyndall Nat. Inst.*, Lee Maltings, Prospect Row, Cork, Ireland.

169. Knudson, A. R., Buchner, S., McDonald, P., Stapor, W. J., Campbell, A. B., Grabowski, S., and Knies, D. L. (1996). The effects of radiation on MEMS accelerometers, *IEEE Trans. Nucl. Sci.*, 43(6), pp. 3122–3126.
170. Lee, C. I., Johnston, A. H., Tang, W. C., and Barnes, C. E. (1996). Total dose effects on microeltromechanical systems (MEMS): Accelerometers, *IEEE Trans. Nucl. Sci.*, 43(6), pp. 3127–3132.
171. McClure, S. S., Edmonds, L. D., Mihailovich, R., Johnston, A. H., Alonzo, P., DeNatale, J., Lehman, J., and Yui, C. (2002). Radiation effects in microelectromechanical systems (MEMS): RF relays, *IEEE Trans. Nucl. Sci.*, 49(6), pp. 3197–3202.
172. Newman, H. S., Ebel, J. L., Judy, D., and Maciel, J. (2008). Lifetime measurements on a high-reliability RF-MEMS contact switch, *IEEE Microw. Wireless Compon. Lett*, 18(2), pp. 100–102.
173. Ehmke, J. C., Goldsmith, C. L., Yao, Z. J., and Eshelman, S. M. (2002). Method and apparatus for switching high frequency signals, U.S. Patent No. 639 167 521.
174. Wairaven, J. A., Soden, J. M., Tanner, D. M., Tangyunyong, P., Cole, E. I., Jr., Anderson, R. E., and Irwin, L. W. (2000). Electrostatic discharge/electrical over-stress susceptibility in MEMS: A new failure model, in *Proc. SPIE*, vol. 4180, pp. 30–39.
175. Mroczkowski, R. S. (1998). *Electronic Connector Handbook*, New York: McGraw-Hill.
176. Machate, M. S. (2003). Joule heat effects on reliability of RF MEMS switches, M.S. thesis, Worcester Polytechnic Inst., Worcester, MA.
177. Kwon, H., Choi, D.-J., Park, J.-H., Lee, H.-C., Park, Y.-H., Kim, Y.-D., Nam, H.-J., Joo, Y.-C., and Bu, J.-U. (2007). Contact materials and reliability for high power RF-MEMS switches, in MEMS 2007, Kobe, Japan, Jan. 21–25, pp. 231–234.
178. Dickrell, D. J., III and Dugger, M. T. (2005). The effects of surface contamination on resistance degradation of hot-switched low-force MEMS electrical contacts, in *Proc. 51st IEEE Holm Conf.*, pp. 255–258.
179. Pothier, A., Blondy, P., Cros, D., Verdeyme, S., Guillon, P., Champeaux, C., Tristant, P., and Catherinot, A. (2002). Low loss ohmic switches for RF frequency applications, in *Proc. 32nd Eur. Microw. Conf.*, Milan, Italy, pp. 1–4.
180. Gaddi, R., Bellei, M., Gnudi, A., Margesin, B., and Giacomozzi, F. (2004). Interdigitated low-loss ohmic RF MEMS switches, in *Proc. NSTI — Nanotechnol.*, vol. 2, pp. 327–330.
181. Ducarouge, B., Dubuc, D., Flourens, F., Melle, S., Ongareau, E., Grenier, K., Boukabache, A., Conedera, V., Pons, P., Perret, E., Aubert, H., and Plana, R. (2004). Power capabilities of RF MEMS, in *Proc. 24th Int. Conf. Microelectron.*, pp. 65–70.
182. Shih, Y. C., Murakami, M., Wilkie, E. L., and Callegari, A. C. (1987). Effect of interfacial microstructure on uniformity and thermal stability of AuNiGeohmic contact to n-type GaAs, *Journal of Applied Physics*, 62(2), pp. 582–590.

183. Wang, B. J., Saka, N., and Rabinowicz, E. (1992). The failure mechanism of low-voltage electrical relays, in *Proc. 38th IEEE Holm Conf. Elect. Contacts*, pp. 191–202.
184. Neufeld, C. N. and Rieder, W. F. (1995). Electrical characteristics of various contact contaminations, *IEEE Trans. Compon., Packag., Manuf. Technol*, 18(2), pp. 369–374.
185. Zangwill, A. (1988). Physics at surfaces, in *Approaches Contacts From Point of View of Surface States and Reconstruction*, Cambridge, U.K: Cambridge Univ. Press.
186. Tabata, O., Tsuchiya, T., Brand, O., Fedder, G. K., Hierold, C., and Korvink, J. G. (2008). Reliability of MEMS: Testing of Materials and Devices, Hoboken, NJ: Wiley-VCH.
187. Hsu, T. R. (2006). Reliability in MEMS packaging, in *Proc. 44th Int. Reliab. Phys. Symp.*, San Jose, CA, IEEE International.
188. Hsu, T. R. (2007). Introduction to reliability in MEMS Packaging, in *Proc. Int. Symp. Test. Failure Anal.*, San Jose, California.
189. Stanimirović, I. and Stanimirović, Z. (2009). Reliability of MEMS, in: *Micro Electronic and Mechanical Systems*, K. Takahata (Ed.), Rijeka: InTech, pp. 177–184.
190. Najafi, K. (2003). Micropackaging Technologies for Integrated Microsystems: Applications to MEMS and MOEMS. *Micromachining and Microfabrication Process Technology VIII*, J. A. Yasaitis, M. A. Perez-Maher, and J. M. Karam (Eds.), *Proc. SPIE*, vol. 4979, SPIE.
191. Rebeiz, G. M. (2003). *RF-MEMS, Theory, Design, and Technology*, John Wiley.
192. Rebeiz, G. M., Entesari, K., Reines, I. C., Park, S.-J., El-Tanani, M., Grichener, A., and Brown, A. R. (2009). Tuning in to RF MEMS, *IEEE, Microw. Mag.*, 10, pp. 55–72.
193. Mulloni, V., Giacomozzi, F., and Margesin, B. (2010). Controlling stress and stress gradient during the release process in gold suspended microstructures, *Sens. Actuators A*, 162, pp. 93–99.
194. Persano, A., Tazzoli, A., Farinelli, P., Meneghesso, G., Siciliano, P., and Quaranta, F. (2012). K-band capacitive MEMS switches on GaAs substrate: Design, fabrication, and reliability, *Microelectron. Reliab.*, 52, pp. 2245–2249.
195. Lisec, T., Huth, C., and Wagner, B. (2004) Dielectric material impact on capacitive RF MEMS reliability, in *Proc. 12th GAAs Symp.*, Amsterdam, September, pp. 471–474.
196. van Spengen, W. M., Puers, R., Mertens, R., and De Wolf, I. (2004). A comprehensive model to predict the charging and reliability of capacitive RF MEMS switches, *Journal of Micromechanical Microengineering*, 14, pp. 514–521.
197. Goldsmith, C. L., Ehmke, J., Malczewski, A., Pillans, B., Eshelman, S., Yao, Z., Brank, J., and Eberly, M. (2001). Lifetime characterization of RF MEMS switches, *IEEE MTT-S Int. Microwave Symp. Dig.* 1, pp. 227–230.

198. Papaioannou, G., Giacomozzi, F., Papandreou, E., and Margesin, B. (2011). Floating electrode microelectromechanical system capacitive switches: A different actuation mechanism, *Applied. Physics. Letters*, 99, pp. 073501 1–3.
199. Persano, A., Tazzoli, A., Cola, A., Siciliano, P., Meneghesso, G., and Quaranta, F. (2012). Reliability enhancement by suitable actuation waveforms for capacitive RF MEMS switches in III–V technology, *Journal of Microelectromechanical. Systems*, 21(2), pp. 414–419.
200. Rebeiz, G. (2003) *RF MEMS, Theory, Design and Technology*, John Wiley.
201. Goggin, R., Fitzgerald, P., Stenson, B., Carty, E., and McDaid, P. (2015). Commercialization of a reliable RF MEMS switch with integrated driver circuitry in a miniature QFN package for RF instrumentation applications, *IEEE MTT-S Int. Microw. Symp.*, pp. 1–4.
202. Malczewski, A., Pillans, B., Morris, F., and Newstrom, R. (2011). A family of MEMS tunable filters for advanced RF applications, in *Proc. Int. Microw. Symp.*, pp. 1–4.
203. Domingue, F., Fouladi, S., and Mansour, R. (2010). A reconfigurable impedance matching network using dual-beam MEMS switches for an extended operating frequency range, in *Proc. Int. Microw. Symp.*, pp. 1552–1555.
204. Belenger, B., Espana, B., Courreges, S., Blondy, P., Vendier, O., Langrez, D., and Cazaux, J.-L. (2011). A high-power Ka-band RF-MEMS 2-bit phase shifter on Sapphire substrate, in *Proc. Microw. Integrat. Circuits Conf.*, pp. 164–167.
205. Daneshmand, M. and Mansour, R. (2011). RF MEMS satellite switch matrices, *IEEE Microw, Mag.*, 12(5), pp. 92–109.
206. Tazzoli, A., Iannacci, J., and Meneghesso, G. (2011). A positive exploitation of ESD events: Micro-welding induction on ohmic MEMS contacts, in *Proc. Electric. Overstress/Electrostat. Discharge Symp.*, pp. 1–8.
207. van Spengen, W. M., Puers, R., Mertens, R., and De Wolf, I. (2004). A comprehensive model to predict the charging and reliability of capacitive RF MEMS switches, *Journal of Micromechanical. Microengineering*, 14, pp. 514–521.
208. Yao, J. J. (2000). R F MEMS from a device perspective, *Journal of Micromechanical. Microengineering*, 10, R9.
209. Rebeiz, G. M. (2003). *RF MEMS: Theory, Design and Technology*, Hoboken, NJ: John Wiley.
210. Lucyszyn (Ed.), S. (2010). *Advanced RF MEMS*, Cambridge: Cambridge University Press.
211. Bouchaud, J. (2010). RF MEMS switches and varactors set to deliver on their promise, iSuppli MEMS Switches Special Report.
212. Love, J. (2011). The race is on for RF MEMS, RF & microwave systems. http://eecatalog.com/rf-microwave/2011/05/23/the-race-is-on-for-rf-mems

213. SPST RF-MEMS switch dc to 12 GHz RMWS101 Radant MEMS datasheet, 2012. http://www.radantmems.com/radantmems.data/Library/Radantata sheet101_1.4.pdf
214. Tyco Electronics Application Note 13C3236, Relay contact life, 2011. http://relays.te.com/appnotes/
215. Grichener, A. and Rebeiz, G. M. (2010). High-reliability RF-MEMS switched capacitors with digital and analog tuning characteristics, *IEEE Trans. Microw. Theory Tech.*, 58(10), pp. 2692–2701.
216. Patel, C. D. and Rebeiz, G. M. (2012). An RF-MEMS switch for high-power applications, in *IEEE MTT-S Int. Microw. Symp. Dig., Montreal*, QC, Canada, pp. 1–4.
217. Rebeiz, G. M. (2003). *RF MEMS Theory, Design, and Technology*, Hoboken, NJ: John Wiley.
218. Zavracky, P. M., McGruer, N. E., Morrison, R. H., and Potter, D. (1999). Microswitches and mircorelays with a view toward microwave applications, *Int. J. RF Microw. Comput.-Aided Eng.*, 9(4), pp. 338–347.
219. Newman, H. S., Ebel, J. L., Judy, D., and Maciel, J. (2008). Lifetime measurements on a high-reliability RF-MEMS contact switch, *IEEE Microw. Wireless Compon. Lett.*, 18(2), pp. 100–102.
220. Sakata, M., Komura, Y., Seki, T., Kobayashi, K., Sano, K., and Horike, S. (1999). Micromachined relay which utilizes single crystal silicon electrostatic actuator, in *IEEE Int. Microelectromech. Syst. Conf.*, Orlando, FL, pp. 21–24.
221. Tazzoli, A., Cellere, G., Autizi, E., Peretti, V., Paccagnella, A., and Meneghesso, G. (2009). Radiation sensitivity of ohmic RF-MEMS switches for spatial applications, in *IEEE 22nd Int. Microelectromech. Syst. Conf.*, Sorrento, Italy, pp. 634–637.
222. Voskoboynik, P., Joos, R. D., Jr. Doherty, W. E., and Goldfarb, R. B. (2006). Low magnetic moment PIN diodes for high field MRI surface coils, *Med. Phys.*, 33(12), pp. 4499–4501.
223. Holm, R. (1967). *Electrical Contacts: Theory and Application*, 4th ed., New York: Springer.
224. McGruer, N., Adams, G., Chen, L., Guo, Z., and Du, Y. (2006). Mechanical, thermal, and material influences on ohmic-contact-type MEMS switch operation, in *IEEE 19th Int. Microelectromech. Syst. Conf.*, Istanbul, Turkey, pp. 230–233.
225. Seki, T., Uno, Y., Narise, K., Masuda, T., Inoue, K., Sato, S., Sato, F., Imanaka, K., and Sugiyama, S. (2006). Development of a large-force low-loss metal-contact RF MEMS switch, *Sensors and Actuators A, Physical*, 132(2), pp. 683–688.
226. Patel, C. D. and Rebeiz, G. (2011). RF-MEMS metal-contact switches with mN-contact and restoring forces and low process sensitivity, *IEEE Trans. Microw. Theory Tech.*, 59(5), pp. 1230–1237.
227. Nishijima, N., Hung, J.-J., and Rebeiz, G. M. (2004). Parallel-contact metal contact RF MEMS switches for high power applications, in *IEEE Int. Conf. Microelectromech. Syst.*, Maastricht, Netherlands, pp. 781–784.

228. Oberhammer, J. and Stemme, G. (2004). Low-voltage high-isolation DC-to-RF MEMS switch based on an S-shaped film actuator, *IEEE Trans. Electron Devices*, 51(1), pp. 149–155.
229. Costa, J., Ivanov, T., Hammond, J., Gering, J., Glass, E., Jorgenson, J., Dening, D., Kerr, D., Reed, J., Crist, S., Mercier, T., Kim, S., and Gorisse, P. (2008). Integrated MEMS switch technology on SOI-CMOS, in *IEEE Solid State Sens., Actuators, Microsyst. Workshop*, Hilton Head, SC, pp. 18–21.
230. Sedaghat-Pisheh, H. and Rebeiz, G. M. (2010). Variable spring constant, high contact force RF MEMS switch, in *IEEE MTT-S Int. Microw. Symp. Dig.*, Anaheim, CA, pp. 304–307.
231. Patel, C. D. and Rebeiz, G. M. (2011). A high power (5 W) temperature stable RF MEMS metal-contact switch with orthogonal anchors and force-enhancing stoppers, in *IEEE MTT-S Int. Microw. Symp. Dig.*, Baltimore, MD, pp. 1–4.
232. Stefanini, R., Chatras, M., Blondy, P., and Rebeiz, G. (2011). Miniature MEMS switches for RF applications, *Journal of Microelectromechanical. Systems*, 20(6), pp. 1324–1335.
233. RF MEMS switch, Omron Electron. Compon. LLC, Schaumberg, IL, 2012. http://www.components.omron.com
234. Radant MEMS switch, Radant MEMS Inc., Stow, MA, 2012. http://www.radantmems.com
235. MEMS Industry Group, MEMS testing standards: A path to continued innovation, Report on MEMS Testing Standards Workshop, Sept. 2011.
236. IEC 62047-5. Semiconductor devices — Micro-electromechanical devices — Part 5: RF MEMS switches, 2011. http://standards.globalspec.com/std/1404606/ds-en-62047-5
237. SEMI MS5, Test Method for Wafer Bond Strength Measurements Using Micro-Chevron Test Structures, 2011. [Online]. Available: http://ams.semi.org/ebusiness/standards/SEMIStandardDetail.aspx?ProductID=211&DownloadID=2442
238. SEMI MS8, Guide to Evaluating Hermeticity of MEMS Packages, 2009. [Online]. Available: http://ams.semi.org/ebusiness/standards/SEMIStandardDetail.aspx?ProductID=1948&DownloadID=1329
239. SEMI MSIO, Test Method to Measure Fluid Permeation Through MEMS Packaging Materials, 2012. [Online]. Available: http://ams.semi.org/ebusiness/standards/SEMIStandardDetail.aspx?ProductID=211&DownloadID=2785
240. JEDEC, Steady-state temperature humidity bias life test, JESD22-A101C, 2009. [Online]. Available: http://www.jedec.org/sites/default/files/docs/22a101c.pdf
241. Fung, C. (2005). Industry study on issues of MEMS reliability and accelerated lifetime testing, in *Proc. IEEE 43rd Annu. Int. Reliab. Phys. Symp.*, pp. 312–316.
242. Miller, S., LaVigne, G., Rodgers, M., Sniegowski, J., Waters, J., and McWorter, P. (1997). Routes to failure in rotating MEMS devices

experiencing sliding friction, in *Proc. SPIE Micromachined Dev. Compon. III*, vol. 3224, pp. 24–30.
243. Challa, V., Rundle, P., and Pecht, M. (2013). Challenges in the qualification of electronic components and systems, *IEEE Trans. Dev. Mater. Reliab.*, 13(1), pp. 26–35.
244. Hu, J., Barker, D., Dasgupta, A., and Arora, A. (1993). Role of failure mechanism identification in accelerated testing, *J. IES*, 36(4), pp. 39–46.
245. Dasgupta, A., Barker, D., and Pecht, M. (1990). Reliability prediction of electronic packages, in *Proc. Annu. Reliab. Maintainab. Symp.*, pp. 323–330.
246. Fuchs, H. and Stephens, R. (1980). *Metal Fatigue in Engineering*, New York: John Wiley.
247. Manson, S. and Hirschberg, M. (1964). *Fatigue: An Interdisciplinary Approach*, Syracuse, NY: Syracuse University Press.
248. Manson, R., Miller, M., Kannard, J., Singleton, M., and Zunino, J. (2007). Development of reliability test guidelines for microelectromechanical systems in military applications, in *Proc. IEEE Sensors Conf.*. pp. 53–57.
249. Olney, A. (2010). Evolving MEMS qualification requirements, in *Proc. IEEE Int. Reliab. Phys. Symp.*, pp. 224–230.
250. Ebel, J., Hyman, D., and Newman, H. (2010). RF MEMS testing — beyond the S-parameters, *IEEE Microw. Mag.*, 8(6), pp. 76–88.
251. JEDEC, Low Temperature Storage Life, JESD22-A119, 2004.
252. Suhir, E., Wong, C., and Lee, Y. (2007). How to make a device into a product: Accelerated life testing its role, attributes, challenges, pitfalls, and interaction with qualification testing, in *Micro- and Opto-Electronic Materials and Structures: Physics, Mechanics, Design, Packaging, Reliability*, New York: Springer.
253. JEDEC, (2008). Application specific qualification using knowledge based test methodology, JESD 94A.
254. Newman, H., Ebel, J., Judy, D., and Maciel, J. (2008). Lifetime measurements on a high-reliability RF MEMS contact switch, *IEEE Microw. Wirel. Comp.*, 18(2), pp. 100–102.

Chien-Ming Huang received the M.S. degree from National Tsing Hua University, Hsinchu, Taiwan. He is currently working toward the Ph.D. degree at the Center for Advanced Life Cycle Engineering, University of Maryland, College Park. His main areas of interest include LIGA fabrication and reliability of electronic packaging. He has over 4 years of experience in the areas related to thermal interface materials, thermal test vehicles, micro-device tests, wafer bumping, and package qualification as a Senior Engineer at Taiwan Semiconductor Manufacturing Company, Taiwan.

Arvind Sai Sarathi Vasan received the B.E. degree in Electronics and Communication Engineering in SSN Institutions, Chennai, India. He received his Ph.D. degree on 2016 in Mechanical Engineering, University of Maryland, College Park. His main areas of interest include prognostics of electronics, diagnostics, and machine learning.

Yunhan Huang received the B.S. degree from Shanghai Jiao Tong University, Shanghai, China. He is currently working toward the Ph.D. degree at the Center for Advanced Life Cycle Engineering, University of Maryland, College Park. His Ph.D. thesis is on the degradation and reliability improvement of radio-frequency MEMS. He has one paper published at the 60th Electronic Components and Technology Conference (ECTC) 2010 and one paper accepted by ECTC 2012. He is the holder of one patent: MEMS Barcode Device for Monitoring Medical Systems at Point of Care. U.S. Patent 2012/0018514 published on January 26, 2012.

Ravi Doraiswami received the M.Tech. (lasers and electro-optical engineering) and M.Sc. degrees (materials science) from the College of Engineering, Anna University, Chennai, India, and the Ph.D. degree in Mechanical Engineering with mechatronics specialization from the Indian Institute of Technology, New Delhi, India. He is currently with CALCE, University of Maryland, College Park, where he is an Assistant Research Scientist. He has over 10 years of experience in the areas related to applied SOC, SIP and SOP systems, design, materials process, fabrication, test, and evaluation and reliability as an Assembly Technology Lead Engineer with Packaging

Research Center, Georgia Institute of Technology, Atlanta. Novel technologies developed by him, to fabricate 5–10-μm-pitch nano nickel–tin flip-chip interconnects, shape-memory organic capacitors, and nanoelectrode systems inside microtubes, have found applications in biosensors, MEMS, microelectronics fabrication, and packaging technologies.

Michael Osterman received the Ph.D. degree in Mechanical Engineering from the University of Maryland, College Park. He is a Senior Research Scientist and the Director of the Center for Advanced Life Cycle Engineering (CALCE) Electronic Products and System Consortium, University of Maryland. He heads the development of simulation-assisted reliability assessment software for CALCE and simulation approaches for estimating time to failure of electronic hardware under test and field conditions. He is one of the principal researchers in the CALCE effort to develop simulation models for the failure of Pb-free solders. In addition, he has led CALCE in the study of tin whiskers since 2002 and has authored several articles related to the tin-whisker phenomenon. Further, he has written various book chapters and numerous articles in the area of electronic packaging. Dr. Osterman is a member of ASME and SMTA.

Michael Pecht (M'83–SM'90–F'92) received the M.S. and Ph.D. degrees in Engineering Mechanics from the University of Wisconsin, Madison. He is the Founder of the Center for Advanced Life Cycle Engineering, University of Maryland, College Park, which is funded by over 150 of the world's leading electronics companies at more than US$6 million/year. He is also a chair professor in mechanical engineering and a professor in applied mathematics with the University of Maryland. He is the chief editor for *Microelectronics Reliability*. He has written more than 20 books on electronic product development, use, and supply chain management and over 400 technical articles. He consults for 22 major international electronics companies, providing expertise in strategic planning, design, test, prognostics, IP, and risk assessment of electronic products and systems.

Dr. Pecht is a Professional Engineer, an ASME fellow, a SAE fellow, and an IMAPS fellow. Prior to 2008, he was the recipient of the European Micro and Nano-Reliability Award for outstanding contributions to reliability research, 3M Research Award for electronics packaging, and the IMAPS William D. Ashman Memorial Achievement Award for his contributions in electronics reliability analysis. In 2008, he was the recipient of the highest reliability honor, the IEEE Reliability Society's Lifetime Achievement Award. In 2010, he was the recipient of the IEEE Exceptional Technical Achievement Award. He served as chief editor of *IEEE Transactions on Reliability* for 8 years and on the advisory board of *IEEE Spectrum*. He is an associate editor for *IEEE Transactions on Components and Packaging Technology*.

Index

3D integration, 18, 32, 42, 46
3D printing, 172, 194
6-axis motion sensor, 75, 78, 79
9-axis motion sensor, 81, 82

accelerometer, 2, 62, 74, 75, 78, 79, 81, 85, 94, 120, 201, 242, 284–289
actuation diaphragm, 175, 177
actuators, 87, 90, 91
adhesive bonding, 141, 144, 145, 151
alginate-PLL capsule, 193
aluminum, 76, 78, 81
anodic bonding, 157, 159–163
asymmetric semi-permeable membrane, 176
Au stud bumps, 49
axis motion sensor, 81

bandwidth-tunable SAW filter, 149
barrier polymer, 177
benzocyclobutene (BCB), 146, 147, 151–153
biodegradable micro-capsules, 190
boron-doped diamond (BDD), 146–148, 166
breathable film, 194
breather valves, 186
BST, 147–149

cellulose acetate, 175
check valves, 186
CMOS process, 73, 75, 76
CMOS-to-MEMS integration, 80
coaxial flow, 190
combination with anodic bonding, 46
consumer electronics, 75, 82
consumer market, 73
consumer products, 73, 75
contacts, 260, 264, 266, 269, 273
contamination, 280, 281, 291, 309, 310, 312, 315
Coriolis force, 75
corona treater, 187
creep, 280, 281, 298, 302–306, 312, 315, 323

daisy chains, 50
deep reactive ion etching (DRIE), 65, 77, 78, 155, 156
device transfer (via-first), 143–145, 152, 153, 155
device transfer (via-last), 143, 144, 149
diaphragm valves, 193
die-level packaging, 120, 123, 124, 127
dielectric breakdown, 306, 307
dielectric welding, 174
digital microfluidic, 100, 108, 109
die-stacking solder bonding, 208
doped silicon pins, 47
double emulsification, 189
dry-film resist, 48

e-CUBES, 46
elastomers, 173
electroosmotic, 106
electrostatic actuators, 88, 90
electrostatic discharge (ESD), 306, 307, 318
electrowetting, 108, 110
embedded orifice, 190
encapsulation, 117–124, 127–129, 134, 135
environmental sensor, 99
Epi-seal, 119–122, 124, 128–131, 133–137
etched trench, 35
eutectic bonding, 60, 76, 78, 81
EWOD, 108–110

failure mechanisms, 280, 287, 288, 290, 305, 311, 312, 319, 321–323
fatigue, 280, 281, 284, 287, 289, 290, 300–303, 305, 312, 322, 323
film transfer, 143, 144, 146, 165
flip-chip packaging, 251, 259
flow sensors, 96, 101
FR4, 85, 86, 94, 101, 102, 108
fully packaged MEMS, 75

γ-Fe_2O_3 nanoparticles, 192
germanium, 77, 78, 81
gyroscope, 1, 13, 74, 75, 78–81, 117, 142, 283–286

hermetic packaging, 241
heterogeneous integration, 141–143, 153, 165
high-temperature tolerant, 48
HISVESTA, 46, 48
hot embossing, 175
hot melt film, 187
hot resistance welding, 174
hybrid integration, 103, 104

impermeable diaphragm, 177
integrated capacitive pressure sensor, 158, 159
interconnect, 251–253, 259, 260, 268, 274
intermediate bonding layer, 181
InvenSense, 75, 76, 78–82

Kapton, 86, 89, 90, 94, 98, 99

Lab-on-PCB, 105, 107
lamination, 182
laser dicing, 125, 126
laser welding, 174
LCP, 86, 92, 93, 95–97, 99
lead zirconate titanate, 144
localized bonding, 181
localized heating, 182, 183, 205, 206, 213, 218, 219
localized heating and bonding, 226–228, 236, 238
low-cost motion sensor, 73
low-temperature co-fired ceramic (LTCC), 141, 161–166

massive parallel EB exposure system, 153, 166
massless lithography, 153
matrix nanocrystal silicon emitter, 153
MEMS encapsulation, 244
MEMS module, 98, 99
Met-Via, 38, 39, 41, 42
Micralyne MicraSilQ^{TM}, 54
MicraSilQ^{TM}, 54
micro drug-delivery system, 179
micro-capsules, 188, 190, 192
micro-encapsulation, 188
micro-heaters, 182
micro-total analysis systems, 105
microelectromechanical systems, 3
microfluidics, 85, 100, 103, 104, 107, 108, 110, 171
microphone, 85, 91, 96–98
microstructure, 85, 87, 283, 291–295, 297–299, 303, 311, 312
microvalve, 106
monolithic integration, 98, 101, 102

motion sensor, 73–75, 78–82
multiclip package (MCP), 197–199

nanoporous Au, 163, 164
Nasiri Fabrication Process, 73
nc-Si, 153, 155–158
NPG, 164

on-demand micro-encapsulation, 192
optical MEMS, 304, 311–313
osmotic micro-actuator, 175
osmotic orthodontic actuator, 194
osmotic pump, 187, 194
oxygen plasma, 180, 181

Parylene, 108, 109
PCB MEMS, 86, 87, 89, 91, 92, 94–100, 110, 111
PCB microfluidics, 100, 104, 107, 110
PDMS, 103, 104, 106, 108, 178, 180, 186, 190
peristaltic pump, 106
phosphocholine, 190
piezoelectric MEMS switch, 149
Planoptik, 46
plasma enhanced bonding, 178
plastic package, 78, 79, 82
plastics-to-plastics bond, 184
plug insertion, 184
PMMA, 103–105
poly(L-lactic acid), 190
polymer micro-fabrication, 172
polymer-based MEMS, 171
polymeric membrane, 171
polymeric microstructures, 172
PolySi, 48
preprogrammed osmotic pumping chip, 184
pressure pump, 106
pressure sensors, 85, 92, 100, 101, 118, 119, 137
pressure-sensitive switch, 187
Process Design Kit (PDK), 58
processing temperatures, 182
PTFE, 85, 86, 94
PZT, 144, 149–151, 166

QFN package, 78

rapid prototyping, 172
release mechanism, 74
reliability, 279, 280, 282–285, 287, 288, 292, 293, 302, 305–307, 309, 311–321, 323, 324
resonator, 117–126, 129, 130, 133–136
RF MEMS Switch, 88, 90, 98
RF-MEMS, 280, 296, 297, 305–309, 313–317, 319, 323

SAW, 147–150, 152, 153, 166
scribe and break, 127
sealing locally, 224
selective laser debonding, 153
selective transfer, 145, 149, 152, 154
self-packaged MEMS, 87, 98
semi-permeable diaphragm, 175
semi-permeable membrane, 179, 187, 194
semi-permeable micro-capsules, 192
shock, 280–286, 304, 310, 312, 318
Si diaphragm capacitive vacuum sensor, 160
Signal to Noise Ratio (SNR), 59
Sil-Via, 35–39, 42
silicon on insulator (SOI), 76
single crystal silicon (SCS), 78
solvent sealing, 174
solvent welding, 175
spin casting, 175
springs, 252–264, 266–271, 273, 274
standards, 279, 280, 313, 317–321
Stanford Epi-seal, 119–121, 124, 128–131
stereolithography, 194
stiction, 280, 283, 285, 286, 290–299, 312, 315, 319, 323
SU-8, 103, 104, 106–108
surface acoustic wave filter, 147
surface micro-machining, 142, 143
surface-treated nanoparticles, 190

tactile sensor network, 151, 152
tactile sensors, 95, 96

temporary embedment, 184
thermal actuators, 87, 90, 91
thermal forming, 187
thermal MEMS switch, 161
thermoplastics, 173
thermosets, 173
thin-film, 253, 254, 271
through glass vias (TGV), 46
through silicon groove (TSG), 151
through silicon insulator (TSI), 35, 36
through silicon via (TSV), 8, 31–36, 38–42, 48, 54, 155–157
through substrate via, 31, 32, 42
time switches, 186
transfer of MEMS to LSI, 142, 143
trilaurin, 190
tunable SAW filter, 149, 150, 166

ultrasonic welding, 174
Under Bump Metallization (UBM), 56

vacuum generators, 186
vapor-permeable thermoplastic polyurethane, 187
vertical chip stacking integration, 203
vibration, 280, 284, 286–289, 312, 318, 322

wafer bumping, 206, 219
wafer saw, 124, 125, 127
wafer-level capping, 3, 5, 6, 12, 25
wafer-level packaging (WLP), 54, 75, 120, 127, 141–143, 157, 158, 161, 162, 164–166
wafer-level transfer, 141–144, 146, 147, 165, 166
wear, 280, 281, 293, 294, 298–302, 316, 319, 323
WLCSP, 219, 221

zero-level packaging, 223, 237